Space Electronics for Deep Space Exploration

This edition first published 2026 by John Wiley & Sons, Inc., 111 River Street, Hoboken, NJ 07030, USA and Scrivener Publishing LLC, 100 Cummings Center, Suite 541J, Beverly, MA 01915, USA

For more information about Scrivener publications please visit www.scrivenerpublishing.com.

Wiley Global Headquarters
111 River Street, Hoboken, NJ 07030, USA

For details of our global editorial offices, customer services, and more information about Wiley products visit us at www.wiley.com.

The manufacturer's authorized representative according to the EU General Product Safety Regulation is Wiley-VCH GmbH, Boschstr. 12, 69469 Weinheim, Germany, e-mail: Product_Safety@wiley.com.

Library of Congress Cataloging-in-Publication Data

ISBN 978-1-394-46645-0

Cover image: Generated with AI using Adobe Firefly
Cover design by Russell Richardson

Set in size of 11pt and Minion Pro by Manila Typesetting Company, Makati, Philippines

Printed in the USA

10 9 8 7 6 5 4 3 2 1

Contents

Preface

Deep space exploration is one of the greatest technological and scientific challenges. In the era of space agencies expanding their missions beyond Earth orbit and private enterprises and research institutions looking to the Moon, Mars, and distant planetary systems, advanced, reliable, and intelligent electronic systems are more important than ever. Space electronics is the cornerstone of any space mission, which includes all technologies required for communication, computation, navigation, sensing, power control and autonomous decision making in the most hostile environment known. Although traditionally very effective for terrestrial use, these electronic systems are facing a number of phenomenological constraints in deep space, including extreme radiation conditions, large and rapid temperature changes, low power supply, long communication delays, and the inability of periodic maintenance and repair. Such restrictions require the evolution of new electronics devices that are radiation hard, energy efficient, intelligent, adaptive, fault tolerant and capable of extended mission endurance without operator intervention.

Space Electronics for Deep Space Exploration, is a comprehensive survey of the technologies which will drive the future of space electronics. It focuses on the latest advances in radiation-tolerant design and design of low-power computing architectures, artificial intelligence, neuromorphic computing, quantum communication, edge computing, advanced sensing, wireless power transfer, additive manufacturing, and autonomous robotics. The technologies are revolutionizing spacecraft design, operation and maintenance, providing unprecedented levels of mission reliability and operational efficiency. The chapters are structured to give the reader basic information and some new ideas on up-to-date space electronics. Starting from the basics of space environments and electronic design problems, the book covers important areas of radiation hardening, intelligent computing systems, secure communications, distributed processing, advanced energy systems, cryogenic electronics, and fault management strategies. Case studies of the James Webb Space Telescope and Mars rover missions provide real-life examples of the implementation of innovative electronic technologies in extreme mission situations. Finally, the book discusses future developments and the future technological development path of interstellar exploration.

The interdisciplinary approach is one of the features of this volume. It unites ideas from electronics engineering and computer science, artificial intelligence, communication engineering, materials science, robotics, and space system design. The book combines theory and application with mission-driven case studies that allow the reader an overall view of the fast-changing field of space electronics. The book is designed for researchers, engineers, scientists, graduate students, space technology professionals and policy makers interested in identifying and participating in the development of advanced electronic systems for future space missions. Readers will find valuable insights and guidance throughout this volume, whether they are designing spacecraft hardware, developing autonomous spacecraft control systems, investigating quantum communication technologies, or exploring novel manufacturing techniques for space applications.

We would like to express our heartfelt thanks to all the authors, reviewers and contributors of this book for their expertise and efforts. The contributions of their collective knowledge and dedication to space technology development have greatly enhanced the quality and scope of this work. We also recognize the innovative contributions of the scientific community, space agencies, and industrial partners around the world in their exploration of space. As the world moves into a new frontier of deep space exploration, intelligent and resilient electronic systems will become more important than ever in breaking technological barriers and supporting sustainable exploration of the cosmos beyond Earth. This book is hoped to be a valuable resource and motivate other research, innovation, and cooperation in the fascinating area of space electronics.

1

Space Electronics: A Primer and the Difficulties of Working in Low Earth Orbit

N. Sathish[1*], Prasanth Aruchamy[2], Rajesh Kumar Dhanaraj[3] and Abdelfateh Kerrouche[4]

[1]Department of Electronics and Communication Engineering, Sri Venkateswara College of Engineering, Sriperumbudur, Tamil Nadu, India
[2]Department of Computer Science and Engineering, Vel Tech Rangarajan Dr. Sagunthala R&D Institute of Science and Technology, Chennai, Tamil Nadu, India
[3]Department of Computer Studies, Symbiosis International (Deemed University), Pune, Maharashtra, India
[4]School of Computing Engineering and the Built Environment, Edinburgh Napier University, Edinburgh, United Kingdom

Abstract
The space electronics is important in facilitating communication, control, sensing, and handling of data in contemporary spacecrafts. Nevertheless, low Earth orbit (LEO) electronic systems design requires special design because space environment is rough and unpredictable. The chapter outlines the basics of space electronics and describes how the behavior of the devices varies in the orbital conditions. It has raised major issues that include exposure to ionizing radiation, flux of charged particles, single-event effects, corrosion of atomic oxygen, and corrosion due to vacuum. Components are also under additional stress because of the rapid thermal cycling that occurs during each orbital pass, and this has an impact on the reliability and the long-term stability. In order to solve these problems, this chapter talks about such strategies in engineering, such as the selection of radiation-hardened devices, shielding, power conditioning, thermal, redundancy, and fault-tolerant system architecture. Practical issues associated with testing, qualification, and space grade component standards are also highlighted. The combination of the theoretical background and practical limitations has helped this chapter give the reader a clear picture of the intricacies that surround the development of reliable electronics in LEO missions. It doubles up as an introduction to students,

**Corresponding author*: n.satyish@gmail.com

N. Sathish, Prasanth Aruchamy, Rajesh Kumar Dhanaraj, Nithya Rekha and Karthik Palani (eds.) Space Electronics for Deep Space Exploration, (1–22) © 2026 Scrivener Publishing LLC

researchers, and engineers who are interested in developing resilient systems that will withstand the harsh working environment in space.

Keywords: Space electronics, low earth orbit, radiation effects, fault-tolerant design, thermal and environmental challenges

1.1 Introduction

The recent surge of space operations, such as Earth observations and communication systems, as well as scientific research and commercial services, has created a substantial pressure on the need to develop reliable, small, and energy efficient electronic systems. This is contrary to terrestrial platforms, which have to work in an environment exposing them to extreme levels of radiation, vacuum conditions, large variations in temperature, and special mechanical stress. The multidisciplinary skills that are needed to design such systems include electronics, material science, thermal engineering, and space mission requirements.

The low Earth orbit (LEO) has become the most used orbital space because it is accessible, cheaper to launch, and can be used in high-speed communication and remote sensing. LEO, however, has its own environmental issues, including the erosion of the atomic oxygen, high thermal cycling, and exposure to the trapped radiation belts [1]. These requirements put very stringent demands on the choice of components, system architecture, and long-range reliability plans. This chapter is a valuable introduction and primer to space electronics and explains the major environmental, electrical and mechanical constraints that dictate system behavior in LEO. It can be used as a starting point toward the realization of why space-grade electronics should be designed differently and why mission performance is largely reliant on a sound electronic design.

1.1.1 Overview of Space Electronics

Space electronics is designated as the special design and assembly of electronic parts with the aim of working in the severe space atmosphere. These systems consist of communication modules, power conditioning units, data handling subsystems, sensors, processing units, and control electronics.

Spaceborne systems have to be able to withstand:

- High-energy radiation that may cause breakdown of devices or cause transient faults.

- Temperatures vary between −150°C in the shade and 120°C in the sun.
- Vacuum is an influence on the transfer of heat and may cause outgassing of materials.
- During the launch, there are mechanical vibrations and shocks.
- Extensive levels of operation without requiring any physical upkeep.

In order to address such requirements, space electronics tends to incorporate radiation-hardened systems, fault-tolerant design, redundant subsystem design, thermologically stable material and strict testing and qualification requirements (e.g., European Cooperation for Space Standardization [ECSS], MIL-STD, NASA).

1.1.2 Importance of LEO Missions

The LEO is the most actively used part of space, with altitude ranges ranging between 160 and 2,000 km in diameter, because it is easy to reach and flexible in terms of missions. LEO satellites are able to take high-resolution images of the earth, aid in weather forecasting, monitoring the environment, and carrying out real-time evaluation of disasters. It is ideal in broadband constellations, IoT connections and scientific experiments and applications that demand rapid transmission of data because the LEO has a lower communication latency than lower orbits. Moreover, the reduced energy to access LEO greatly lowers the cost of launching satellites which now provides universities, startups, and other organizations the opportunity to launch small satellites and CubeSats to conduct research, technology demonstration, and commercial purposes [2].

Despite such benefits, the LEO operation presents a number of engineering problems which directly affect the electronics design. During their orbits, satellites undergo a rapid change of temperature, high concentrations of charged particles in places like South Atlantic anomaly, and constant exposure to atomic oxygen, which tends to damage materials and surfaces. Orbit stability is also influenced by atmospheric drag and high-precision systems of attitude and power management are needed. These environmental stress factors render LEO a very significant yet challenging environment in which strong, radiation-resistant, and thermally stable electronic systems are a must to the reliability and persistence of the mission.

1.2 Fundamentals of Space Electronics

The basics of space electronics entail the principles, design practice, and component technology needed to assure dependable functioning in the severe circumstances of space. In comparison with terrestrial systems, space electronics need to operate without service during long periods in the mission without being subject to the radiation, vacuum, low or high temperatures, or mechanical forces. These limitations lay stress on the thorough choice of components, healthy circuit design, severe qualification, and special reliability engineering. The inherent knowledge of semiconductor behavior, space grade component requirements, and the fundamentals of power and signal integrity requirements are the foundations of building resilient electronic architectures of orbital missions [3].

1.2.1 Semiconductor Device Behavior in Space

When subjected to the orbital environment, semiconductor devices experience a large change in performance. Energy particles, protons, electrons, and heavy ions, which are of high energy, have the potential of penetrating structures of the devices and induce degradation or temporary malfunctions. Important applications are total ionizing dose (TID) that causes the gradual alteration of threshold voltages and causes leakage currents to increase and single-event effects (SEE) in which the strike of one particle can trigger bit flips, latch-up, or even burnout. Space temperature cycling also has an impact on carrier motion and junction behavior, whereas the vacuum environment has an impact on heat dissipation, as well as reliability of the packaging. Consequently, designers have to consider the variation of the electrical parameters, apply protective circuitry and apply predictive models to guarantee stable behavior of the device throughout the mission lifecycle [4].

1.2.2 Space-Qualified Components and Standards

Space missions are based on components which are under strict quality, reliability, and environmental tolerance criteria. Qualified space parts are subjected to stringent screening and radiation test, thermal cycling, and burn-in to guarantee the uniform behavior even when subjected to extreme conditions. Agencies like NASA, ESA, and military organizations have standards, such as MIL-STD-883, ECSS-Q-ST-60, and NASA EEE-INST-002, that outline acceptable material, fabrication process, and testing procedure. These standards assist in categorizing the components into levels like commercial-off-the-shelf (COTS), radiation-tolerant and radiation-hardened devices. Space-grade parts can be more reliable, but are more expensive and have longer lead times and thus tradeoffs between performance, availability, and risk are an important aspect of mission planning.

1.2.3 Power, Signal Integrity, and Reliability Basics

The reliability of any spaceborne electronic system is based on the power management and signal integrity. Satellites in LEO go through the sun and the eclipse often, resulting in variable power supplies and thermal conditions. Power systems should hence be characterized by constant regulation, sound energy storage as well as defenses against voltage spikes, transients caused by radiation and deterioration of batteries. On the same note, signal integrity is necessary considering that circuits are prone to noise, electromagnetic interference, and bit errors caused by radiations. The techniques are shielding, filtering, high-resilience grounding, and the use of differential signaling. Reliability engineering provides additional protection of redundancy, fault detection, error-correcting code, and graceful-degradation. Combined these measures guarantee that electronic subsystems still operate properly even in environments that are disrupted and where the mission time is long [5].

1.3 The Low Earth Orbit Environment

The LEO environment poses a set of physical, thermal, and radiation issues, which have a direct impact on the effectiveness, maintainability and reliability of electronic systems. The satellites in the LEO orbiting between 160 and 2,000 km experience varying temperatures, significant radiation of particles, and high concentrations of atomic oxygen, as well as the material effects in the vacuum. Such environments can dramatically change electrical properties, material degradation and cause mechanical loading of spacecraft base. To develop electronic systems that are capable of surviving such stresses in long working life cycles, it is important to understand LEO environment. All environmental factors should be taken into consideration when selecting the components, circuit design, thermal management, packaging, and the system architect of the system [6].

These environmental conditions in LEO are dynamic as opposed to being static. As an example, satellites orbit every 90 to 120 minutes, or rather they are constantly in the direct sunlight and then in total darkness. These high-rate cycles add thermal gradients, mechanical deformation and electrical variations. Likewise, the concentrations of the charged particles are not homogeneous and the South Atlantic Anomaly is one such region that exposes satellites to greater background of radiation. The complete knowledge of these LEO properties assists engineers in coming up with extremely strong, fault tolerant designs which reduce the risk of spaceborne electronics.

1.3.1 Radiation Types and Their Effects

One of the greatest serious threats to electronic systems is radiation in the LEO, since it has a direct effect on the performance and life cycle of semiconductor devices. There are three sources of high-energy particles that hit LEO satellites, which are solar winds, trapped particles, and Van Allen belt, and high-energy cosmic rays. In the near-earth radiation, protons and electrons prevail and they are capable of penetrating through layers of semiconductor causing charge accumulation and structural defects. The net effect of the long-term exposure is called TID that leads to shift in threshold voltage, augmentation in leakage currents, diminished gain and general device decay. SEEs are also another significant category [7]. A single charged particle impacting a sensitive node in a device can cause transient or permanent faults including single-event upsets (bit flips), latch-up (high current conditions), burnout, or gate rupture. These effects may cause disruption of the missions or a total failure of the subsystem. Non-ionizing particles cause a further degradation of materials through displacement damage dose (DDD), which occurs as a result of the atoms on the crystal lattice being displaced. Engineers can overcome these threats by using radiation hardened parts, shielding techniques, redundancy, error correcting codes, as well as architectural fault tolerance in order to ensure reliability over the mission period.

1.3.2 Atomic Oxygen Interaction

Another significant environmental issue in LEO is atomic oxygen (AO), particularly at the altitudes of 200-700 km, where it is concentrated most of all. AO is formed by dismantling the molecular oxygen, which is present in the atmosphere, under the influence of ultraviolet solar radiation and is highly reactive and able to oxidize and erode exposed materials. Of particular susceptibility to degradation by AO are polymers, composites, thermal blankets, and protective coating. Mass loss is not the only effect of the oxidation process but surface roughening may also result, which may change the thermal properties and lower the efficiency of insulation layers. In the case of electronic systems, it is AO degradation which impacts the insulation of cables, connector material and external surfaces which may cause short circuiting or mechanical breakdown of the system. AO may contaminate and make the lenses and filters less transparent in optical systems. In order to reduce such risks, spacecraft designers use AO-resistant coating like silicon dioxide, aluminized Kaplon, and protective metallic films. In LEO, AO exposure testing should be carried out over a long period and material selection should be completed to guarantee long-term system performance.

1.3.3 Vacuum and Outgassing Issues

Space is a vacuum, which means that convective heat transfer is not possible and, therefore, spacecraft must only use conduction and radiation to manage heat. In the absence of proper design, electronic parts may overheat or suffer point hot spots. There is also outgassing caused by vacuum conditions in which gases trapped by materials, adhesives, and coatings are released. Such vapors may settle on adjacent surfaces, including optical sensor, thermal radiator, or high-voltage circuit and contaminate, lower the efficiency, and may cause arcing. Guidelines used to select materials like low-outgassing criteria established by NASA and ESA make spacecraft components to be of high-quality safety and performance standards. Mitigation of contamination issues is usually done by use of out gassing resistant materials, proper venting, and pre-launch bake out.

1.3.4 Temperature Extremes and Thermal Cycling

Satellites in LEO undergo a high rate of thermal cycling due to their orbital period which is a transition between daylight and the shadow of the earth every 45 to 60 minutes. This leads to temperatures ranging between −150°C and +120°C. These fluctuations introduce a lot of mechanical stress on solder joints, printed circuit boards, connectors, and structural elements that may result in fatigue, delamination or cracking. Semi-conductors are also sensitive to temperature; changes may impact the carrier velocity, reference levels, and analog signal precision. Radiators, multi-layer insulation (MLI), heat-pipes, and thermostatically controlled heaters are all necessary as thermal control systems to ensure that the operating temperatures are kept at safe levels and to prevent failure of the electronic components because of thermal effects.

1.3.5 Microgravity and Mechanical Stresses

Even though microgravity has no direct destructive effect on electronics, it affects fluid dynamics, thermal, and mechanical stability. No convection changes the cooling dynamics, and it is important to pay special attention to thermal design, to avoid hotspots. The major sources of mechanical stresses, however, are during launch, electronics are exposed to severe vibration, shock loads, and acoustic energy. Those forces may break the connectors, crack the solder joints, or destroy delicate circuits prior to the spacecraft even escaping into orbit. Microgravity may cause problems in fluids or moving components, such as pumps, or attitude control components, post-launch. Strict environmental tests such as tests of vibration, shock and thermal vacuum tests are thus necessary to promote reliability over time.

1.4 Challenges for Electronic Systems in LEO

There are a set of environmental and operational challenges that affect electronic systems working in Low Earth Orbit which differ greatly as compared to what can be found on earth. The radiation exposure, thermal cycling, atomic oxygen erosion, vacuum conditions and microgravity are all combined to affect the device performance and must be carefully taken into account during design and qualification. Also, LEO missions tend to be highly limiting in size, weight and power usage, complicating even more the optimization of subsystems. These issues may affect semiconductor stability and material stability up to power stability and communication integrity. These stressors are important in understanding how to design electronics that will be able to operate and survive faults over the whole lifetime of the mission [8].

Besides environmental stress, specific aspects of the missions like the small size of the satellite architecture, lack of shielding, and the COTS components also add to the problem of reliability. Smaller satellites, though notably CubeSats, often have little radiation or thermal protection. These platforms therefore are more prone to electronic anomalies, abrupt failures and deterioration of system performance. It is necessary to have radiation mitigation measures, sound thermal design, fault tolerant architectures, and trusted communication subsystems to tackle these issues. In the subsections below, the most important issues impacting electronics in LEO have been described [9].

1.4.1 TID and Displacement Damage

TID is the cumulative impact of long-term exposure to Ionizing radiation. The charges in LEO accumulate overtime through protons and electrons in the semiconductor oxides, insulating layers, shifting threshold voltages to levels higher than normal, leaking more than normal, and decreasing the gain of transistors. Throughout the time these changes will cause timing errors, impaired analog performance and even circuit failure. On the other hand, Displacement Damage takes place when high-energy particles or particles move atoms within the lattice of a material, especially when applied to devices like photodiodes, solar cells or bipolar transistors. This translates to less efficiency, noise as well as poor signal response. Both TID and displacement damage must be mitigated with one of the mitigation strategies being radiation-hardened components, shielding, and derating to ensure long-term stability [10, 11].

1.4.2 Single-Event Effects

SEEs are the situation when a single charged particle hits a sensitive part of an electronic device resulting in an immediate perturbation. Such effects include soft (single-event upsets, or SEUs) or hard (single-event latch-up, or SEL; single-event burnout, or SEB), or cause permanent damage to components. SEEs especially pose a challenge to digital electronics, power devices and memory arrays. Mitigation methods can be seen in terms of error correcting codes (ECC), redundancy, current limiting circuits and radiation hardened systems that tend to reduce sensitive volumes in the structure of semiconductor devices.

1.4.3 Material Degradation and Surface Erosion

The degradation of materials in LEO is caused due to the exposure to atomic oxygen, ultraviolet light, thermal stress, and micrometeoroids. Atomic oxygen in particular is highly destructive, and is able to react with polymers, coatings, and thermal materials to bring about erosion, embrittlement and roughness of surface. With time, these degradations have the ability to affect insulations, thermal blankets and connector housings. UV radiation can additionally increase the rate of chemical degradation, whereas the exposed surfaces can be physically damaged by the impact of micrometeoroid. To secure electronics, it is necessary to use AO-resistance coating, strong materials and multilayer protective structures to ensure long-term stability [12].

1.4.4 Power System Instability

LEO Power systems are unstable because of the changing sunshine, battery life, radiation related failures and temporary loads. Due to the frequent interchange of sunlight and eclipse between satellites, there is a rapid fluctuation in power availability. Solar panels should endure loss of efficiency due to radiation, batteries are subjected to thermal stress and charge-discharge cycles which cause wear in the long term. Transient faults can also be caused by radiation to power converters or regulators. Serious design of power systems needs efficient energy storage, redundancy, robust regulation and protection circuits to manage variation of voltage spikes and load variation.

1.4.5 Communication and Electromagnetic Disturbances

The problems facing communication subsystems include Doppler shifts, intermittent contact with ground stations and signal attenuation. Moreover, receiver sensitivity and data integrity can be affected by electromagnetic interference (EMI) by onboard electronics or by the solar activity. The charged particles have the ability to cause noise on communication circuits hence diminishing the link reliability. To ensure stable communications lines, shielding, filtering, error-correction coding, and grounding of the circuits are necessary. In high-frequency systems, it is also important that the antennas are perfectly aligned and thermal stability is maintained.

1.5 Design Strategies for LEO Electronics

Electronic systems to be designed to LEO must be developed using a multidisciplinary approach taking into account radiation effects, temperature extremes, vacuum environment conditions, power surge, communications, and mechanical forces. Space systems cannot be easily repaired or replaced like on earth therefore reliability must be designed in at all levels- semiconductor devices to system architecture [13]. LEO is especially complicated since systems work in the magnetic field of the Earth but still undergo the severe radiation of the Sun, flux of charged particles, and the exposure to the atomic oxygen. In addition, repeated thermal cycles are due to the high frequency of orbital day–night cycle, about 90 minutes, that exertes stress on materials and solder joints. These issues require a very complex collection of design solutions incorporating radiation mitigation, thermal control, shielding, fault tolerance, redundancy, and sound system-level design. The subsections that follow elaborate on the key strategies employed in order to provide long-term survivability and optimum performance of electronics within the LEO conditions. Figure 1.1 presents an overview of the LEO environment, the major effects on electronics, and the major mitigation measures mentioned above.

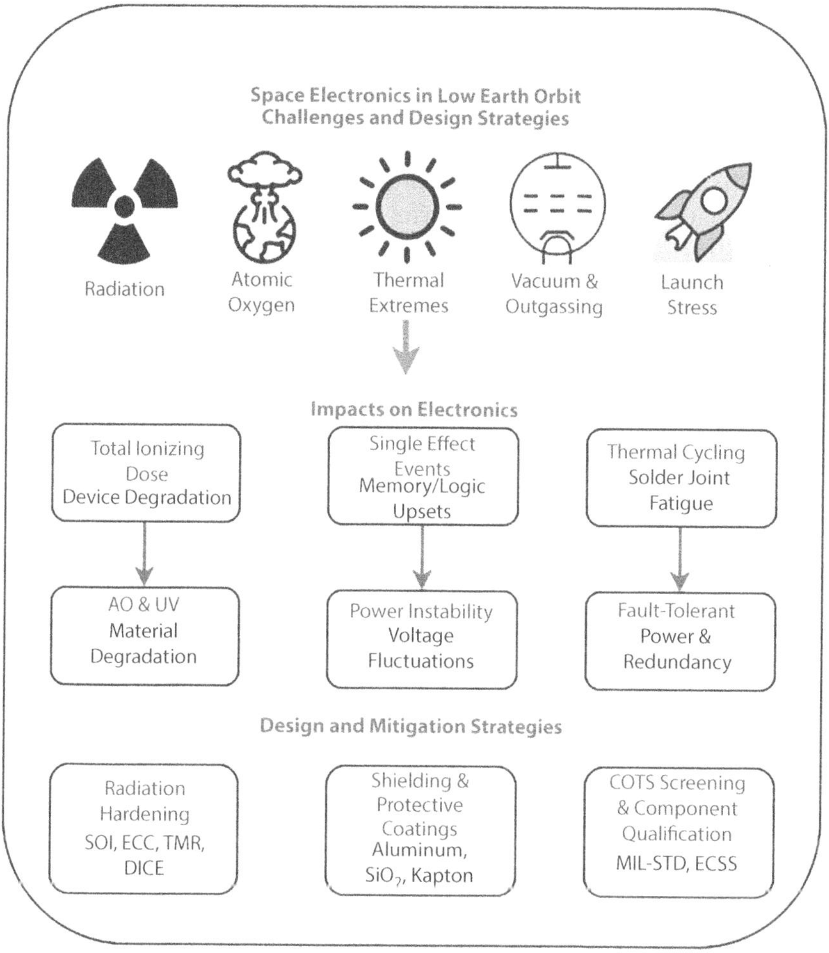

Figure 1.1 Space electronics in low earth orbit: challenges and design strategies.

1.5.1 Radiation Hardening Techniques

Radiation hardening is used to make sure that electronic devices can work under the particle environment of LEO that contains protons, electrons, heavy ions, secondary radiation, among others. The inflicted damage by high-energy particles may be TID damage, displacement defect, or SEEs including upsets, latch-ups, and burnout. Radiation-hardening techniques start at the device level and include Silicon-on-Insulator, enclosed-layout transistors, and optimized doping to minimize charge collection. Hardened flip-flops, ECC, triple modular redundancy (TMR), and dual interlocked storage cells can also be used as a circuit-level protection against transient faults. Watchdog timers, memory scrubbing, and fault-monitoring software at the system level identify and fix errors. It is also important to select radiation-tolerant or hardened components, which meet the requirements of MIL-STD and ECSS standards. Through the use of a combination of device, circuit, and system level giving LEO electronics the opportunity to ensure data integrity and long-term reliability even when subjected to moderate levels of radiation [14].

1.5.2 Shielding and Protective Coatings

Electronics are shielded and coated to provide protection against LEO threats like radiation and AO, ultraviolet (UV) radiation and impacts of micrometeoroids. Radiation shielding is typically made of aluminum, tantalum or polyethylene, occasionally with graded-Z configurations in order to minimize the production of secondaries. Silicon dioxide (SiO_2), aluminum oxide (Al_2O_3), and diamond-like carbon are used as AO-resistant coating to stop erosion of the polymers, thermal blankets, and insulations. UV-resistant films and paints minimize degradation and embrittlement of materials, and conductive coats eliminate electrostatic charges. Whipple shields and reinforced buildings offer micrometeoroid and orbital debris protection and consequently indirectly protect electronics. Mass, protection is balanced with the mass in an effective shielding with the help of modeling tools to optimize the thickness and location. Combined, the shielding and the coatings are used to supplement radiation hardening and thermal management, to guarantee the long-term durability and stable functioning of the LEO environment which is quite challenging [15].

1.5.3 Thermal Control Methods

The thermal control ensures that electronic components do not operate above safe operating temperatures even during the extreme day-night cycles of LEO, which may be as cold as −150°C and as hot as +120°C. Passive techniques comprise of MLI, reflective coating, radiators, and heat pipes to control conduction and radiation heat transfer. High-power systems take care of accurate temperatures with active techniques, such as electric heaters, thermoelectric coolers, and pumped loops. Thermal modeling is used to maintain components within the limits of design without influencing much thermal stress. Specially designed material coefficients of thermal expansion, can be used to alleviate mechanical fatigue, and the combination of thermal and radiation shielding is used to stabilize the system. Through efficient thermal control, overheating, freezing, solder joint fatigue, and sensor drift is avoided and in the end, the electronics will work consistently and reliably throughout the mission life of the satellite.

1.5.4 Power Management and Fault Tolerance

The management of power in LEO guarantees the continuous power supply during sunlight and in time of the eclipse and reduces the incidences of failure due to radiation or due to factors referred to as load transients. Solar arrays can be used with battery storage to provide power, and the lithium-ion batteries are popular due to their high energy density. Power conditioning units are used to control the voltage, offer protection against surges, and maximum power point tracking (MPPT) is used to maximize efficiency. The fault-tolerant designs have redundant buses, current limiting, load shedding, and battery management systems to avoid thermal runaway or unplanned shutdowns. Solar cells and protection shields are radiation-hardened, thereby remaining efficient. The autonomous algorithms check the consumption of energy, assign power to vital subsystems and adjust to the changing environment conditions. LEO electronics can attain consistent and continuous operation even under unstable conditions and environmental stresses through the incorporation of redundancy, monitoring, and predictive management [16, 17].

1.5.5 Redundancy and Robust System Architecture

Mission continuity is guaranteed through redundancy and sound architecture such that when failures occur to spacecraft they do not have catastrophic effects on the spacecraft. Redundancy may be component, subsystem or system, active-standby or cold-standby. TMR and ECC guard against momentary radiations in the computation and memory. Presentation of local failures Local failures are not propagated through the modular system architecture. The autonomous recovery, fault detection and monitoring allow an anomaly to be corrected in real time. Graceful degradation enables non-critical subsystems to get shut down without effecting critical functions. Hardware can be reconfigured, e.g. logic based on FPGA or software-defined radios, which increases flexibility and life expectancy. Resilience is also increased by secure communication and fault-tolerant networking. The combination of redundancy and strong design principles makes it possible to build LEO systems, which withstand radiation, thermal, and operational conditions and are reliable over the long term to be effective and successful in their missions [18].

joints fatigue, connector loosening, components expansion mismatches, and material degradation. Long-term thermal cycles may also expose hidden problems in adhesives, surface finish, or composite constructions, which may not work in orbit.

Vibration and shock testing imparts the mechanical loads during the launch process. To detect loose loosely connected components, mechanical connections, or defects in assembly, shaker tables are used to expose the components to high-frequency vibrations and abrupt shocks. This measure is to make sure that sensitive electronics, sensors, and connectors remain intact as the dynamic launch environment takes place. Humidity and pressure tests can be added as well to ensure outgassing is a concern or degradation by moisture occurs before subjecting the components to vacuum. ESS is supplemented by pre and post stress functional and electrical testing. Voltage, current, signal integrity, and operational performance measurements are made to ensure that the hardware is still operating within design limits. The faults identified during ESS enable the engineers to apply corrective actions like rework, part replacement, or design changes, which minimizes the chances of early failures in the life cycle after being put in space. ESSS will speed up possible failure modes by subjecting the system to thermal, mechanical, and electrical stress tests, which, otherwise, would go unnoticed. This screening process is normally used with sensitive subsystems, such as power electronics, flight computers, communication modules and payload sensors. ESS is standardized as per MIL-STD and ECSS standards, so the results can be repeated, reliable and documented. Comprehensively, ESS is a proactive measure to make the spacecrafts more reliable, validate the design robustness, and give an assurance that the electronic systems have the capability to withstand the LEO environmental challenges without any mission critical failures.

1.6.3 Standards (MIL-STD, ECSS, NASA Guides)

The use of accepted space standards is essential in maintaining high quality, reliability, and safety in LEO electronics. MIL-STD-883 defines testing procedures of microelectronics which encompasses radiation, vibration, thermal cycling as well as electrical performance requirements. ECSS offers quality assurance guidelines, testing, and environmental compliance guidelines to missions in Europe. Electrical, mechanical, and environmental spacecraft hardware verification procedures are given in NASA standards including NASA-STD-8739 [24]. These standards determine the choice of components, screening procedures, radiation qualification, and documentation procedures. Interoperability, reduced design risk, and certification of launch approval are ensured by compliance. Using these guidelines, designers are able to methodically test electronics, use best practices and have a high degree of certainty that the systems will be capable of supporting mission needs as well as enduring the harsh LEO environment [25].

1.7 Recent Space Electronic Developments

The recent advances in space electronics have been fuelled by the growing need of economy, high performance, miniaturized systems with the capability of operating under severe conditions. Material, fabrication, and processing developments have allowed smaller satellites and CubeSats to do what used to only be performed by large spacecraft. The adoption of COTS components and close screening and hybrid radiation-tolerant methods has also spurred the development of satellites by making them both faster and cheaper. At the same time, the onboard data processing, autonomous control, and communication are changing due to trends in miniaturized electronics, low-power computing, and AI-assisted systems. These innovations favor LEO missions especially since they can deploy the constellations quickly, process the data in real-time, and improved scientific measurements. Large scale applications, including Earth observation networks and Internet-of-Space-Things and AI-based mission planning, are enabled as well by these developments. Here, the discussion of the main fields of innovation of the modern space electronics is presented.

1.7.1 Innovations in Small Satellites and CubeSat Electronics

One of the biggest emergent trends in space electronics is the small satellites, especially CubeSat. They have developed over the last 10 years to be more than mere educational and technology demonstration platforms and become full-fledged scientific, commercial and communications satellites. New technologies encompass a high degree of integrated electronics modules incorporating power, computing, and communication capabilities in very small form factors, usually less than 1U to 12U. Minimized size and weight features radiation-tolerant microcontrollers and miniaturized attitude control sensors, low-power, and high-efficiency solar arrays allow reliable operation in LEO despite the size and weight constraints. High-power management controllers with smart battery management and MPPT maximization of the energy efficiency and modular electronics buses ease the assembly and integration. Mission programs, such as NASA CubeSat Launch Initiative (CSLI), ESA QB50, and commercial constellations (Planet Labs), indicate that CubeSats can be useful in carrying out important scientific measurements and Earth observation activities. New materials, radiation shielding, and communication technology are also being tested in these platforms leading to the further development of compact or resilient space electronics [26, 27].

1.7.2 COTS Components Adopted in Space

Adoption of COTS is one of the key drivers of the recent advancements in space electronics. Historically, space systems were based on radiation-hardened components designed on a case-by-case basis, which were not only costly, slow, and in short supply but also difficult to develop. COTS devices, which were originally intended to be used on land are currently being modified to be used in the orbital environment by means of screening, derating, and system-wide mitigation techniques [28]. They have such benefits as low cost, short development times, high levels of availability, and exploitation of cutting-edge commercial semiconductor technology. Newer CubeSat and small satellite programs are increasingly based on COTS microprocessors, sensors, FPGAs, and communication modules, with protective coatings or shielding or error correction to reduce the threat of radiation. Examples are commercial imaging satellites and LEO constellations prototypes whereby hybrid designs are used, a combination of radiation tolerant COTS and hardened components are used to provide performance comparable to fully custom designs. Such a strategy has transformed space electronics by making it easier to enter the market and making it possible to quickly demonstrate new technology missions [29].

1.7.3 Novel Miniaturized and AI-Based Space Electronics

The current trends of miniaturized and AI-sensitive electronics are bringing changes in the way the satellites process data, control subsystems, and communicate autonomously with their surroundings. Miniaturization has also made it possible to have high-density circuits, low-power processors, and compact communications systems that can be included in very small satellites and still perform as well as or better than the larger platform they replaced in size. Real-time image processing, anomaly detection, predictive maintenance, autonomous navigation, and mission planning are also onboard AI applications [30]. As an example, it is possible to have AI algorithms on low-power radiation-tolerant FPGAs or system-on-chip (SoC) architectures to enable spacecraft to recognize features in Earth observation images, optimize communications plans, and dynamically change power or attitude control without ground control. These capabilities will minimize the latency, maximize mission efficiency, and also allow constellations of small satellites to work together. Combined with miniaturized sensors and edge computing, AI-assisted electronics is a paradigm shift, where LEO satellites are capable of performing complex scientific and operational operations without requiring much power or mass [31].

1.8 Conclusion and Future Directions

The LEO electronics have experienced a massive development, which is as a result of the increasing demand of reliable, compact and high-performance systems, which can work under extreme space conditions. This chapter has also identified the problem of radiation, atomic oxygen, thermal extremes, microgravity, and mechanical stresses and the design solutions that have been developed to reduce these problems as radiation hardening, shielding, thermal control and fault-tolerant power systems, and robust system architectures. Small satellites, CubeSats, COTS integration, and AI-assisted electronics are just some of the recent trends that show the transition towards inexpensive, small, and autonomous platforms. In the future, it is projected that there will be further miniaturization enhanced, more physics-tolerant semiconductors, adaptive AI-based onboard processing, and power-efficient power management. The ability of missions will continue to increase with the development of new technologies such as in-orbit reconfigurable electronics, novel thermal metamaterials, and constellations of satellites working together. These innovations combined will help LEO electronics to keep on increasing mission reliability, cutting down development costs and operating autonomously leading to new scientific, commercial, and exploration opportunities in space.

AI disclosure statement: The author confirms that no artificial intelligence (AI)-generated content or AI-assisted writing tools were used in the preparation of this chapter (including summaries and the abstract). All text, analysis, and conclusions are the result of the author's original work.

References

1. Reddy, Y.M., Raj, V.H., Thethi, H.P., Gupta, S., Maan, P., Ghani, R.H., Low Earth Orbit (LEO) satellite networks: A new era in global communication, in: *10th IEEE Uttar Pradesh Section Int. Conf. on Electrical, Electronics and Computer Engineering (UPCON)*, Gautam Buddha Nagar, India, pp. 1563–1568, 2023.
2. Crisp, N.H., Roberts, P.C.E., Livadiotti, S., Oiko, V.T.A., Edmondson, S., Haigh, S.J., *et al.*, The benefits of very low earth orbit for earth observation missions. *Prog. Aerosp. Sci.*, 117, 100619, 2020.
3. Gutiérrez, O., Prieto, M., Perales-Eceiza, A., Ravanbakhsh, A., Basile, M., Guzmán, D., Toward the use of electronic commercial off-the-shelf devices in space: Assessment of the true radiation environment in low earth orbit (LEO). *Electronics*, 12, 4058, 2023.
4. Baumann, R. and Kruckmeyer, K., *Radiation handbook for electronics*, Texas Instruments, Dallas, TX, USA, 2016.
5. Yue, P., *et al.*, Low earth orbit satellite security and reliability: Issues, solutions, and the road ahead. *IEEE Commun. Surv. Tutorials*, 25, 3, 1604–1652, 2023.
6. Pardini, C. and Anselmo, L., Evaluating the impact of space activities in low earth orbit. *Acta Astronaut.*, 184, 11–22, 2021.
7. Youn, S., Nam, U., Kim, S., Park, W.-K., Kwak, J., Sohn, J., *et al.*, Monitoring the radiation environment in high-latitude low-earth orbit during an ascending phase of the solar cycle (July 1, 2023 to July 31, 2024). *Acta Astronaut.*, 2025.
8. Jun, I., Garrett, H., Kim, W., Zheng, Y., Fung, S.F., Corti, C., *et al.*, A review on radiation environment pathways to impacts: Radiation effects, relevant empirical environment models, and future needs. *Adv. Space Res.*, 2024.
9. Adams, J.H., Silberberg, R., Tsao, C.H., Cosmic ray effects on microelectronics. *IEEE Trans. Nucl. Sci.*, 29, 1, 169–172, 1982.
10. Samwel, S., Hady, A.A., Mikhail, J.S., Ibrahim, M., Hanna, Y., Studying the total ionizing dose and displacement damage dose effects for various orbital trajectories, in: *Proc. First Middle East-Africa Regional IAU Meeting*, pp. 55–58, 2008.
11. Kocaman, B. and Harmandali, M., Total ionizing dose (TID) and displacement damage (DD) analyses for different space missions. *AIP Conf. Proc.*, 2700, 1, 020042, 2023.
12. Verker, R., Keren, E., Refaeli, N., Carmiel, Y., Bolker, A., David, D., *et al.*, Measurements of material erosion in space by atomic oxygen using the on-orbit material degradation detector. *Acta Astronaut.*, 211, 818–826, 2023.
13. El Hachimi, Y., Daghouri, A., Mahmoudi, H., Ouhammam, A., El Hani, S., A framework for sizing, designing, and testing of electrical power system for nanosatellites: A case study of a 3U university CubeSat. *e-Prime – Adv. Electr. Eng. Electron. Energy*, 12, 101018, 2025.

14. Poivey, C., Radiation hardness assurance for space systems, NASA Goddard Space Flight Center, 2002.
15. Finckenor, M.M., Materials for spacecraft, NASA Technical Memorandum NASA/TM-2003-212479, NASA Marshall Space Flight Center, 2003.
16. Shakoor, U., Alayedi, M., Elsayed, E.E., Comprehensive analysis of CubeSat electrical power systems for efficient energy management. *Discover Energy*, 5, 9, 2025.
17. Mejía Escobar, M.A. and Algora, C., Surveying the potential of flexible and high-specific-power photovoltaic assemblies and arrays for space applications. *Joule*, 9, 11, 102194, 2025.
18. Hogan, S.L., Effective fault management guidelines, Digital and Integrated Circuit Electronics Department, Electronics and Engineering Subdivision, 2009.
19. Gordo, P., Frederico, T., Melicio, R., Duzellier, S., Amorim, A., System for space materials evaluation in LEO environment. *Adv. Space Res.*, 66, 2, 307-320, 2020.
20. Kameda, T., Nagata, A., Kimura, Y., Imai, R., Shrestha, P., Kimura, K., *et al.*, Space environment evaluation and low-earth-orbit demonstration of a communication component with a commercial transceiver integrated circuit. *Aerospace*, 9, 280, 2022.
21. Torasso, A., Diram, R., Kowaltschek, S., Design and test results of a rad tolerant space qualified CVG, in: *DGON Inertial Sensors and Systems (ISS)*, Braunschweig, Germany, pp. 01-16, 2022.
22. Nealy, J.E., Cucinotta, F.A., Wilson, J.W., Badavi, F.F., Dachev, T.P., Tomov, B.T., *et al.*, Pre-engineering spaceflight validation of environmental models and the 2005 HZETRN simulation code. *Adv. Space Res.*, 40, 11, 1593-1610, 2007.
23. Yeritsyan, H.N., Sahakyan, A.A., Grigoryan, N.E., Harutyunyan, V.V., Arzumanyan, V.V., Tsakanov, V.M., *et al.*, Space low earth orbit environment simulator for ground testing materials and devices. *Acta Astronaut.*, 181, 594-601, 2021.
24. Dijks, J., de Jong, S., Menicucci, A., Akay, I., Systematic review of engineering and testing approaches for radiation hardness assurance in commercial space avionics. *Acta Astronaut.*, 238, A, 1354-1366, 2026.
25. Aydes, S. and Oznalbant, Z., Quality assurance and standards in space projects, in: *10th Int. Conf. on Recent Advances in Air and Space Technologies (RAST)*, Istanbul, Türkiye, pp. 1-6, 2023.
26. Bouzoukis, K.-P., Moraitis, G., Kostopoulos, V., Lappas, V., An overview of CubeSat missions and applications. *Aerospace*, 12, 550, 2025.
27. Sajjad, W., Shafique, A., Mahmood, R., Designing of reliable, low-power, and performance-efficient onboard computer architecture for CubeSats. *IEEE J. Miniatur. Air Space Syst.*, 5, 2, 59-72, 2024.
28. Brunetti, G., Campiti, G., Tagliente, M., Ciminelli, C., COTS devices for space missions in LEO. *IEEE Access*, 12, 76478-76514, 2024.

29. Cho, I.-C., Tan, C.M., Baruah, N.A., Shabir, A., Singh, H.T.T., Chao, T.-C., *et al.*, Degradation analysis of commercial off-the-shelf (COTS) electronics system under gamma radiation. *Results Eng.*, 25, 103802, 2025.
30. Carrasco-Casado, A., Shiratama, K., Kolev, D., Trinh, P.V., Ishola, F., Fuse, T., *et al.*, Development and space-qualification of a miniaturized CubeSat's 2-W EDFA for space laser communications. *Electronics*, 11, 2468, 2022.
31. Li, B. and Ne, R., A novel miniaturized dual-layer frequency selective surface. *AEU – Int. J. Electron. Commun.*, 130, 153580, 2021.

2

Hardening Electronics against Space Radiation: Strategies and Technologies

Manibharathi D.[1]*, Deepthi P.[1], Dhinakaran M.[1] and Mahalakshmi R.[2]

[1]*Department of Electronics and Communication Engineering, Government College of Engineering, Salem, Tamil Nadu, India*
[2]*Department of Electrical and Electronics Engineering, University College of Engineering, Nagercoil, Tamil Nadu, India*

Abstract

Deep space missions put electronic systems in harsh radiation conditions that can severely degrade the performance, cause faults, or cause outright system failure. The requirement of resilient, radiation-resistant electronics increases when space exploration missions come far enough to the point where spacecraft are no longer shielded by the magnetic field of the Earth. The chapter is a general overview of the mechanism with which radiation influences semiconductor devices, and they are total ionizing dose (TID), single-event effects (SEE), and displacement damage. It discusses the underlying design principles and the design strategies used to reduce these effects, both at the material level (innovation and design of circuits) and in the shielding of circuits and fault tolerance of the system as a whole. The focus is put on the traditional hardening methods and on the emerging technology, such as the use of commercial off-the-shelf (COTS) components in space projects. Using case studies and recent developments, the chapter describes the changing environment of space electronics, and provides some insight into future trends in building strong structures that can withstand the extreme radiations in deep space.

Keywords: Hardening electronics, total ionizing dose, space electronics, cosmic rays, radiation, shielding strategies

**Corresponding author*: manibharathi@gcesalem.edu.in

N. Sathish, Prasanth Aruchamy, Rajesh Kumar Dhanaraj, Nithya Rekha and Karthik Palani (eds.) Space Electronics for Deep Space Exploration, (23–52) © 2026 Scrivener Publishing LLC

2.1 Introduction

Electronics are at the heart of each contemporary space mission, either in the form of navigation computers in satellites or sensing components on scientific collections. The onboard electronics reliability is becoming a determining factor in the overall success of the mission as the spacecraft operations are getting more autonomous, and the missions are getting further into space. Space hardware is required to work in such a setting where it does not have natural protection of atmosphere and magnetic field like on earth. This puts electronic devices into the wide range of energetic particles, which can be used to degrade materials, corrupt logical states, and in the worst cases destroy components [1, 2].

This difficulty is further aggravated with the increasing reliance on large-scale semiconductor technologies. The integrated circuits of modern times have lower supply voltages, have smaller node geometries, and transistors are densely packed. Although these are useful in improving the speed and power efficiency of their devices, these architectural enhancements also expose devices to disruptions by radiations. This means that radiation-resistance design is no longer a requirement only for deep-space probes, but is also equally important to satellites in low-Earth orbit, communication constellations, navigation system and crewed spacecraft [3].

Radiation hardening can thus be considered a vital science in the space engineering industry. It contains a set of design philosophies, material selections, circuit methods and system-level approaches that interact to keep the system operational under the influence of deleterious radiation. It is not only to ensure that the individual parts are not damaged but also that the whole system will remain predictable through the entire mission life [2, 4]. To meet this objective, there should be an appreciation of radiation environment, an appreciation of how the energetic particles can interact with the semiconductor material, and an appreciation of both preventive and corrective options that can be offered to the designers.

This chapter summarizes the view of the strategies and technologies involved to harden the electronics against the space radiation. It starts by describing the radiation environment one would experience in various orbital regimes with a further explanation of the physical processes by which radiation interacts with electronic devices. Later sections discuss on the system and circuit level mitigation mechanisms, material and shielding development. The other qualification processes that were most common are also reviewed in the chapter, and real-life examples of how radiation-hardened electronics have been successfully utilized to accomplish a mission. The combination of these aspects seeks to provide the researchers, designers, and students with a technical basis of achieving strong electronics that can survive in the harsh environment of space.

2.2 Space Radiation Environment

Space electronic systems are used in an environment that is fundamentally different as compared to that of the surface of the earth. The lack of an atmospheric shield enables the spacecraft hardware to be hit by a large number of energetic particles. This knowledge is the initial move towards developing electronics that will survive a prolonged mission period when it comes to radiation. Despite the variation of the intensity and the composition of the radiation depending on the altitude, orbit, and the solar activity, the space environment can be well characterized under three broad categories, namely galactic cosmic rays, solar energetic particles (SEPs), and trapped radiation belts [5, 6] as represented in Figure 2.1.

2.2.1 Galactic Cosmic Rays

Galactic cosmic rays (GCR) are outside the solar system and are made up of mostly high-energy protons, alpha particles, and heavier ions. Their energies can reach GeV range and they can easily penetrate spacecraft shielding. GCR flux is quite steady with time but changes with the solar cycle; strong outbursts of solar activity suppress GCR intensity because of the increased strength of solar magnetic fields [6, 7]. Although the GCRs are not as common as the solar particles, they are of great concern since the charge that is deposited by a single heavy ion has enough power to make the electronic equipment malfunction or cause permanent damage.

2.2.2 Solar Energetic Particles

The solar flares and coronal mass ejections create SEPs. Such events cause the acceleration of protons and heavier ions to high energies comparable to cosmic rays in some cases. The bursts of SEP are rare and unforeseen, but once they have happened, the particle flux will soar several hours or days. SEP bursts can cause an unexpected stress on electronics on spacecraft particularly those on interplanetary missions or at high altitude, resulting in transient anomalies or in extreme cases, catastrophic failures [6, 7]. SEP statistics are usually considered in mission planning to calculate shielding requirements and onboard operating margins.

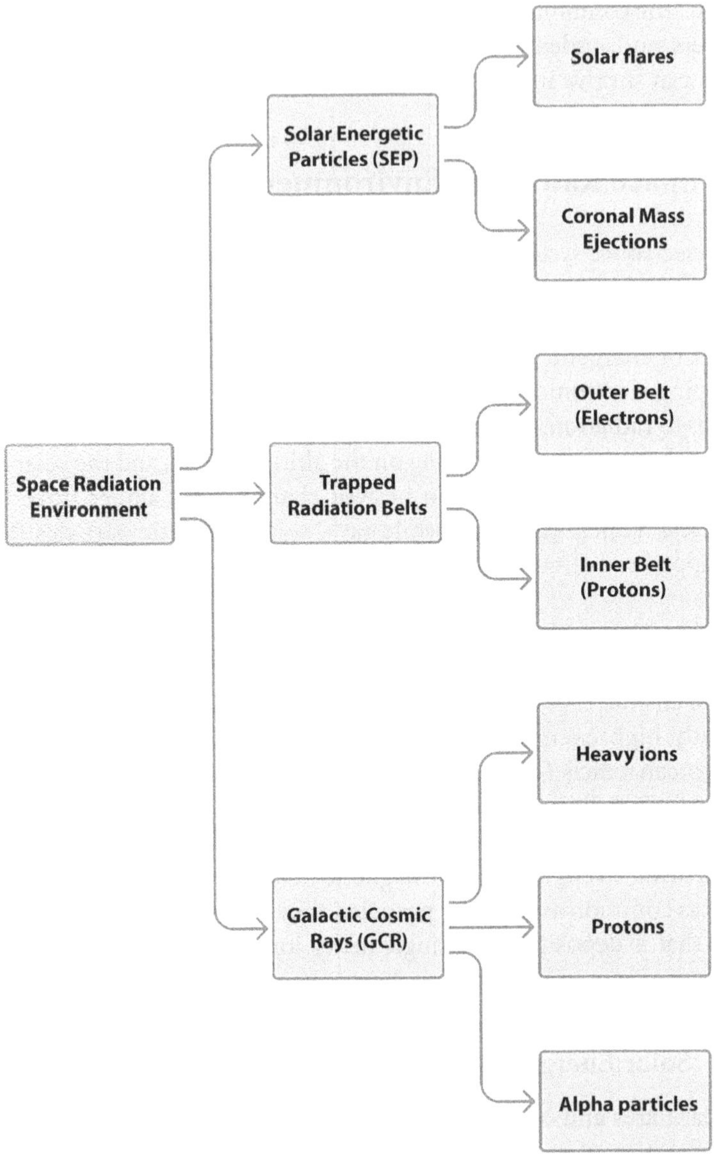

Figure 2.1 Classification of the space radiation environment affecting spacecraft electronics.

2.2.3 Trapped Radiation Belts

The Van Allen radiation belts are formed around the earth as the charged particles are trapped by the geomagnetic field. The protons are mostly located in the inner belt whereas the outer belt is dominated by electrons.

Satellites in a medium Earth orbit (MEO), geostationary orbit (GEO) or highly elliptical orbit often pass over high-density particle regions. Even the spacecrafts in low-Earth-orbit experience even more radiation on their way through the South Atlantic Anomaly [8], where the inner belt is nearest to the surface of the Earth. In the case of electronic systems, extended exposure in these areas may be an added factor to cumulative radiation dose.

2.2.4 Types of Particles and Energies

Protons and electrons, heavy ions, and secondary particles produced during the interaction with spacecraft structures are included in the radiative environment. The particles have varying interactions with materials [9, 10]. Protons and heavy ion may leave charge on their tracks or push atoms off the lattice or the high-energy electrons can produce bremsstrahlung photons that contribute to the total radiation burden. The energy spread, between a couple of keV and a couple of hundred MeV—or even GeV in the cosmic rays-case—renders shielding and prediction models complicated.

2.2.5 Mission Radiation Exposure Dependent on the Mission

Mission profile has a strong effect on radiation levels. Low-Earth orbit spacecraft are exposed to moderate levels of trapped radiation and comparatively lower levels of the cosmic ray. GEO missions or deep space missions are more affected by GCR and susceptible to solar events. The cumulative dose is also dependent on the duration of the mission; long planetary missions or multi-year communication satellites have to be able to sustain a significantly higher cumulative exposure. Therefore, orbital-specific radiation models like AE8/AP8, AE9/AP9/SPM, and CREME96 are used to determine probable flux and dose by designers [9, 10].

2.2.6 Radiation Transport and Shielding Concerns

When energetic particles are traveling through spacecraft material, they either become less energetic, spread out, or produce secondary particles. This knowledge of these interactions is necessary in the determination of the degree of penetration of radiation into electronic components. Radiation transport software, e.g., GEANT4 or FASTRAD, is used to determine shielding thickness and detect possible hotspots. Shielding is, however, limited by masses in spacecraft design [11]. This is why this involves a middle ground approach consisting of moderate shielding and device and system-level hardening.

2.3 Radiation

Radiation can cause damage to electronic equipment. When the electronic components are in space, they are subjected to particles that are energetic enough to change the property of materials and disorient electrical behaviour. Such interactions may result in the gradual degradation or the unexpected failure, depending on the type of the particle, energy, and the technology of the device. Radiation effects are typically divided into two major groups: cumulative effects and single-event effects (SEE) [12, 15] as given in Figure 2.2. These mechanisms need to be understood in order to choose the proper hardening strategies.

2.3.1 Total Ionizing Dose

This is the long-term charge accumulation on the insulating layers (oxides, interfaces etc.) used. The trapping of electrons and holes created by radiation in these areas cause the shifts in transistor threshold voltages, changes in leakage currents and timing margins. CMOS technologies can exhibit TID as a raised static power, reduced gain, or even functional failure once bias points have gone out of acceptable range. Interfaces Deep-submicron processes have a lower resistance (and thus thinner oxides) but interface traps may still be a problem [13, 50].

2.3.2 Damage Dose Displacement Dose

The displacement damage occurs when excited particles dislodge atoms in their lattice positions. This causes defects which serve as recombination centers or trap carriers, which decreases the lifetime of minority-carriers and influences photodiodes, solar cells and bipolar transistors. Devices that are used in imaging systems or communication links may lose sensitivity

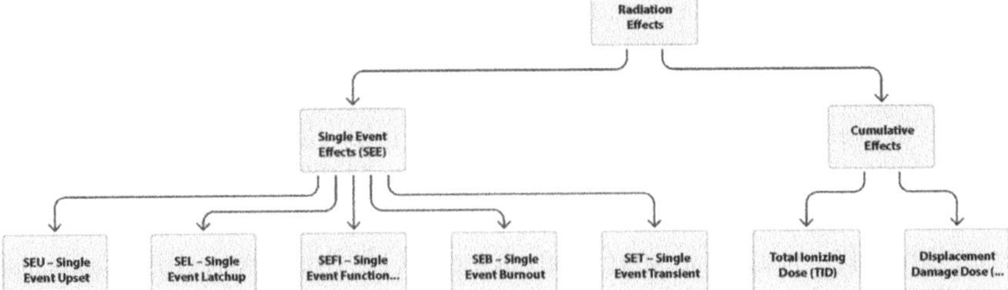

Figure 2.2 Classification of radiation effects in electronic devices.

or gain due to the build-up of displacement defects [14]. Since the magnitude of the damage varies depending on the type of particle and the material composition, it is normal to find that mission planners are focusing on DDD exposure as well as the ionizing dose.

2.3.3 Single-Event Effects

Single-event effects are the effects that are produced by the discharge of one charged particle into a device which produces localized charge. Their effects ranges from benign transient errors to irreparable damage [12, 16].

I. Single-Event Upset
A single-event upset (SEU) is a logic or memory bit flip that occurs due to transient charges collection. Although the equipment is not damaged physically, the corrupted data may pass into the system without being fixed.
II. Single-Event Latchup
Parasitic thyristor structure can be excited by an energetic ion in CMOS circuits into a low-impedance connection between supply rails. Single-event latchup (SEL) events may cause excessive current and may also permanently destroy the device unless the power is switched off fast.
III. Single-Event Burnout
MOSFETs or IGBTs can also fail when subjected to a heavy voltage and a high level of energy ions. single-event burn (SEB) is a vital issue in power conditioning units of spacecraft.
IV. Single-Event Transient
These are brief spikes of voltages on an analog or digital path. Even though short, single-event transients (SETs) can cause timing disturbance or produce false control circuit triggers.
V. Single-Event Functional Interrupt
A single-event functional interrupt (SEFI) is the occurrence of a particle strike that interrupts a complete functional unit, e.g., an ADC, PLL, or configuration controller. Recovery typically requires resetting or reinitializing the affected module [16, 17].

2.3.4 Cumulative vs. Prompt Effects

Gradually developing cumulative effects (e.g., TID and DDD) can be estimated using mission dose estimates. Conversely, SEE are probabilistic phenomena and need to be stochastic. These two forms of effects have to be looked into jointly because a device that is resistant to prolonged contacts with ionizing doses might still be susceptible to a single energetic ion [18, 19].

2.3.5 Impact of Technology Scaling

The scale effect of technology is identified based on scaling factors which are directly connected on the economies of scale on the technology market. The development of semiconductor scaling has resulted in smaller geometries, lower operating voltages and greater densities. The trends tend to decrease stored charge in nodes, which makes certain technologies less susceptible to charge-induced upsets. Scaling, however, also brings its own sensitivities: thinner oxides have higher chances of interface damage, and tight structures create the probability of multiple-bit upsets due to a single event [18, 19]. New types of transistors, including FinFETs, SOI devices, and 3D architectures, have varied radiation characteristics, and have to be characterized separately.

2.4 Radiation Hardening Methods—System Level

System-level hardening concentrates on structural and operational methods that guarantee the trustworthy operation even in the occurrence of disturbances by radiations on respective parts. These approaches offer resilience to firmware, architecture, redundancy, and mission planning instead of depending on dedicated hardware only. They are particularly useful with complicated spacecraft were developing all parts to be radiation hard through manufacture of radiation-hardened processes might be unfeasible because of cost, performance, or mass penalties [20, 21].

2.4.1 Architecture-Level Fault Tolerance

Fault-tolerant architecture remains operating correctly when some of the subsystems possess temporary or permanent faults. Layered fault-handling mechanisms, such as watchdog timers, periodic integrity checks, and modular isolation, are common in spacecraft computing platforms. Critical functions can be designed in blocks and thus in case one of the paths fails then the failure does not spread to other parts of the system [22]. The methodology decreases the chance of a one-event upset or latchup expanding to a system failure.

2.4.2 Redundancy Strategies

Redundancy is also among the most common measures of hardening the system level. The principle is simple: copy or duplicate important modules to ensure that in the event one of them suffers, the rest will be able to replace them.

I. Triple Modular Redundancy
Triple modular redundancy (TMR) follows three identical processing or logic paths the outputs of which are constantly compared by a majority voter. A failure in one of the paths, e.g. an SEU is covered by a coincidence of two paths. TMR finds common application in flight computers, FPGAs, and control loops that are involved with software that needs to be run continuously.
II. Dual Modular Redundancy
Dual modular redundancy (DMR) uses two copies of a module and identifies inconsistency between them. DMR cannot fix any mistake on its own, but it can give early alerts of something wrong, allowing corrective action to be taken by resets or switching to safe-mode.
III. N-Modular and Hybrid Redundancy
To trade off power consumption against reliability, large spacecraft can use a combination of various types of redundancy such as cold spares, warm spares, and dissimilar redundancy. Dissimilar redundancy is where the same function is implemented with dissimilar hardware or software to reduce the chance of common-mode failures [23, 24].

2.4.3 Error Detection, Correction, and Mitigation Algorithms

Bit flips due to radiation may frequently be countered algorithmic means. Error-correcting codes (ECC) like Hamming, BCH, or Reed-Solomon codes, can be used to guard the memory and data transmission channels against a single-bit error, and also against multi-bit errors. Data integrity is also increased with the help of cyclic redundancy checks (CRC) and checksums in regard to packets and tele commands [25]. Software systems can minimize the effect of transient errors on the flow of computations through the use of techniques like temporal filtering and instruction duplication. Figure 2.3 shows the flow of Error detection and correction for radiation-induced faults.

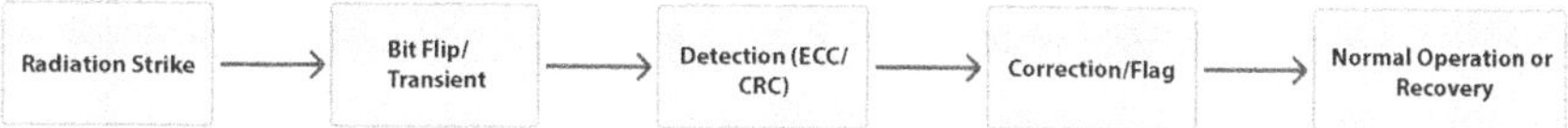

Figure 2.3 Error detection and correction flow for radiation-induced faults.

2.4.4 Operational Margins and System Derating

Derating is the process of running the components with lower electrical, thermal, and environmental values than the maximum ones. The radiation effects on components can be mitigated by reducing the supply voltages, temperatures, or currents. Operational margins are also designed into spacecrafts, and the worst-case radiation conditions are estimated, and enough performance headroom is built. This will reduce the likelihood that the parameters changed by TID will turn functional at the end of a mission.

2.4.5 Radiation-Aware Design Flow

A radiation tolerant system needs to coordinate hardware, software, and mission planning decisions. The simulation frameworks enable the engineers to plan the anticipated dose of the radiation and assess the potential reaction of various subsystems [20]. The consideration of radiation at early stages of the design flow will lead to an increased number of redesigns at the late stage and allow optimal combination of shielding, the choice of devices, and redundancy. More and more interdisciplinary designing groups incorporate mechanical, electrical, and software views in order to reach a balanced solution.

2.5 Radiation Hardening Techniques—Circuit and Device and Level

Although system level methods improve the general resilience, fundamental resilience of electronics to radiation starts at the device and circuit level. The techniques in this category include changing the structure of transistors, the material used in devices, circuit topology, and layout styles in order to suppress charge collection, unwanted conduction paths, and reduced sensitivity to cumulative damage. These are done such that even minor building blocks of an integrated circuit will act in a predictable manner in a radiation-immersed environment [25].

2.5.1 Hardened-by-Design Techniques

Hardened-by-design (HBD) techniques are techniques that modify the circuit or layout to lower the vulnerability, without a special manufacturing process. Their application is common since they strike equilibrium between performance, area, as well as compatibility with commercial fabrication lines [26].

I. Enclosed Layout Transistors

The enclosed layout transistor (ELT) is a geometric modification in which the gate is in the form of a closed loop around the active area. This form removes the traditional drain-edge at which the charge build up may cause leaked paths under ionizing radiation. ELTs can be particularly useful in analog circuits and sensor interfaces where leakage stability is an issue over mission times.

II. Guard Rings

Guard rings are implanted areas surrounding vulnerable devices in order to gather charge created by ion strikes prior to it spreading to active transistor nodes. Guard rings suppress parasitic conduction by offering a controlled route of charge dissipation, with the effect of lowering the chance of latching up or transient disruption.

III. Transistor Sizing and Node Capacitance

Increasing the size of devices or the capacitance required to change the state of a logic device increases the charge necessary to change a logic state. Even though this method increases the area used, and might decrease the switching speed, it is effective at decreasing the SEU rates in digital circuits.

IV. Current Limiting and Latchup Prevention

Current-limiting resistors can be used in circuits, the power rails segmented or protective clamps can also be used to limit the current surge during a latchup event. These steps reduce thermal stress and provide system controllers with time to start power cycling in case it is needed [26, 27].

2.5.2 Hardened-by-Process Techniques

Hardened-by-process (HBP) methods are based on changes to the process of semiconductor. Such techniques usually have high radiation resistance, but must be fabricated in special radiation-hardened foundries [30].

I. Silicon-on-Insulator Technologies

Silicon-on-insulator (SOI) isolate devices with a buried oxide layer, which has a great effect on the reduction of charge collection volume. Such isolation eliminates latch up and reduces SEEs. Fully Depleted SOI (FD-SOI) technology further extends the resistance by reducing parasitic conduction.

II. Doping and Isolation Engineering
Doping profiles, wells with a deeper depth and more isolation trenches are some of the techniques used to contain charge generated during particle strikes. These processes are typical in rad-hard CMOS processes, and they are applied to preserve the stability of the threshold in the presence of ionizing radiation.
III. Thick Oxides and Radiation Tolerant Materials
Thicker gate oxides are used in certain applications to enhance TID tolerance, or in other cases, to use other dielectric materials. Though thicker oxides slow down the speed of devices, they are reliable in applications of analog and power devices where they are exposed to high radiation [30].

2.5.3 Radiation Response of Emerging Devices

Other new architecture devices include FinFETs, multi-gate transistors, and 3D-stacked devices, which have various behaviors when exposed to radiation. FinFETs, such as these, have lower charge collection rates as their narrow channels and better electrostatic control, but with a multi-fin structure make them sensitive to multi-node upsets. Three-dimensional ICs have further issues of inter-layer interactions and thermal control, but have the advantage of interconnect reduction and lower node capacitance [27–29].

2.5.4 Layout-Level Hardening

Radiation susceptibility is highly dependent on the physical layout of transistor and interconnects. Techniques include:

- Spatial isolation of overlapping nodes in order to prevent overlapping disturbances by the same particle.
- Minimization of charge diffusion by a shallow trench isolation.
- Metal routing techniques of minimizing interconnection among sensitive regions and aggressor nodes.

These measures can be used together with circuit-level redundancy and enhance robustness without modifying devices fabrication.

2.5.5 Modeling and Simulation Tools

The workings of the devices when they are subject to radiation are predictable by using a framework of simulations that include particle interactions and charge transport. Technology computer-aided design (TCAD) is able to simulate ion impacts at transistor granularity, and offers information on charge deposition, transient currents, and breakages. TCAD is used alongside circuit-level simulations to verify the effectiveness of hardening strategies to be used prior to fabrication by designers [31]. This will reduce the number of iterations of prototypes and reduce the time spent in the development of space-qualified electronics.

2.6 Radiation Hardening Materials and Technologies

The radiation tolerance does not depend only on the design of the circuit, but the selection of the materials and technology of the devices. Materials affect the semiconductor lattice interaction of energetic particles and charge gathering, as well as structural flaw building-up over time. Mission to higher-radiation orbit or deep space destinations has made material engineering a key foundation of hardened electronics development [32]. In this part, material systems and technological platforms, which can be critical in radiation resilience, are presented.

2.6.1 SOI Technologies

The use of SOI-based electronics has been a new choice in many space applications due to its intrinsic isolation. With SOI wafers, the substrate is separated by an oxide layer between the active silicon layer and the substrate. The architecture greatly restricts the amount of charge that can be accumulated in a strike by a particle. Consequently, SOI devices are less affected by SEEs and intrinsically immune to latchup since the parasitic thyristor path is effectively switched off. Fully depleted SOI (FD-SOI) devices further reduce floating-body effects as well as enhance the stability of threshold voltage with TID [33, 34].

2.6.2 Compounds Semiconductors: Gallium Nitrate and Silicon Carbide

Gallium nitrate (GaN) and silicon carbide (SiC) are wide-bandgap materials, which are characterized by their robustness in high-temperature and high-electric field environments. These properties have various benefits to radiation spaces.

- GaN devices, in particular, high-electron-mobility transistors (HEMTs) are highly resilient to displacement damage and can be operated with high performance even after being subjected to high levels of protons.
- SiC devices exhibit low leakage current variations under ionizing dose and are well suited for high-power conditioning stages onboard spacecraft.

Since the crystalline bonds of these materials are stronger than that of silicon, they tend to be less susceptible to the effects of the radiation. They are currently finding more applications in power modules, RF amplifiers and propulsion control electronics [35, 36].

2.6.3 Advanced Composite Shielding Materials

Although semiconductor materials define the hardness of devices, the structural materials of the surrounding also affect the radiation levels. Best practices in modern spacecrafts have involved the use of lightweight composites including aluminum-lithium alloys, carbon-based laminates and hydrogen-rich polymers which enhance shielding efficiency without unnecessary additional mass. Hydrogen containing materials, especially, are efficient against protons and aid in curtailing the formation of second-generation particles by heavy ions. Currently studies being done on layered shields and graded-Z structures, the materials with different atomic numbers are combined to provide optimum absorption and scattering [37].

2.6.4 Space Grade Electronics Novel Materials

New materials are under investigation to drive the radiation resilience boundaries:

- Although they are in the formative stages, diamond-based electronics boast of exceptional thermal conductivity and high ionising radiation resistance.
- Synthetic two-dimensional materials such as graphene can be used as radiation tolerant because of the low density of defects and the special properties of charge transport,

but the fabrication of devices on the basis of such materials is complicated.

- Amorphous oxide semiconductors, which are found mainly in sensor interfaces and display electronics demonstrate stable performance with ionizing dose due to an ionizing dose-independent conduction mechanism.

These materials can be used to supplement the traditional silicon technologies in future missions with extreme needs of durability [38, 39].

2.6.5 Radiation/Radiation Resilience Packaging Technologies

The aspect of packaging is not given much consideration; however, it has a significant impact on the space reliability of devices. Radiation has the capability to change the properties of encapsulants, solder joints and interconnects, creating mechanical or electrical degradation. To address this:

- To reduce outgassing and to protect against the effects of contamination and humidity, and this is achieved with the help of ceramic and hermetic packages.
- Low-Z encapsulants decrease the process of secondary particle creation upon impact with the high-energy ions.
- Underfills and adhesives are radiation-tolerant making them stable with cumulative dose.

Such sophisticated packaging strategies as flip-chip bonding and 3D stacking are becoming increasingly significant as well. As much as these architectures have performance benefits, there should be a consideration of the charge spreading between layers and thermal behavior during exposure to radiations [40].

2.7 Shielding Strategies

The most direct and old method of covering the radiation that is received by sensitive electronics is shielding. Though by itself, the shielding cannot prevent all radiation, particularly the very high-energy cosmic ion, but is significant in reducing the cumulative dose, and in reducing effects of lower-energy particles. Shielding design should be able to compromise between mass constraints, mechanical capability, thermal capability and integration with other spacecraft subsystems. New approaches, in the present day, involve a combination of material science, geometry optimization, and the new active protection [41].

2.7.1 Passive Shielding

Passive shielding is based on the use of physical materials that absorb or diffract the incident radiation. Aluminum has traditionally been the preferred material of spacecrafts because it has an appropriate strength to weight ratio and extended flight record. Nevertheless, aluminum is capable of forming secondary particles when struck by high-energy ions, and this can limit its usefulness in certain missions.

Other passive shielding materials available are:

- The polymers that are effective against protons based on the fact that the hydrogen atoms decrease the formation of secondary particles are polyethylene and other hydrogen-rich polymers.
- Multi-layers laminates. In multi-layers, the layers of various atomic numbers allow the charging particles as well as the secondary radiation to be moderated.
- Graded-Z shields, that use successive layers of grading materials with low-Z to high-Z to gradually decelerate the particles and minimize the products of spallation.

Passive shielding works best in the low-Earth orbit, or in the conditions where the proton and electron flux are predominating [42–44].

2.7.2 Active Shielding Concepts

Active shielding describes the technologies that involve the use of plasma barriers or electromagnetic field to intercept and decelerate the incoming charged particles. These strategies are still in experimental or conceptual phase, but they are being investigated as the passive shielding in long-duration deep-space missions where the mass constraints would make passive shielding impractical.

There are two concepts under investigation:

- In a magnetic shielding, high-current or superconducting coils are used to generate magnetic fields that repel charged particles.

- The method of using grid structures or charged surfaces to shield ions by repelling them with electric fields is called electrostatic shielding.

The active shielding technologies have engineering issues, including electricity requirements, the weight of the supporting equipment, and the inability to keep the fields stable in space. However, they are good long-term prospects of high-radiation missions [46].

2.7.3 Hybrid Shielding Approaches

Hybrid shielding involves the combination of both passive and active to bring higher efficiency. An example is that lightweight polymers can be used as outer coating of a spacecraft and the active devices inside can be used to further safeguard the spacecraft in the event of intense sunflare. The other hybrid solution is to take structural elements and add shielding functionality- such as fuel or water tanks as radiation-absorbing layers to essential electronics.

There are also other concepts of hybrids:

- Deployable shields, which are extendable at the high radiation times.
- Variable configuration shields in which moveable components are shifted to align with the direction and intensity of radiation.

The strategies are trying to maximize the mass with the ability to offer flexible protection throughout various stages of the missions [45].

2.7.4 Optimization and Lightweight Spacecraft Materials

The design of the modern spacecraft is becoming more and more dominated by the use of composite materials in order to decrease the mass of the spacecraft and maintain its structural integrity. A lot of such materials that include carbon fiber reinforced polymers (CFRP), aluminum-lithium alloys, etc. provide competent performance and tolerable radiation attenuation when intelligently incorporated into the spacecraft structure.

The optimization of shielding is frequently based on the simulation tools of particle transport at complex geometries. Engineers can also consider radiation exposure significantly by changing wall thickness, material distribution and electronic module location without large mass addition [46, 47]. It should also be placed where it is needed; such as putting electronics behind fuel tanks, battery assemblies or other naturally massive objects can offer a lot of shielding without the added mass.

2.8 Testing and Qualification Standards

Radiation hardening can only be successful when it is reinforced by intensive testing and qualification. As the true space radiation environment cannot be perfectly simulated on the ground, engineers use special ground-based laboratory equipment and standardized test procedures to test the behavior of equipment under ionizing dose, displacement damage, and SEEs. These tests help in giving a certainty that the electronic components will be able to endure the amount of radiation that will be present in the target mission profile. These evaluations are guided by standards given by organizations like NASA, ESA, and the U.S. Department of Defense which make the space industry consistent [48].

2.8.1 Ground-Based Testing Facilities

Globally, there are numerous accelerator laboratories and test centers which aid in the radiation qualification:

- Proton accelerators, which are used to recreate trapped belt protons and solar particle conditions.
- Heavy-ion accelerators, which mimic the action of galactic cosmic rays and high linear energy transfer (LET) events that result in severe single-event actions.
- Electron irradiators are used in measuring displacement and dose effects in materials and semiconductor structures.
- As a measure of the effects of TID, gamma irradiators, usually with cobalt-60 sources.

Devices can be subjected to a series of tests to get a variety of interactions of particles depending on the needs of the mission [49, 50]. Characterization of pre-irradiation, incremental exposure and post-irradiation are often part of test campaigns.

2.8.2 Standards and Qualification Guidelines

A number of agencies have elaborate guidelines outlining the conditions of tests, dose rate, reporting forms, and acceptability requirements:

- MIL-STD-883: This is commonly used in the microelectronic equipment, including test methods of TID and latchup, as well as test methods of other radiation phenomena.
- MIL-STD-750: Specializes in power semiconductor devices, the burning out and avalanche mode.
- NASA EEE-INST-002: Gives the instructions on the choice, screening, and qualification of electronic components to be used in space missions.
- ECSS-Q-ST-60-15C (European Cooperation Space Standardization): Instills procedures of radiation assessment of elements in ESA missions.
- The JEDEC JESD57 specifying techniques of SEE characterization.

These standards enable reproducibility and enable the comparison of results of other laboratories with high reliability [51–53].

2.8.3 Accelerated Radiation Testing

Since the mission can last years or decades, it is necessary to test it faster. Greater dose rates are employed to take the anticipated lifetime exposure in manageable duration. Nonetheless, the acceleration should be done very cautiously, excessively high dose rates may result in non-representative degradation processes. Engineers also test dose-rate dependence and annealing effects so that accelerated results can be predictable to be as accurate as possible in a real mission situation [54].

Single-event effects test typically utilizes a variety of ions, each of a different LET, which enables investigators to identify threshold LET and cross-section curves. These measures contribute to the forecasting of the rates of the intended orbit being upset with the assistance of the environmental models.

2.8.4 Data Analysis and Qualification Criteria

The analysis of the test results is crucial as the test is. Engineers analyze:

- Ionizing dose parameter drift,
- Leakage current changes,
- Functional disruptions,
- Dependence on LET cross-section,
- Worst-case operating post irradiation margins.

The qualification normally demands that all parameters should be within acceptable limits throughout the entire mission dose with an allowance towards uncertainties. In the case of critical subsystems, e.g. power control units, flight computers as well as communication links, redundancy and derating is included on the basis of testing results.

Reliability modeling tools also take input of test data and predict the likelihood of failures during the mission life. These studies influence selection of the components, defensive decisions and fault-tolerant strategies at the system level [55].

2.9 Cases and Practical Applications

The radiation hardened electronics have developed over a period of decades under the lessons of work in operating spacecraft. Investigating actual applications can also be useful in demonstrating how the principles in the previous sections can be applied to practice within the engineering environment. This part indicates the representative processors, subsystems on spacecrafts, and experiences in missions, which have influenced the present-day solutions in radiation hardening.

2.9.1 Radiation-Hardened Processors

I. RAD750 Processor

BAE Systems has developed the RAD750, the most popular radiation-hardened processor used in both scientific and deep-space missions. It is based on the PowerPC 750 architecture, and it adds the hardened-by-process fabrication, enclosed-layout transistors, and robust error-correcting memory systems to it. Its tolerance has been shown to have a number of hundred kilorads of the TID and ability to withstand a single event. It has been used in space missions like NASA Mars Reconnaissance Orbiter and Juno spacecraft which have offered long term reliability with high radiation exposure [56, 57].

II. RHBD Processors and Microcontrollers

Radiation-hardened processors designed using radiation-hardened by design (RHBD) have become popular because they are cheaper and offer better performance. Such architectures are based on circuit-level forms of non-specialised fabrics, including redundancy in registers, hardened flip-flops and current-limiting structures. They show that with very common process technologies, reasonable radiation tolerance can be obtained through attention to design [58].

2.9.2 Radiation-Tolerant FPGAs

FPGAs are becoming popular in the spacecraft avionics due to their flexibility. Radiation-tolerant families include radiation-tolerant devices like Microchip RTG4 and Xilinx space-grade devices:

- Hard disk configuration memory,
- Triple-module redundant fabric alternatives,
- Flash-based or antifuse architectures Flash-based and antifuse Latchup-immune technologies.

These gadgets enable designers to incorporate tailored digital logic with dependability that can be used in long-term missions [59, 60].

2.9.3 Spacecraft and Subsystems Using Radiation-Hardened Electronics

I. Satellites in Geostationary Orbit

The electron flux on communication satellites in GEO is steady. Their payload control units, transponders and power system utilize a combination of hardened semiconductors, shielded package and system level correction of error. With this combination, service lifetimes of more than 15 years are possible and 24/7 operational.

II. Science Mission in High-Radiation Space

Van Allen Probes, Parker Solar Probe, and Europa Clipper missions are missions that work in extremely hostile radiation environments. These spacecrafts contain high-reliability components that are checked by an extended radiation screening. As an example, the electronics of the Europa Clipper will have to stand orders of magnitude greater doses than the normal Earth-orbiting satellites. The avionics system is designed to combine shielding, redundancy and hardened elements, which are specific to the extreme radiation belts of Jupiter.

III. Small Satellites and Commercial Missions

Radiation-tolerant methods are being used even in small satellites and commercial constellation systems, traditionally limited in cost. They can not necessarily operate with all rad-hard components, but system-level precautions, including error correction, watchdog systems and selective shielding, can allow it to operate reliably in low-Earth orbit. Their success proves that radiation conscious design may be achieved within a very wide range of mission budgets [61, 62].

2.9.4 Lessons Learned on Mission Failures

Radiation effects have been associated with failures which are instructive:

- A number of early spacecrafts have suffered power outages due to latchup before latchup protection was a normal practice.
- Single-event upsets which cause memory corruption have resulted in unexpected reboots, improper execution of commands or loss of data on orbit.
- Solar particle storms have resulted in burnout of high-power components (e.g. MOSFETs and IGBTs).

These events underscore the importance of thorough testing, sufficient shielding and layered mitigation. More to the point, they show that one technique will never work with radiation effects; effective designs are characterised by smart selection of materials, circuit hardening, resilience at the architecture level, and operational protection [63].

2.10 Radiation-Hardened Electronics Future Trends

The increasing numbers of new generations of radiation-resistant electronics are necessitated by progress in the field of space exploration, longer mission durations, and growing autonomy in spacecraft. The conventional hardening methods albeit inseparable could not be entirely satisfactory to the future missions to the moon, Mars, asteroids, and deep space conditions. With the future developments in semiconductor technologies, researchers are looking into new paths that integrate material innovation, intelligent design and new computing paradigms. The following section describes the emerging trends in the field of radiation-hardened electronics that are likely to continue defining the coming decade [65].

2.10.1 Self-Healing Systems and Fault Prediction with the Help of AI

Machine learning enables novel opportunities of real-time surveillance and prediction of anomalies caused by radiations. Through the examination of telemetry streams, sensor outputs and internal state transitions, AI algorithms are capable of detecting small patterns that foretell the presence of faults [64, 65]. These prediction models allow:

- Preemptive identification of degradation,
- Adjustment of operating points, and
- Recovery or reconfiguration of an automated fashion prior to a failure.

In the avionics of the future, self-healing architectures might be implemented, in which hardware modules automatically isolate faulty areas and reroute capability, thereby enhancing long-term reliability with no human assistance.

2.10.2 Adaptable and Reconfigurable Electronics

Radiation-tolerant FPGAs and other reconfigurable systems are likely to become more important in the next-generation spacecraft. These devices permit:

- On-orbit controls to counteract aging or drift due to TID,
- Reconfiguring of logic resources on-demand, and
- Installation of new capabilities on longer missions.

Dynamic tunable analog interfaces, adaptive power converters, reconfigurable sensor electronics, etc. will also give more flexibility to the missions. These systems convert the more traditional radiation hardening process to a more mission and dynamic process [66, 67].

2.10.3 Quantum and Cryogenic Systems

Scientific missions and deep-space communications are of interest in quantum sensors, superconducting circuits, and cryogenic electronics. Their radiation behavior is not similar to other conventional technologies:

- Superconducting devices may be resistant to displacement damage, but vulnerable to magnetic disturbances.
- The quantum communication items are based on the concept of coherence in photons, which requires the development of new shielding and fault repair schemes.
- Cryogenic CMOS studied to infrared astronomy can have different TID and SEE behaviors because the carrier behavior in low temperature is different.

Even in its infancy, these technologies will need to be rethought in terms of hardening methods in the age of maturity [68].

2.10.4 Material Engineering and 3D Integration

Material engineering in electronics Material engineering will play a major role in the production of radiation-tolerant electronics in the future:

- High-power and high-temperature applications are likely to switch to wide-bandgap semiconductors like GaN, SiC, and diamond.
- Two-dimensional materials can potentially be used in sensors and power-minimal circuits with a high radiation resistance *in vitro*.
- 3D-stacked ICs have potential benefits of compact and more performance electronics, but generate new charge spreading, thermal gradient, and inter-layer single-event issues.

Study is taking place on the customization of material interfaces, optimization of the layer thicknesses and enhancement of through-silicon *via* (TSV) designs to sustain high levels of radiation [69, 70].

2.10.5 Co-Design and Digital Twins on a System Level

The growing sophistication of spacecrafts demands systems thinking designs in which electronics, mechanical structure, thermal, and shielding are designed in a co-optimized manner. Virtual replicas of spacecraft hardware models (so-called digital-twin models) allow engineers to model the effects of radiation transport, device degradation, and system-wide behavior over space missions (long mission) on a virtual platform [71, 72]. This combined method enables the identification of vulnerabilities earlier and more effective trade-off between mass, power, and reliability constraints.

2.11 Conclusion

Radiation hardening is a vital facilitator to dependable space electronics. The interaction of energetic particles with semiconductor devices is a very important design constraint as spacecrafts venture into more demanding parts of the solar system and begin to operate over extended periods of time. The chapter has discussed the environment of space radiations, the physical processes governing the behavior of electronic components, and the wide range of methods of addressing these issues.

Radiation hardening needs a multi-layered solution in terms of system-level redundancy and algorithmic fault corrections, device-level adjustments, material engineering, and shielding optimization. There is no one solution to this question that will provide absolute immunity, but a well-thought algorithm of measures can be taken to guarantee a guaranteed performance even in the conditions of severe radiation exposure. The experience of space missions in the real world has shown that these methods, properly implemented, allow very reliable performance during the mission periods in the years.

In the future, the future of radiation-tolerant design will be likely to be transformed by such technologies as adaptive electronics, novel semiconductor materials, AI-based diagnostics, and 3D integration. They can enable future spacecrafts to do even better with all the performance being supported by the rigorous testing and the qualification requirements that they provide, although they retain the robust nature required of spacecrafts.

After all, radiation hardening is not a discipline that stands the test of time; it changes every time there is a new development in semiconductor technology and every new mission need. Further interdisciplinary studies, including materials science, circuit design, systems engineering, and space physics, are the only way to ensure that electronics can be relied upon to work in one of the most hostile environments on the planet.

AI disclosure statement: The author confirms that no artificial intelligence (AI)-generated content or AI-assisted writing tools were used in the preparation of this chapter (including summaries and the abstract). All text, analysis, and conclusions are the result of the author's original work.

References

1. Wertz, J.R., Everett, D.F., Puschell, J.J., *Space Mission Engineering: The New SMAD*, Microcosm Press, Hawthorne, CA, 2011.
2. Johnston, A.H., Radiation effects in advanced microelectronics technologies. *IEEE Trans. Nucl. Sci.*, 45, 1339–1354, 1998.
3. Ferlet-Cavrois, V., Massengill, L.W., Gouker, P., Single-event transients in digital CMOS—A review. *IEEE Trans. Nucl. Sci.*, 60, 1767–1790, 2013.
4. Messenger, G.C. and Ash, M.S., *The Effects of Radiation on Electronic Systems*, Springer, New York, 1997.
5. Xapsos, M.A., *et al.*, Space radiation environment models. *IEEE Trans. Nucl. Sci.*, 54, 2294–2307, 2007.
6. Simpson, J.A., Elemental and isotopic composition of the galactic cosmic rays. *Annu. Rev. Nucl. Part. Sci.*, 33, 323–382, 1983.
7. Reames, D.V., Particle acceleration at the Sun and in the heliosphere. *Space Sci. Rev.*, 90, 413–491, 1999.
8. Van Allen, J.A. and Frank, L.A., Radiation around the Earth to a radial distance of 107,400 km. *Nature*, 183, 430–434, 1959.
9. Ginet, G.P., *et al.*, AE9, AP9, and SPM: New models for specifying the trapped energetic particle and space plasma environment. *Space Sci. Rev.*, 179, 579–615, 2013.
10. Tylka, A.J., *et al.*, CREME96: A revision of the cosmic ray effects on micro-electronics code. *IEEE Trans. Nucl. Sci.*, 44, 2150–2160, 1997.
11. Agostinelli, S., *et al.*, GEANT4—a simulation toolkit. *Nucl. Instrum. Methods Phys. Res. A*, 506, 250–303, 2003.
12. Holmes-Siedle, A. and Adams, L., *Handbook of Radiation Effects*, Oxford University Press, Oxford, 2002.
13. Schwank, J.R., *et al.*, Radiation effects in MOS oxides. *IEEE Trans. Nucl. Sci.*, 55, 1833–1853, 2008.
14. Srour, J.R., Marshall, C.J., Marshall, P.W., Review of displacement damage effects in silicon devices. *IEEE Trans. Nucl. Sci.*, 50, 653–670, 2003.
15. Dodd, P.E. and Massengill, L.W., Basic mechanisms and modeling of single-event upset in digital microelectronics. *IEEE Trans. Nucl. Sci.*, 50, 583–602, 2003.
16. Johnston, A.H., Latchup in integrated circuits. *IEEE Trans. Nucl. Sci.*, 43, 494–504, 1996.

17. Ferlet-Cavrois, V., *et al.*, Radiation effects in deeply scaled CMOS technologies. *IEEE Trans. Nucl. Sci.*, 60, 1953–1969, 2013.
18. Gouker, P., *et al.*, Single-event effects in advanced FinFET technologies. *IEEE Trans. Nucl. Sci.*, 65, 377–384, 2018.
19. Baumann, R.C., Radiation-induced soft errors in advanced semiconductor technologies. *IEEE Des. Test Comput.*, 22, 258–266, 2005.
20. NASA, EEE Parts Selection, Screening, Qualification, and Derating, NASA EEE-INST-002, 2018.
21. Avizienis, A., Laprie, J.-C., Randell, B., Landwehr, C., Basic concepts and taxonomy of dependable and secure computing. *IEEE Trans. Dependable Secure Comput.*, 1, 11–33, 2004.
22. Lyons, R.E. and Vanderkulk, W., The use of triple-modular redundancy to improve computer reliability. *IBM J. Res. Dev.*, 6, 200–209, 1962.
23. Kastensmidt, F.L., Sterpone, L., Carro, L., Reorda, M.S., On the optimal design of TMR architectures. *IEEE Des. Test Comput.*, 22, 486–495, 2005.
24. Hamming, R.W., Error detecting and error correcting codes. *Bell Syst. Tech. J.*, 29, 147–160, 1950.
25. Cai, C., Chi, Y., Cai, L., Radiation effects of advanced electronic devices and circuits: Recent progress and challenges. *Electronics*, 14, 2896, 2025.
26. Seung, M., Choi, J.-G., Choi, W.-Y., Kwon, I., Radiation-hardened-by-design CMOS preamplifier with improved total ionizing dose tolerance. *Nucl. Eng. Technol.*, 56, 621–629, 2024.
27. Pandey, I., Gupta, K., Kumar, V., Kamat, S.V., A MOSFET-based single-event latch-up detection and self-recovery circuit for spacecraft electronics. *Microelectron. Reliab.*, 152, 114713, 2024.
28. Chi, Y., Cai, C., Zhang, L., Radiation effects and mitigation in SOI and advanced CMOS technologies. *Electronics*, 13, 1073, 2024.
29. Gouker, P., Black, D.A., Schrimpf, R.D., Radiation effects in FinFET-based technologies: A review. *IEEE Trans. Nucl. Sci.*, 69, 1021–1032, 2022.
30. Black, D.A., *et al.*, Single-event effects in three-dimensional integrated circuits. *IEEE Trans. Nucl. Sci.*, 69, 725–734, 2022.
31. Zhang, Y., Liu, J., Wang, H., TCAD-based modeling of single-event transients in nanoscale CMOS technologies. *Microelectron. Reliab.*, 147, 114401, 2023.
32. Fleetwood, D.M., Radiation effects in a post-Moore world. *Microelectron. Reliab.*, 76–77, 3–11, 2017.
33. Schwank, J.R., *et al.*, Radiation effects in SOI technologies. *IEEE Trans. Nucl. Sci.*, 50, 522–538, 2003.
34. Ferlet-Cavrois, V., *et al.*, Total ionizing dose and single-event effects in advanced FD-SOI technologies. *IEEE Trans. Nucl. Sci.*, 69, 845–854, 2022.
35. Pearton, S.J., *et al.*, Radiation effects in GaN materials and devices. *J. Appl. Phys.*, 86, 1–78, 2019.
36. Kimoto, T. and Cooper, J.A., *Fundamentals of Silicon Carbide Technology*, 2nd ed., Wiley-IEEE Press, Hoboken, NJ, USA, 2023.

37. Wilson, J.W., *et al.*, Shielding Strategies for Human Space Exploration, NASA Conf. Publ. 3360, 2018.
38. Isberg, J., *et al.*, High radiation tolerance in single-crystal diamond detectors. *Science*, 297, 1670–1672, 2002.
39. Neudeck, P.G., *et al.*, Extreme environment electronics. *Proc. IEEE*, 109, 1726–1756, 2021.
40. Johnston, A.H., Radiation effects in advanced microelectronics packaging. *IEEE Trans. Nucl. Sci.*, 67, 1373–1385, 2020.
41. Durante, M. and Cucinotta, F.A., Physical basis of radiation protection in space travel. *Rev. Mod. Phys.*, 83, 1245–1281, 2011.
42. Wilson, J.W., *et al.*, Issues in space radiation protection. *Acta Astronaut.*, 68, 319–332, 2011.
43. Guetersloh, S.B., *et al.*, Polyethylene as a radiation shielding material in space. *Nucl. Instrum. Methods Phys. Res. B.*, 252, 319–332, 2006.
44. Spillantini, P., *et al.*, Shielding from cosmic radiation for interplanetary missions: Active and passive methods. *Radiat. Meas.*, 42, 14–23, 2007.
45. Tripathi, R.K., *et al.*, Hybrid radiation shielding concepts for deep space missions. *Acta Astronaut.*, 170, 1–10, 2020.
46. Norbury, J.W. and Townsend, L.W., Nuclear data for space radiation. *Adv. Space Res.*, 64, 1817–1832, 2019.
47. Badavi, F.F., *et al.*, Optimization of spacecraft radiation shielding using Monte Carlo simulations. *Acta Astronaut.*, 198, 350–360, 2022.
48. Johnston, A.H., Radiation testing and qualification of electronics for space environments. *IEEE Trans. Nucl. Sci.*, 60, 2054–2073, 2013.
49. Srour, J.R. and Palko, J.W., Displacement damage effects in irradiated semiconductor devices. *IEEE Trans. Nucl. Sci.*, 60, 1740–1766, 2013.
50. Oldham, T.R. and McLean, F.B., Total ionizing dose effects in MOS oxides and devices. *IEEE Trans. Nucl. Sci.*, 50, 483–499, 2003.
51. U.S. Department of Defense, MIL-STD-883: Test Method Standard, Microcircuits, 2019.
52. U.S. Department of Defense, MIL-STD-750: Test Methods for Semiconductor Devices, 2020.
53. ECSS, ECSS-Q-ST-60-15C: Radiation Hardness Assurance, European Cooperation for Space Standardization, 2019.
54. Fleetwood, D.M., Dose rate effects in MOS devices. *IEEE Trans. Nucl. Sci.*, 60, 1706–1730, 2013.
55. Baumann, R.C., Radiation-induced soft errors: A reliability perspective. *IEEE Trans. Device Mater. Reliab.*, 5, 305–316, 2005.
56. BAE Systems, RAD750 Radiation-Hardened PowerPC Processor Product Datasheet, BAE Systems, 2022.
57. Johnston, A.H., Radiation effects and reliability of microprocessors for space applications. *IEEE Trans. Nucl. Sci.*, 61, 1726–1735, 2014.
58. Kastensmidt, F.L., *et al.*, Design strategies for radiation-tolerant microprocessors. *IEEE Des. Test.*, 31, 6–17, 2014.

59. Quinn, H., Radiation effects in reconfigurable FPGAs. *IEEE Trans. Nucl. Sci.*, 62, 2432–2449, 2015.
60. Microchip Technology Inc., RTG4 Radiation-Tolerant FPGA Datasheet, 2023.
61. Xapsos, M.A., *et al.*, Radiation effects and mitigation strategies for spacecraft avionics. *IEEE Aerosp. Electron. Syst. Mag.*, 30, 4–13, 2015.
62. Swartwout, M., CubeSat mission success rates and reliability trends. *J. Small Satell.*, 9, 541–558, 2020.
63. Normand, E., Single-event upset at ground level. *IEEE Trans. Nucl. Sci.*, 43, 2742–2750, 1996.
64. Borkar, S. and Chien, A.A., The future of microprocessors. *Commun. ACM*, 54, 67–77, 2011.
65. Zhang, Y., *et al.*, Machine learning-based fault prediction for safety-critical electronic systems. *IEEE Trans. Reliab.*, 72, 812–825, 2023.
66. Quinn, H., Reconfigurable computing for radiation-tolerant space systems. *IEEE Aerosp. Electron. Syst. Mag.*, 33, 30–41, 2018.
67. Rollins, N., *et al.*, On-orbit reconfigurable computing for long-duration space missions. *Acta Astronaut.*, 208, 45–55, 2023.
68. Holmes, S.N., *et al.*, Radiation effects in cryogenic CMOS for space applications. *IEEE Trans. Nucl. Sci.*, 68, 845–853, 2021.
69. Radebaugh, R., Cryogenic electronics for space applications. *Proc. IEEE*, 109, 1810–1824, 2021.
70. Black, D.A., *et al.*, Radiation effects and mitigation in 3D integrated circuits. *IEEE Trans. Nucl. Sci.*, 69, 1215–1226, 2022.
71. Glaessgen, E. and Stargel, D., The digital twin paradigm for future NASA and U.S., Air Force vehicles, in: *AIAA Infotech@Aerospace Conf.*, 2012.
72. Tuegel, E.J., *et al.*, Reengineering aircraft structural life prediction using a digital twin. *Int. J. Aerosp. Eng.*, 4561873, 2020.

3

Low-Power and High-Efficiency Computing Architectures for Space Missions

T. Sathya[1*], M. Parimala Devi[2], G. Raja[3] and Sathishkumar V. E.[4]

[1]*Department of Computer Science and Engineering, Kongu Engineering College, Erode, Tamil Nadu, India*
[2]*Department of ECE, Velalar College of Engineering and Technology, Erode, Tamil Nadu, India*
[3]*Department of Mechanical Engineering, Kongu Engineering College, Erode, Tamil Nadu, India*
[4]*School of Computing and Artificial Intelligence, Faculty of Engineering and Technology, Sunway University, Sunway City, Selangor, Malaysia*

Abstract

Low-power, high-efficiency computing architectures are essential for humans to starts for even more ambitious space missions. Every watt of electricity, gram of weight, and line of computation must be optimized for life and performance in resource-constrained environment of space. With a focus on architectures that accomplish high computational performance under demanding power and thermal limitations, this chapter offers a forward-looking examination of the state of space grade computing systems. Space-borne systems must function with constrained energy sources and withstand adverse conditions like radiation exposure, extremely high temperatures, and communication delays, in contrast to terrestrial computing platforms. The chapter examines new neuromorphic and RISC-V-based architectures, radiation-hardened processors, and reconfigurable logic devices (such FPGAs), all of which are designed to provide reliable performance

**Corresponding author*: sathyamce@gmail.com; ORCID: https://orcid.org/0000-0002-6417-2544
M. Parimala Devi: ORCID: https://orcid.org/0000-0003-3955-1339
G. Raja: ORCID: https://orcid.org/0000-0002-7442-6720
Sathishkumar V. E.: ORCID: https://orcid.org/0000-0002-8271-2022

N. Sathish, Prasanth Aruchamy, Rajesh Kumar Dhanaraj, Nithya Rekha and Karthik Palani (eds.) Space Electronics for Deep Space Exploration, (53–82) © 2026 Scrivener Publishing LLC

with low energy costs. In order to enable edge computing on satellites, landers, and autonomous space probes, particular focus is given to heterogeneous computing systems that integrate general-purpose CPUs, low-power GPUs, and AI accelerators. The article offers a thorough analysis of how the future of intelligent space systems is being shaped by the confluence of mission-specific resilience, high-efficiency computation, and low-power design. The need for the development of robust, flexible, and effective computing architectures is essential in order to progress toward self-governing space travel for inevitable missions.

***Keywords*:** Low power computing, RISC-V-based architectures, FPGA, energy-efficient processors, space missions

3.1 Introduction

Space exploration faces numerous resource constraints that predominantly impact the space mission planning and execution. Although there are many constraints that play a critical role, power constraints become more pronounced in deep space missions, where solar panels are less effective. The restricted storage and payload capacity of spacecrafts need optimal utilization of space. The multifunctional equipments are to be optimized within stringent weight limitations. Communication latencies deteriorate increasing distance from Earth, requiring autonomous systems for instant decision making. Efficient and low power computing capabilities are essential for space missions because of distinctive limitations and problems that occur with functioning in extra-terrestrial settings. Spacecraft and satellites possess constrained power resources, predominantly depending on solar, panels, and batteries. This indicates the need for implementation of energy-efficient computing systems. These systems must execute complex computations, should manage extensive data, and regulate essential operations in space while utilizing minimum power. In addition to this, the adverse conditions of space, characterized by extreme temperature and radiation, necessitate a durable and dependable computing system capable of handling all environmental challenges. Efficient computing facilities extended mission duration, improved payload capacity, and enhances data processing capabilities. It results in more productive and economical space exploration initiatives.

Space-grade computing systems have advanced considerably since the early days of the exploration of space. During the 1960s, the Apollo Guidance Computer (AGC) was among the initial digital computers utilized in spacecraft, using integrated circuits and user interface. With technological advancements, radiation-hardened CPUs became crucial for enduring the severe conditions of space. In 1980s, the advancement of fault-tolerant systems and redundancy methods are implemented to improve reliability. During 1990s and 2000s, commercial off-the-shelf (COTS) components were adapted for space applications, which act as economical alternatives to modified hardware. In recent days, the utilization of Field Programmable Gate Arrays and Multicore processors facilitates enhanced and adaptable computing capabilities for space missions.

3.2 Literature Survey

Linling Kuang *et al.* [1] explained about the integration of communication and computing capabilities across various satellite types. It includes low-Earth and medium-Earth orbital missions, which emphasize on the need for power efficient networks in space. Grisafi [2] in his work proposed a design and evaluation multi-core hardware which aimed at secure, low-power computing architecture need for space missions. Kuang [1] proposed different techniques for achieving power efficiency at the processor core level, which is critical for the constrained environments of space missions. Wang [3] observes trends in high-performance computing (HPC) energy consumption. Wang discusses strategies to enhance energy efficiency which is necessary for space computing frameworks. Guimarães *et al.* [4] explore the positioning of computational resources on satellites for processing huge data, addressing the unique challenges in space environments. Bersuker *et al.* [5] discuss the advantages of neuromorphic computing architectures, which can perform complex calculations with lower power requirements, making them suitable for space applications. Muralidhar *et al.* [6] focus on architectural techniques to decrease static and dynamic dissipation of power in computing systems. Kaul [7], in an article, explores various approaches for optimizing power, performance, and area in digital design. It can significantly impact space mission computing efficiency. Zhao *et al.* [8] introduce an innovative on-board intelligent computing and data processing architecture designed for space applications. Yu *et al.* [9] confer the potential of neuromorphic computing as a low-power, high-efficiency alternative for AI-based applications in space exploration. Weston *et al.* [10] proposed reviews and advancements in power technologies for small spacecraft, emphasizing the importance of energy density for cost-effective launches. Bussemaker and Firchau [11] demonstrate the application of system architecture optimization techniques specifically for space mission design. Wallace *et al.* [12] present a communication framework for deep-space missions that integrates multi-level AI for enhanced performance. Kuang *et al.* [1] analyze different strategies for enhancing energy efficiency in space systems. Wang *et al.* [3] outline the principles for low-power computing systems which overcome the challenges in space environments. Kim *et al.* [13] explore the advancement of radiation-hardened processors that can offer low power consumption for space applications. Liu *et al.* [14] focus on methods for optimizing energy consumption in computing systems used in space missions. Zheng *et al.* [15] discuss innovative power management strategies for enhancing the efficiency of space computing systems. Goodwill *et al.* [16] inspect how machine learning techniques can be applied to improve low-power computing in space missions. Wang *et al.* [3] observe various strategies for optimizing the performance of computing architectures used in space missions. Geda and Tang [17] discuss the potential of hybrid computing architectures that combine varied processing models for space applications. Jones *et al.* [18] review the use of low-power FPGA architectures explicitly designed for the constraints of space missions. Shahid *et al.* [19] explore energy harvesting methods that can be integrated into space computing systems to enhance power efficiency. Erosy *et al.* [20] examine how effective thermal management strategies can improve power efficiency in space computing systems. Underwood *et al.* [21] discuss the design and implementation of low-power computing systems for autonomous spacecraft operations. Wu *et al.* [22] appraise various strategies for enhancing power efficiency in sensor networks installed in space environments. Liu *et al.* [14] discuss the technical challenges faced when implementing low-power architectures in space missions. Wang *et al.* [3] explore evolving trends and technologies that could shape the upcoming of low-power computing in space missions.

3.3 Basics of Power/Energy Efficiency Techniques

Quantified energy in joule is one of the most significant measurements in every computer system but especially in space, wearable, and mobile systems where energy usage is directly linked with battery life. The measurement now becomes relevant in large systems. Energy usage is one of the world's biggest running expenses, and as such, decreasing it is essential for every computing system. The quantity at which energy is released is called power. Watts (W), or Joules per second, is the unit of power. Power density, a related statistic, is power per unit area. By reducing the area, power can be minimized but it is a challenging task, when there is no system for active cooling in space. The energy needed to run an instruction is termed as energy-per-instruction (EPI). Another important metrics for expressing the aggregate energy-performance of loads is the energy-delay product (EDP). It combines energy and performance. As EDP is based on both energy consumption and runtime of the workload, it will rise if either of them rises. Thus, low EDP is advantageous.

3.3.1 Power Consumption and Dissipation

Average power (Equation 3.1) is defined as the addition static and dynamic power

$$P_{avg} = P_{dynamic} + P_{leakage} \tag{3.1}$$

$$P_{dynamic} = ACV^2 f \tag{3.2}$$

Dynamic power (Equation 3.2) depends on clock frequency f, supply voltage V, node capacitance C (function of wire lengths), and switching activity factor A. Minimizing switching activity, clock frequency, capacitance, and supply voltage reduces dynamic power and performance. Dynamic power is observed during switching, while static power is ongoing. Sub-threshold leakage power, dissipated by a transistor with an off gate, is the most significant leakage component. Transistors leak current even at voltages less than the threshold voltage because of their weak switching behaviour. Leakage power becomes the leading cause of power consumption in small geometries. In today's and tomorrow's fabrication schemes, microprocessors spill 20% to 40% of their power budget. Low temperature effect compensation (LTEC) compensates for low/high temperatures. Clock gating minimizes hardware unit activity when it is idle, conserving energy. Aside from power dissipation, supply voltage is influenced by the clock frequency f. Generally, higher clock frequencies demand higher supply voltage.

3.3.2 Techniques Adopted for Energy Efficiency

Apparently, reducing supply voltage, activity factor, and frequency of operation can reduce power consumption. Clock, power, and dynamic voltage, frequency scaling are used to achieve this.

I. Clock and Power Gating

Clock gating eliminates the use of clock signal when the system is idle, providing additional logic. The trick is to shut off circuitry so flip-flops do not need to change states, which increases dynamic power. Suppressing unnecessary activity that has no impact on calculation is the objective. It can be done at any level, ranging from functional units, structures, and subsystems, i.e., memory, I/O, CPU to flip-flops and circuits. More complex functional units employ power Gating, or gated Vdd, which reduces functional unit voltage.

II. Managing Dynamic Power in Computing

At the system level, there are two main power-saving approaches: race-to-halt and crawl-to-idle. Race-to-halt means running tasks at maximum speed, finishing them quickly, and then switching to low-power or shut-down modes to save energy. This method reduces static power by keeping tasks short. Crawl-to-idle, on the other hand, runs tasks more slowly, using less power over a longer time, which helps extend battery life. Modern systems often use a combination of both methods for better efficiency. These are system level power management techniques and thermal management techniques:

(1) Management of idle power: Idle power management does nothing effectively. Idle power management is employed on all circuit layers, such as processors, memory controllers, hard disks, network processors, IO subsystems, and interconnects. Operating system entities, such as device drivers, observe scheduler load, interrupts, and other heuristics to maximize system idleness using software and hardware coordination.

(2) Active power management: It is system efficiency in terms of energy during active workloads. Dynamic frequency scaling (DFS) can be used, but a number of system components run at fixed voltage. Different OS components like kernel, device drivers, and firmware manage DVFS, depending on workload, hardware utilization, and others.

(3) Duty Cycle: Power cycle or duty cycle is the ratio of time that a signal or system is up. Duty cycling within computing systems and logic blocks can cycle devices on/off in terms of workload, usage counters, etc.

(4) Race-to-idle: Execute the system at its maximum possible frequency and voltage while it is operating to complete the task at hand in as little time as possible, and then set it to sleep or its lowest possible active mode.

(5) Crawl-to-halt: Being at the highest state all the time might not be the best, based on the workload or phases of the workload. In IoT/wearable devices, some parts of the system are maintained at reduced running states to preserve battery life and ensure maximum usage. Although the workload or system can be slower, battery life could be extended where feasible.

3.3.3 Power Management Guidance through Software

With time, computer software has created sophisticated, tunable CPU and subsystem power and performance plans. Operating systems, such as Windows and Linux, utilize architectural interfaces to track CPU, interconnect, and peripheral idleness and invoke device idle states for the management of power. The Linux CPU idle subsystem tracks CPU and system idle time, adhering to interrupt and QOS demands to initiate CPU and SOC low power states with the MWAIT x86 instruction. The ARM processor has a similar interface called WFI. Advanced power management regulations for CPU, GPU, and peripherals are a part of contemporary operating systems.

3.3.4 Moore's Law

Moore's Law has driven a continuous evolution in device design, process technology, circuit innovations, and micro-architecture, effectively doubling the number of transistors on a chip roughly every 18 months. This exponential growth has spurred advancements and development of highly compact form factors. Notably, Gordon Moore foresaw a slowdown as CMOS technology come close toed its physical and practical limits.

Other principles have also shaped computer system development. In 1974, Robert Dennard observed that as transistors shrink, their power density remains roughly constant. The primary aim of reducing transistor dimension is to decreases the overall power consumption. The linear dimensions of a transistor enable the accommodation of four times as many transistors within the same area. If both the voltage and current are simultaneously reduced by half, the power drops by factor of four. Its enables high performance without exceeding the same power. Under these conditions, utilizing low cost, smaller transistor could be constructed and operated which consumes less energy. During the 21st century, chip designers made advanced processor architectures, memory hierarchies and exploitation of instruction-level parallelism (ILP).

Instruction level parallelism (ILP) is achieved through many ways based on application. Scientific computation and graphics programs can have large ILP, but cryptography might not. Micro-architectural techniques for exploiting ILP are as follows:

(1) Instruction pipelining: Decreases Clocks-per-instruction (CPI) by overlapping several instructions partially.

(2) Superscalar execution, Very Long Instruction Word, EPIC: These employ a number of execution units to execute instructions concurrently. Superscalar machines enable many instructions to be executed per clock cycle by sending them to different processor execution units.

3.4 Energy Efficiency Attainment through System Level Techniques

The techniques explore higher layers of the software stack, such as firmware, the operating system, and applications leverage architectural and micro-architectural strategies, to enhance energy efficiency across the system. These techniques are applied to tailor performance and power consumption according to workload requirements and system limitations. Since system-level energy optimization is closely tied to the underlying hardware, examples from ARM and x86 platforms are used to illustrate these approaches.It is necessary to start looking at these systems' system-level techniques and then how software maximizes energy efficiency with them [23].

3.4.1 Arm System Architecture

Figure 3.1 illustrates a hierarchical ARM system architecture, labeled as "System Type A" at the top. The system consists of three components, each are represented by Chiplet type A, Chiplet type b and Chiplet type C as shown in Figure 3.1. All the three chiplets are interconnected through standardized interfaces. The overall system architecture indicates a complex system where different components work together to ensure flexibility in system configuration.

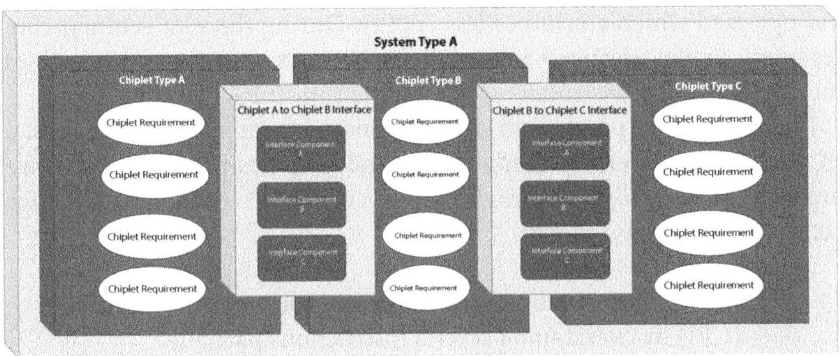

Figure 3.1 System architecture of ARM processor.

I. Clock Gating, Latency, Power Breakdown

The Wait-For-Idle (WFI) instruction in ARM processors interrupts the CPU clocks. The majority of ARM cores provides clock gating the L2 cache, debug logic, and other parts through instructions provided to co-processor. Latent mode shuts down the CPU and the cache controller but not cache memory. Cache RAMs can be configured in low-power retention to retain their content but not operate. This power-saving mode shuts down the cache masters without performance impact because of cache invalidation/flushing. Power gating a core necessitates context reset during resume [23]. ARM-based systems can have numerous core clusters that share L2. When there is power down of all the CPU cores in a cluster, snoops from cache are disabled and the L2 cache is power gated. DVFS is controlled by system control processors (SCPs) that handle clocks and voltage regulators.

II. DVS/DVFS/AVFS

Many ARM SoC providers provide a boost mode for CPU over-clocking and an absolute maximum sustained frequency. A single voltage rail for all (heterogeneous) cores within a cluster is the standard configuration for SMP and HMP systems incorporating multiple (heterogeneous) cores. Implementations per core voltage rails are uncommon since they are design intensive. Certain SoCs employ per-core clock lines for core frequency regulation using glue logic to synchronized voltage rail voltage across common ones.

ARM11 introduces an intelligent energy manager (IEM) estimates the minimum voltage required for operation. The Adaptive Voltage Frequency Scaling (AVFS) system utilizes on-chip sensors in order to monitor the system characteristics in a closed-loop manner. Data received from these sensors allows the IEM to reduce voltages below the standard values thus improves the energy efficiency.

ARM BIG.LITTLE platforms use SoCs with clusters of high-performance and power-efficient cores. Mobile devices frequently shift between heavy and light workloads. Efficient job allocation across these heterogeneous clusters optimizes performance and power consumption. DynamIQ combines high-performance cores with high-efficiency cores in memory-coherent clusters. A CPU cluster, with global memory and optimized intruder management, manages task relocation between big and little cores which enables energy saving. It also facilitates the exchange of data among cores, hence minimizing memory traffic and total power consumption.

III. Device PM/Power Domains

ARM SoCs use multiple voltage domains to independently power the system and support DVFS. Voltage regulators are arranged hierarchically, allowing software to identify idle nodes that do not need clock or power. This hierarchical structure of voltage and clock domains allows unused IP blocks to enter power collapse. Device drivers take care of the hardware clocks and power, while the Operating System controls system power.

IV. Shared Low-Power States in ARM SoCs:

- **S2R:** All functional units are turned off apart from for wake-up logic while operating in this mode.
- **Low Power Audio:** SoCs have a low-power audio to save energy during activities like music playback with the screen off. The audio SRAM, DRAM and DMA controller remain active, while the CPU or DSP updates audio data intermittently.
- **Low Power Display:** ARM-based devices offer several low-power modes.

3.4.2 Power Management of Intel x86

Fine-grained control over device and system level power management is offered by Intel x86 SoCs. Power managers in operating systems convert the platform into various power modes according to user-defined power configurations. In Intel SoCs, the OS and firmware coordinate to manage power for the CPU, devices, subsystems, and the entire system. Without impacting user experience or system operation, the power management architecture aggressively shuts down unused subsystems.

Figure 3.2 shows an integral part of Intel x86 power management is the Advanced Configuration and Power Interface (ACPI) system. Being an open standard, it can be used by operating system to control the usage of power. It can be used to find, configure devices and manage the system events. Power management can be precisely done by using ACPI, which grants the OS control over the power state of individual components. Because of this, the energy efficiency and battery life of mobile devices can be improved. The ACPI system manages temperature, which act as an interface for setting hardware. Since it has plug-and-play functionality, it can able to handle events at both system and device levels. Because of ACPI-implemented power management, Intel x86 maximizes system performance by utilizing minimal power.

Multiple updates to the platform's hardware and software have made this possible:

- **On-die clock and power gating:** Subsystems, controllers, fabrics, and peripherals use on-die clock and power gating to eliminate power wastage.

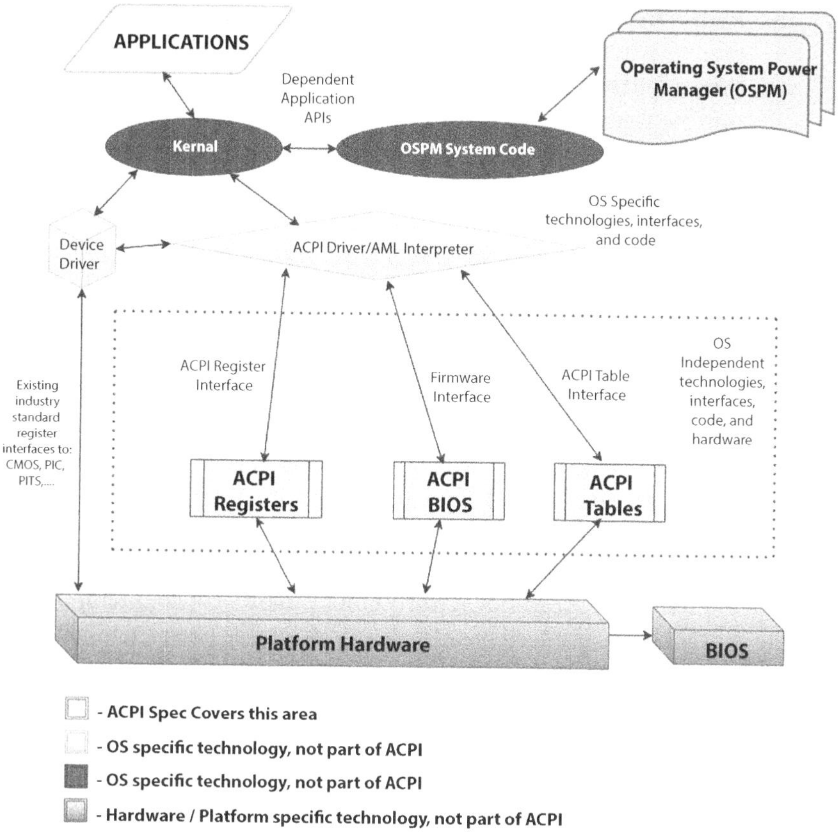

Figure 3.2 Advanced configuration and power interface (ACPI) system.

- **CPU C-states:** These represent the low-power or idle modes of the CPU cores. Each Cx state saves increasing amounts of power. C1 stops instruction execution automatically. C3 clears caches and disables processor clocks, while C6 disables the CPU core by withdrawing voltage. These transitions are started and monitored by operating systems.
- **CPU P-states:** Px is defined as fixed voltage-frequency operating points called as performance states. P-state transitions use architectural registers and hardware-software coordination. It balances CPU power efficiency and performance.
- **Subsystem active-idle states:** Operating systems and sub system or device drivers manage these states for finer component control.

- **Platform idle states:** Platform idle states increase energy savings when devices are idle.
- **Microcontrollers for power management:** Power distribution between the north (CPU, GPU) and south (peripherals) is managed by dedicated microcontrollers. They control voltage rails, device state transitions, and system wake-ups.
- **Integrated Voltage Regulators (IVR):** Many Intel SoCs design has a three-tier CPU hierarchy containing of the core, module, and package. During in-package consolidation, the system can schedule tasks to perform on one or both modules using all four cores.

Intel CPUs manage processor performance *via* OS-controlled techniques like DVFS or P-states. ACPI tables inform the operating system of available P-states, which CPU frequency governors in Linux and Android regulate based on workload demands.

Intel Atom cores also support turbo frequencies, which work similarly to the improve feature in ARM SoCs. Turbo mode allows cores to temporarily operate above their guaranteed base frequency as long as power, temperature, and current constraints are respected. This capability leverages the fact that the highest certified operating frequency is typically determined using conservative worst-case conditions, leaving room for opportunistic performance gains.

I. Low-Power Systems

Transient low power system states supported by Intel SoCs:

(1) S0ix: SOC is wide shallow idle.
(2) S0ix-Display: Periodic wake of display controller to supply material, with the remainder of SOC turned off.
(3) S0ix-Audio: Low-power SOC minus audio block.
(4) S0ix-Sensing: Low power SOC minus sensor hub in order to enable pedometer and other low-power sensing modes.

Figure 3.2 illustrates the architecture and components of the Advanced Configuration and Power Interface (ACPI) system. At the top level, the applications interacting with the kernel through dependent application APIs. The kernel communicates with the Operating System Power Manager (OSPM) system code, which interfaces with multiple OSPM components. The ACPI Driver/AML Interpreter acts as a bridge between OS-specific and OS-independent technologies. Below the ACPI Driver, the three key ACPI components are shown: ACPI Registers, ACPI BIOS, and

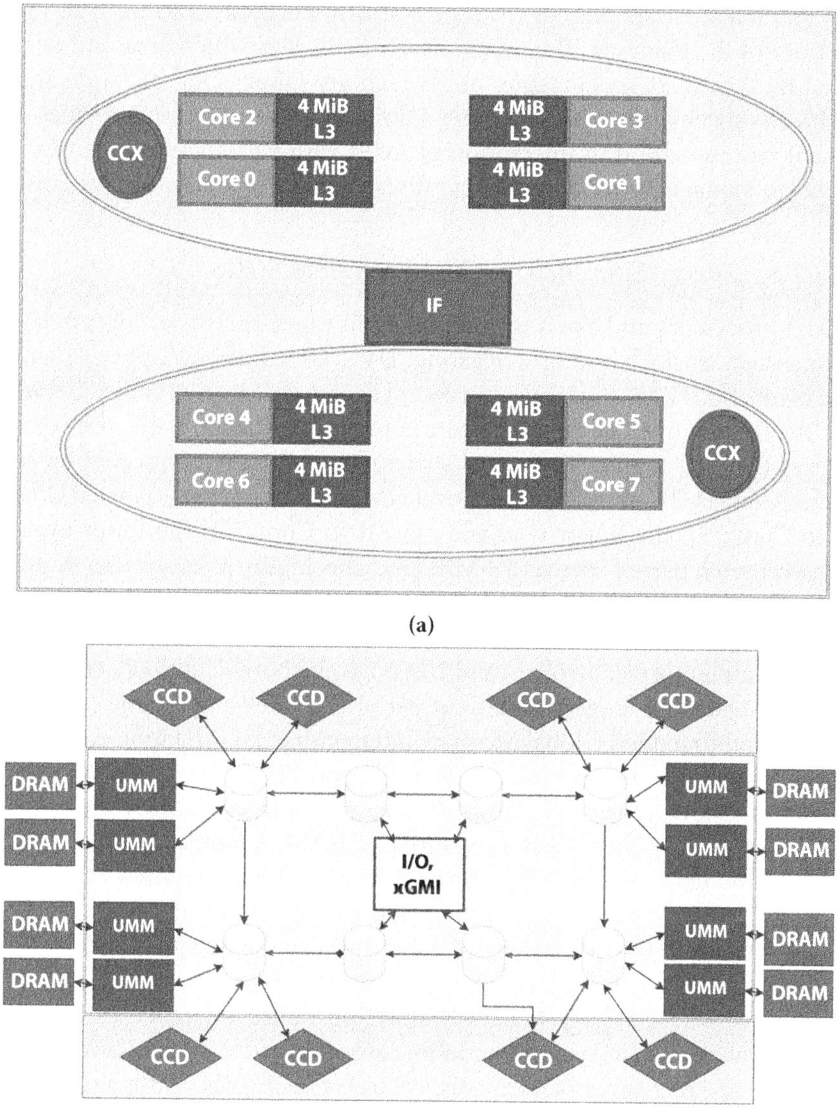

Figure 3.3 (a) Core complex die (CCD) with core complexes (CCXs). (b) Block diagram of I/O die with memory controller, attached memory, IF switches (yellow), CCD and repeaters.

ACPI Tables. These interact with the platform hardware and BIOS at the bottom of the diagram. The image clearly delineates which areas are covered by the ACPI specification and which are OS-specific or hardware/platform-specific technologies. This comprehensive diagram provides a visual representation of integration of ACPI with various system components to manage power and configuration in modern computing systems.

3.4.2.1 Energy Efficiency in AMD Processors/SOCS

AMD processors and SoCs include different types energy saving mechanisms such as clock and power gating, DVS, DFS, link-level power management and DVFS. A summary of recent advancements is provided below.

The first Figure 3.3a depicts a core complex die (CCD) with core complexes (CCXs). It shows two oval structures with CCX connected by an infinity fabric (IF) interface. Each oval consists of four cores (labeled Core 0 to Core 3 in the upper oval and Core 4 to Core 7 in the lower oval). Between each pair of cores is a 4 MiB L3 cache. Figure 3.3.a shows a multi-core processor's cores and cache symmetrically arranged and interconnected between die halves. Figure 3.3.b shows a processor's architecture, including memory controllers and I/O ports. It shows eight core complex dies (CCDs) are arranged around the periphery of the chip. In the center is an I/O and Extended Global Memory Interconnect (xGMI) unit. Adjacent eight yellow cylinders represents the interconnects or data pathways. The diagram also includes eight unified memory manager (UMM) units and eight dynamic random access memory (DRAM) modules on the outer edges. This Figures 3.3a and 3.3b shows the complex integration of multiple CCDs with memory controllers and I/O interfaces. It highlights the scalability and interconnectivity of modern processor designs.

The power monitor uses onboard models to detect workload-induced power changes and voltage, frequency, and temperature consequences. Models get calibrated per SOC version/model for consistent performance. Power management includes three performance controllers: GEAPM, CBB, and MBB. IP-level power management also employs Locally Efficient APM (LEAPM).

3.5 Various Space Architectures for Low-Power, High-Efficiency Space Missions

The following architectures/approaches for low-power, high-efficiency space missions, and their trade-offs are employed

3.5.1 NASA's High Performance Spaceflight Computing

NASA is developed flight computing platform, high-performance spaceflight computing (HPSC), led by the Jet Propulsion Laboratory, for missions beyond 2040 as shown in Figures 3.4a and 3.4b. It addresses the

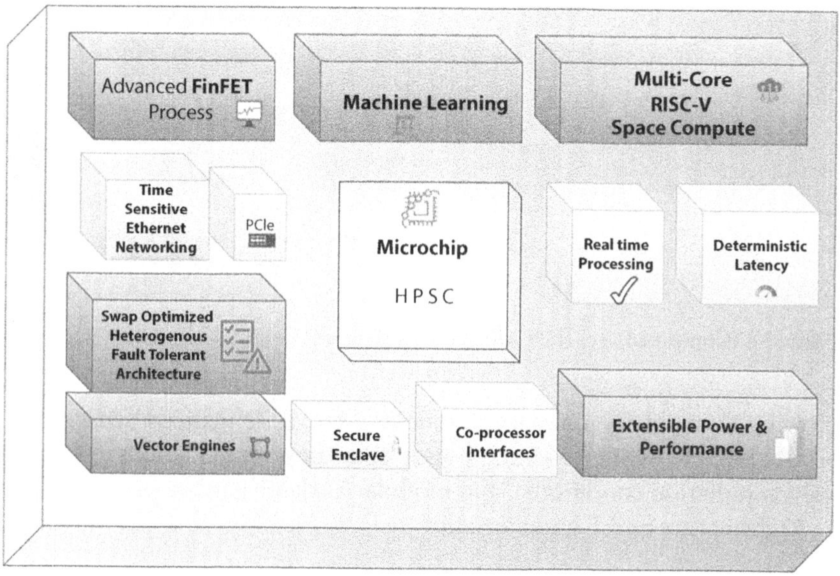

(a)

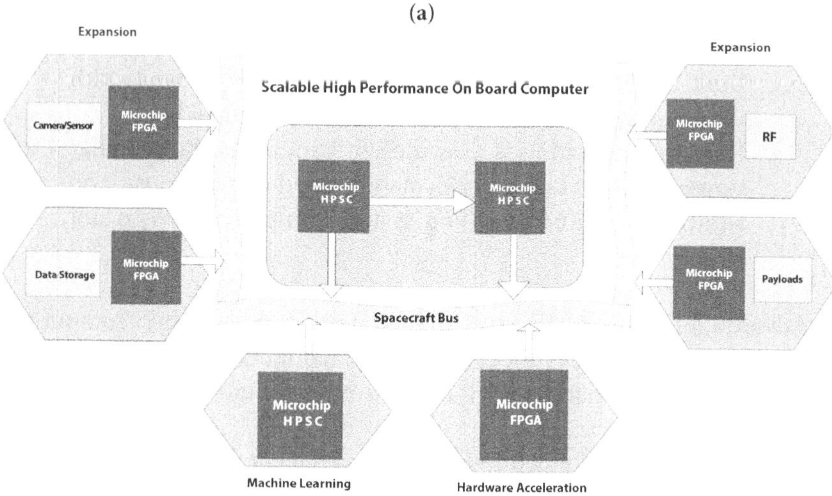

(b)

Figure 3.4 (a) HSPC architecture. (b) HSPC enabling extensive space system. (*Continued*)

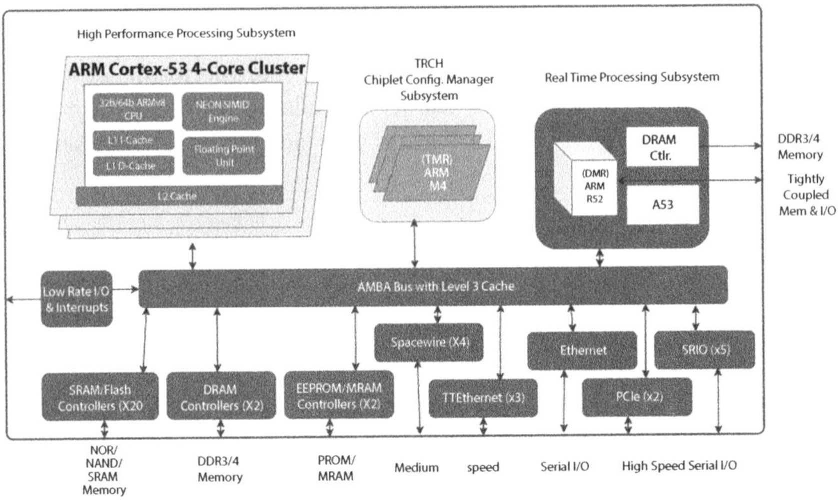

(c)

Figure 3.4 (Continued) (c) HSPC chiplet architecture.

growing demand for onboard autonomy, AI/ML, enhanced sensor data processing, etc. while retaining power efficiency, dependability (particularly in radiation conditions), and modular/scalable performance.

Architecture and Characteristics

- A multicore, radiation-hardened, fault-tolerant System-on-Chip (SoC) architecture.
- Built-in TSN Ethernet switch supports high-bandwidth interconnects and reduced data mobility overheads.
- Power optimization is a key design aim, including granular power modes, scaling performance based on power use, and shutting down or throttling system components when not in use.

Advantages: This architecture offers significantly higher computing capacity than older space processors. It also maintains power efficiency for spacecraft which is mandatory for real-time computation, autonomy, or sensors needed for space.

Drawbacks: The architecture may be less efficient compared to cutting-edge commercial chips due to process node scaling lag. Obtaining high Technology Readiness Level (TRL) for harsh space settings is difficult and costly. When implementing networked on-chip, weight, heat management, and latency are trade-offs.

The HSPC Chiplet Architecture, as depicted in Figure 3.4c , is a sophisticated system-on-chip design that integrates various components to achieve high performance computing. The core of ARM Cortex-53 consists of 4-core cluster which consists of a 32b/64b ARMv8 CPU. It has L1 I-Cache, L1 D-Cache, Floating Point Unit, and L2 Cache. This cluster unit forms the primary processing unit of the system. The architecture also has AMBA bus with Level 3 Cache as the interconnect,links various subsystems. It also has TRCH chiplet Config Manager Subsystem for real time processing with DRAM,DMR ARM R52, and A53 processors. The bus in Figure 3.4c also interconnects all data in bidirectional way to multiple controllers and interfaces, such as SRAM/Flash Controllers, DRAM Controllers, EEPROM/MRAM Controllers, Spacewire, TTEthernet, PCIe, Ethernet, and SRIO.

This modular design allows for efficient data flow between the high-performance processing core, memory systems, and various I/O interfaces, enabling the system to handle a wide range of computing tasks while maintaining real-time processing capabilities and high-speed connectivity.

3.5.2 Neuromorphic Computing and Sensing in Space

For efficiency, neuromorphic computing architectures resemble biological neural networks in structure, function, or both [24]. This is attractive for space since power, size, and resilience are constrained and pattern recognition, sensory processing, etc. are bursty and tolerant to approximation computation or error [23].

Neuromorphic Computing Architecture and Characteristics

- The approach involves sparse, event-driven computation using neurons/spiking neural networks instead of always-on, timed digital logic. Idle power consumption drops dramatically [24].
- Consider integrating neuromorphic sensors with neuromorphic compute to reduce data transport and computation waste by processing only when "something interesting" occurs. Neuromorphic systems provide intrinsic noise/approximation resilience, benefiting radiation-prone areas by careful design.

- **Advantages:** Low power consumption for vision, detection, and classification workloads, particularly when sparsely activated. High energy efficiency per inference or event can outperform standard CPUs for specific jobs.
- **Drawbacks:** Difficulty in mapping general purpose jobs due to lack of optimized techniques for spiking or approximate computing Radiation impact and reliability remain issues. Also, the tooling, libraries, compilers etc., is less established than for traditional architectures.

Neuromorphic computing architecture is built to imitate how biological neural networks work. Compared to traditional von Neumann model, neuromorphic computing model processes information in a unique way. In these systems, data is exchanged through spikes, similar to neurons communicates to brain. The main part of the architecture consists of network of neurons and synapses that handle both computation and memory. There is a clear division between processing and storage found in conventional systems. Figure 3.5 illustrates key differences between neuromorphic and traditional von Neumann architectures used in processors. Furthermore, neuromorphic architectures use spiking neural networks for programming and spikes for communication, as opposed to binary instructions and data in traditional systems. The timing in von Neumann architectures is synchronous and used clock-driven approach. But the timing in neuromorphic systems is asynchronous and event-driven. These characteristics helps neuromorphic systems to potentially achieve higher energy efficiency. Its performance for certain types of computations, particularly those involving pattern recognition and sensory processing is also better compared to traditional processor.

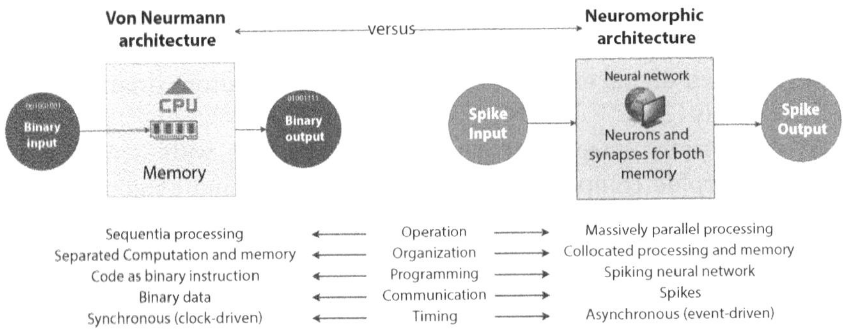

Figure 3.5 Neuromorphic computing architecture.

3.5.3 SPACECUBEX: A Hybrid and Heterogeneous CPU-FPGA/Digital Signal Processing Design

In order to meet space constraints and handle high throughput requirements from Earth-observing sensors, hyperspectral imaging equipments, heterogeneous systems combine general-purpose processors with accelerators which constitute FPGA and DSP [25].

Architecture and characteristics:
SpaceCubeX is an Earth Science mission-specific hybrid CPU/FPGA/DSP architecture. Its CPUs handle control, coarse operations, and OS, whereas FPGA or DSP co-processors expedite high-speed, fine-grain processes like compression and signal processing. Accelerators can accomplish more work per watt than central processing units (CPUs) for certain algorithms, and the system can be designed to shut down or run at less power when it is not in use.

Advantages: Accelerators offer substantial efficiency benefits in data throughput while implementing compression and filtering. Heterogeneous architectures balance CPU flexibility and FPGA/DSP efficiency.

Drawbacks: Co-designing hardware and software increases design complexity. Verification is harder, especially in radiation. Added thermal/mass management difficulties. Accelerators may not be beneficial for uncertain or variable mission requirements.

SPACECUBEX is a heterogeneous/hybrid CPU-FPGA/DSP architecture designed for space applications is shown in Figure 3.6, it consists of multiple layers:

This heterogeneous computing environment is structured into four distinct layers. The Physical System forms the foundation, with the Host Evaluation System serving as the base for testing and the Host Operating System providing multi-platform support (e.g., Jira Software, Apple, Android). The System comprises a diverse set of computational platforms like Multi-Core Platform for general-purpose processing, the FPGA Platform for hardware acceleration, and the DSP Platform for specialized signal processing. All are interconnected by transaction-level modeling (TLM) interfaces. The Bare Metal/OS Layer ensures low-level system control and management. Finally, the top layer is composed of the Benchmark/App segment which executes applications and benchmarks. It leverages the capabilities of the underlying heterogeneous architecture. This architecture combines the strengths of different computing paradigms and creates a flexible and powerful system suitable for various space computations.

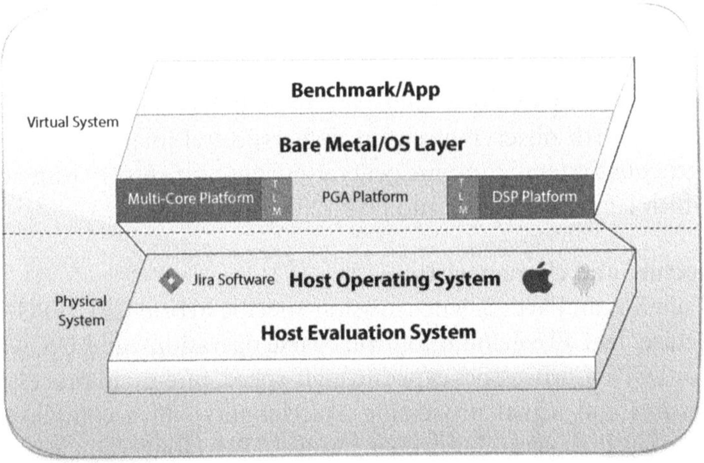

Figure 3.6 Run-time architecture of the SpaceCubeX.

Since the processor allows heterogeneous operation, it allows efficient task allocation and processing based on the specific requirements of space.

3.5.4 Multiprocessor SoCs Fault Tolerance on the Software Side

Single event upsets, transient failures, etc. are more likely in space with contemporary multiprocessor SoCs, which are more power efficient and denser. A solution is to use many layers of software for fault detection, rectification, and mitigation, rather than solely depending on hardware. This allows low-power hardware with great reliability.

The NASA Dependable Multiprocessor shown in Figure 3.7 involves three stages: coarse-grained thread-level lockstep or self-validation, FPGA reconfiguration, and mixed criticality. Using MPSoCs (popular in embedded/mobile markets) with software wrappers, monitoring, and repair procedures to enhance mission reliability.

Dynamic adaptability allows for low power mode or idle operation, and fault handling during flight. This architecture is comparatively cheaper than entire hardware and it has triple modular redundancy. It is efficient in weight and size and provide better power-budget performance. Associated software makes upgrades easier. Its drawbacks include Complex software, probable failure due to improper fault tolerance, lag in detection/recovery, and necessity for radiation testing and validation of the integrated system.

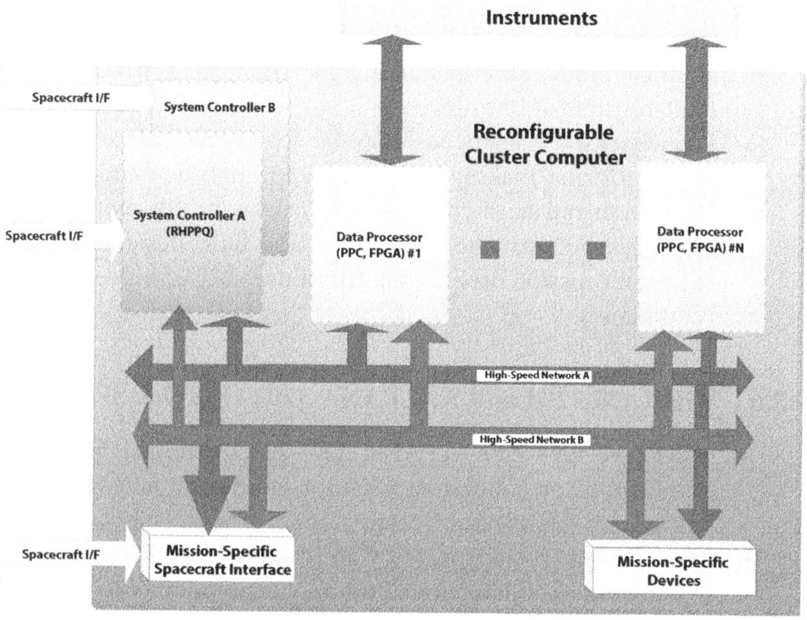

Figure 3.7 NASA-dependable multiprocessor.

3.5.5 Resilient Affordable CubeSat Processor and Power-Efficient Adaptive Computing

Tiny spacecraft (CubeSats, tiny sats, swarms) have limited power, every gram and watt counts. Hence designs must be efficient, durable, and typically created with "off the shelf" or cheaper components added for space. Resilient affordable cubeSat processor (RACP) and power-efficient adaptive computing (PEAC) are relevant projects.

I. Architecture and Characteristics

- RACP features an ARM SoC, radiation-tolerant microcontroller, and non-rad-hard-by-design FPGA. It functions with health monitoring and a flexible power usage range. Its peak power range starts from ~0.53 to ~8.7 W depends on clock and workload.
- PEAC uses a low-power always-on domain for essential functions for processing intensity. Power scheduling, variable voltage/clock scaling, etc., are special features available in boards and chips/firmware/software.

- **Advantages:** Ideal for small/swarm missions with limited mass, volume, cost, and power. Flexible power scaling and adaptive modes save lots of energy. RACP performs and is fault-resistant at low power.
- **Drawbacks:** Performance may be lower than large mission platforms, making it unsuitable for high-demand jobs. Using non-rad-hard components is risky; health monitoring helps but may not match full rad-hard designs. Ray exposure/mission lifespan may limit usability or necessitate redundancy.

3.5.6 VIKRAM 3201 and KALPANA 3201 (ISRO & SCL)

- VIKRAM 3201 & KALPANA 3201 are developed by the Semiconductor Laboratory (SCL), Chandigarh, and ISRO's Vikram Sarabhai Space Centre (VSSC).
- These 32-bit microprocessors are designed specifically for use in space and launch vehicle applications. Vikram 3201 is a chip that was entirely manufactured in India [26]. Based on IEEE 1754, the SPARC V8 RISC Kalpana 3201 CPU is made to work with open-source tool chains.

I. Efficiency/Low-Power Relevant Details

It is fabricated at SCL using 180 nm CMOS technology. Although it may not have the highest density, using a moderately mature process node can aid with dependability, faster qualification, and cheaper cost.

They are capable of operating in hostile environmental circumstances (e.g., temperature, vibration, etc.). Because of this the mentioned architecture has resilient and have minimal wasted overhead.

Advantages: They allow for more onboard processing than older, simpler processors, use 32-bit with floating-point, are capable of operating at higher levels of language, and so on, significantly increasing India's self-sufficiency in the field of space electronics. On top of that, having local resources like tool chains and simulators reduces overhead and could potentially save energy that would otherwise be wasted on inefficient development or mismatches [5].

Drawbacks: Since they are fabricated using 180 nm, they will not be able to achieve the same power-to-performance ratio. Furthermore, additional specialized accelerators or heterogeneous architectures may still be required for extremely compute-intensive workloads like as artificial intelligence, machine learning, and image processing.

II. Shakti/IRIS (IIT Madras & ISRO)

- The Shakti series is an open-source processor effort (RISC-V ISA) that comes from IIT Madras [27].
- IRIS is a particular chip in that series that is intended for use in space applications; it has capabilities that include bespoke modules like CORDIC units and watchdog timers as well as fault-tolerant memory.

Details on Efficiency and Low Power Consumption:
Using RISC-V for space missions, the design provides more flexible, allowing for the removal of unnecessary ISA or module components to save space and power. Fault tolerance in memories, lowers the need for extra error correction or excessive guard bands in space, which could make things run more smoothly. The IRIS is presently under development for command and control operations, among other purposes. Such processes generally require reliability over maximal throughput; therefore, they frequently utilize low-power modes and simpler pipelines, among other strategies.

Advantages: Because of the open design, Indian institutions have the ability to modify or scale down the design to meet their requirements. In addition, it contributes to the development of indigenous capabilities. The possibility of minimal standby power or efficient power gating arises if the latter is implemented.

Drawbacks: There is a high probability that it is not yet sufficiently developed to qualify for situations with severe radiation. Furthermore, the environment for space-grade certification can be slow to develop, and in many instances, systems based on RISC-V still require sophisticated testing.

Figure 3.8 shows India's aerospace semiconductor milestone and its four main components. First is IRIS Chip Development, a collaboration between IIT Madras and ISRO. The SHAKTI Microprocessor (RISC-V) powers this indigenous space controller. India's space technology self-reliance is advanced by this breakthrough. The second and third parts cover government efforts. The National Policy on Electronics (2019) envisions an ESDM hub and the INR 76,000 Cr Semicon India Programme. Modified semiconductor fab schemes with 50% financial assistance, EMC 2.0, 100% FDI in Electronics Manufacturing, and PLI incentives for domestic manufacturing are further attempts. This third government initiative, coordinated by IIT Madras, aims to create a secure indigenous processor ecosystem utilizing RISC-V (DIRV). The fourth component, "Make in India & Atmanirbhar Bharat," highlights manufacturing, with SCL Chandigarh

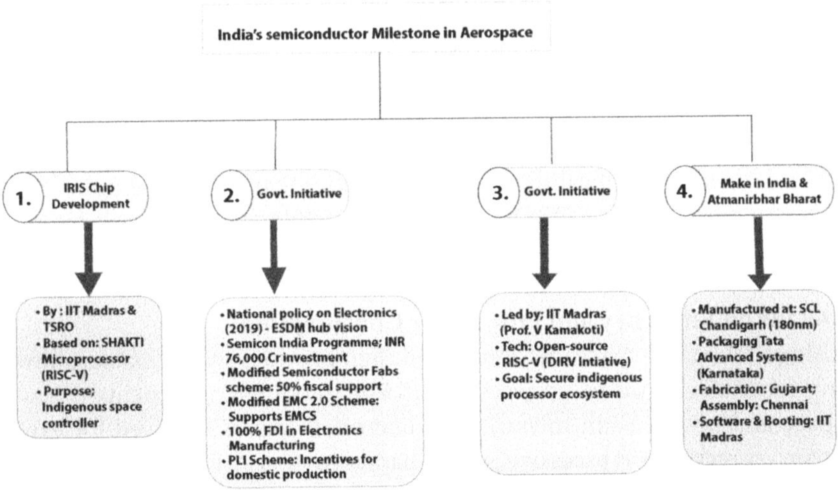

Figure 3.8 India's semiconductor milestone in space.

(180 nm) producing, Tata Advanced Systems packaging, Gujarat fabrication, Chennai assembly, and IIT Madras software and booting.

3.6 Comparison

These processors reflect a spectrum of trade-offs between power, performance, cost, and resilience. For example:

- HPSC and SpaceCubeX are ideal for high-performance missions but come with complexity and cost.
- RACP and PEAC shine in small satellite missions where power and size are constrained.
- Vikram and Shakti chips represent India's growing self-reliance in space-grade computing. Table 3.1 shows the comparison of different high-performance architectures based on origin, price, uniqueness with its advantages and disadvantages.

Table 3.1 Gives comparison of different high-performance architectures based on origin, price, uniqueness with its advantages and disadvantages.

Name/ architecture	Make/origin	Price (Est.)	Need in space missions	Uniqueness/ innovation	Advantages	Disadvantages
HPSC (High Performance Spaceflight Computing)	NASA/Jet Propulsion Lab	High	Autonomous operations, Artificial Intelligence/Machine Learning, sensor data processing	Radiation-hardened multicore System on Chip with TSN Ethernet	Fault-tolerant, scalable, high computing capacity, modular design	Lag in process node scaling, expensive qualification, thermal challenges
Neuromorphic Computing and Sensing	ESA/Research Labs	Medium	Event-driven tasks, vision, detection, classification	Imitate biological neurons, sparse event-driven logic	Ultra-low power, robust to noise, efficient for bursty workloads	Limited general-purpose use, small ecosystem, radiation concerns
SpaceCubeX (Hybrid CPU-FPGA-DSP)	NASA	High	High-throughput tasks (suited for hyperspectral imaging)	Heterogeneous architecture with accelerators	Efficient for parallel tasks, flexible, scalable	Complex design, verification challenges, thermal/mass overhead
MPSoC with Software Fault Tolerance	Various applications, space, academic	Medium	Fault-resilient computing with lower hardware redundancy	Multi-stage software-side fault tolerance	Lower cost, adaptable, efficient, software-based resilience	Complex software, latency in fault recovery, validation needed

(Continued)

Table 3.1 Gives comparison of different high-performance architectures based on origin, price, uniqueness with its advantages and disadvantages. (*Continued*)

Name/ architecture	**Make/origin**	**Price (Est.)**	**Need in space missions**	**Uniqueness/ innovation**	**Advantages**	**Disadvantages**
RACP (Resilient Affordable Cubesat Processor)	NASA/ TechPort	Low to Medium	CubeSats, small satellites, swarm missions	Combines ARM SoC + FPGA + radiation-tolerant microcontroller	Flexible power scaling, fault resilience, low power	Lower performance, risk from non-radiation hard parts
PEAC (Power Efficient Adaptive Computing)	NASA/ TechPort	Medium	Adaptive computing for small spacecraft	Fine-grained power domains, always-on and on-demand logic	Excellent power scheduling, scalable, efficient for variable workloads	Limited compute capacity, complexity in power arrangement
Vikram 3201 & Kalpana 3201	ISRO & SCL (India)	Low	Launch vehicle control, onboard computation	Indigenous 32-bit microprocessors, SPARC V8 RISC	Robust, qualified for harsh conditions, Make-in-India	Older 180nm node, limited for AI/ ML or high-performance tasks
Shakti/IRIS	IIT Madras & ISRO	Low to Medium	Command/control, fault-tolerant embedded systems	RISC-V based, customizable, open-source	Adaptable, low standby power, domestic capability	Not fully space-qualified yet, ecosystem still maturing

3.6.1 Processor Suitability by Mission Type

Table 3.2 describes the list of various processors and its suitability based on different space missions types.

Table 3.2 Lists the suitable processors based on the type of mission that is launched.

Mission type	Recommended processor(s)	Suitability
Lunar Rover/ Planetary Exploration	HPSC, SpaceCubeX	High autonomy, AI/ML capability, fault tolerance, real-time sensor fusion
Earth Observation Satellite	SpaceCubeX, MPSoC with Fault Tolerance	High-throughput imaging, parallel processing, moderate fault resilience
CubeSat/Swarm Missions	RACP, PEAC	Low power, cost-effective, scalable, fault-resilient for distributed systems
Launch Vehicle Control	Vikram 3201, Kalpana 3201	Proven reliability, robust under extreme conditions, indigenous support
Deep Space Probes	HPSC, Neuromorphic Computing	Radiation-hardened, adaptive learning, low power for long-duration missions
Onboard AI/Edge Computing	HPSC, Shakti/IRIS	AI-ready, customizable, scalable, open-source flexibility
Experimental/ Research Payloads	Neuromorphic, Shakti/IRIS	Innovative architectures, ideal for testing new paradigms

3.7 Conclusion

The comprehensive analysis of advanced space-grade computing architectures, the path forward for ambitious space exploration is presented in this chapter. In order to overcome the extreme limitations of the space environment, it is necessary to combine neuromorphic and RISC-V designs with strong radiation-hardened CPUs and reconfigurable field-programmable gate arrays (FPGAs). By adopting heterogeneous computing systems that flawlessly integrate diverse processors enable critical edge computing capabilities for autonomous space missions. Ultimately, the success of long-duration crewed missions and autonomous space flight depends on our capacity to consistently design and implement highly adaptable, robust, and energy-efficient computer platforms that can withstand the difficulties of deep space. The chapter convincingly establishes that achieving low-power, high-efficiency computing architectures is not merely an optimization goal, but a fundamental prerequisite for humanity's sustained and ambitious venture into space. Constraints on power budget, mass, and the harsh, radiation-laden environment are forcing a change from traditional computer paradigms on Earth to specialized systems designed for use in space.

AI disclosure statement: The author confirms that no artificial intelligence (AI)-generated content or AI-assisted writing tools were used in the preparation of this chapter (including summaries and the abstract). All text, analysis, and conclusions are the result of the author's original work.

References

1. Kuang, L., *et al.*, Space Computing Power Networks: Fundamentals and Techniques. *Engineering*, 2025.
2. Grisafi, M., On the importance of software-hardware co-design of Trusted Computing architectures and services. 2025.
3. Wang, G., *et al.*, High-performance On-orbit Intelligent Computing and Real-time Services for Remote Sensing Satellites Based on Large-scale Computing Power in Space. *IEEE Access*, 2025.
4. Guimarães, C., *et al.*, A Survey on Satellite Computing: Connecting the Dots Between Networks and Applications. *IEEE Commun. Surv. Tutorials*, 2025.
5. Bersuker, G., Mason, M., Jones, K.L., Neuromorphic computing: The potential for high-performance processing in space, Game Changer, pp. 1–12, 2018.
6. Muralidhar, R., Borovica-Gajic, R., Buyya, R., Energy Efficient Computing Systems: Architectures, Abstractions and Modeling to Techniques and Standards. *ACM Comput. Surv.*, 54, 11s, 236, 2022.
7. Kaul, S., Low-Power Design for Test Techniques in Sub-7nm Technologies. *J. Comput. Sci. Technol. Stud.*, 7, 5, 930–937, 2025.
8. Zhao, W., *et al.*, A Novel Space Intelligent Computing and Data Processing Architecture--The Spacecraft Payload Health Management Unit (SPHMU). *Front. Astron. Space Sci.*, 12, 1657487.
9. Yu, A., Woo, S., Ahn, H., Toward transforming space exploration with artificial intelligence neuromorphic computing. *Eng. Appl. Artif. Intell.*, 154, 111055, 2025.

10. Weston, S.V., *et al.*, State-of-the-art small spacecraft technology, 2025.
11. Bussemaker, J.H. and Firchau, T., System architecture optimization: an example application to space mission planning: JH Bussemaker, T., Firchau. *CEAS Space J.*, 1–9, 2025.
12. Wallace, J., *et al.*, A High-Performance 5G/6G Framework for Deep-Space Missions with Multi-Level AI, *2025 IEEE 11th International Conference on Space Mission Challenges for Information Technology (SMC-IT)*, IEEE, 2025.
13. Kim, S.-J. and Jo, S.-H., A Radiation-Hardened Low-Power SRAM with Enhanced Write Capability for Space Applications. *Appl. Sci.*, 15, 13, 6988, 2025.
14. Liu, Y., *et al.*, Computing over Space: Status, Challenges, and Opportunities. *Engineering*, 2025.
15. Zheng, Y., *et al.*, A novel sun-shading design for indoor visual comfort and energy saving in typical office space in Shenzhen. *Energy Build.*, 328, 115083, 2025.
16. Goodwill, J., Wilson, C., MacKinnon, J., Current AI technology in space, in: *Precision Medicine for Long and Safe Permanence of Humans in Space*, pp. 239–250, Elsevier, Academic Press Publisher, Greenbelt, Maryland, United States, 2025.
17. Geda, M. and Tang, Y.M., Adaptive hybrid quantum-classical computing framework for deep space exploration mission applications. *J. Ind. Inf. Integr.*, 44, 100803, 2025.
18. Jones, S., *et al.*, ANGEL: An Architecture for Simplified AI/ML Implementation on Low-Power FPGAs for Small Satellites, 2025.
19. Shahid, H., *et al.*, Effectiveness of energy harvesting systems subjected to flow-induced vibrations in confined spaces. *Renewable Sustainable Energy Rev.*, 210, 115183, 2025.
20. Ersoy, K., Review of electronic cooling and thermal management in space and aerospace applications. *Eng. Proc.*, 89, 1, 42, 2025.
21. Underwood, L., *et al.*, Astra--Autonomous Satellite Technology for Resilient Applications: A Technology Demonstrator, in: *39th Annual Small Satellite Conference*, 2025.
22. Wu, Q., *et al.*, Optimizing Collaborative Space–Air–Ground Earth Observation: A decision-making perspective. *IEEE Geosci. Remote Sens. Mag.*, 2025.
23. de Jesus, B.C. and Jorge, R., Evaluating and optimising the AArch64 ecosystem for HPC, 2025.
24. Rudge, Z.A., *et al.*, Memristor-Based Neural Network Accelerators for Space Applications: Enhancing Performance with Temporal Averaging and SIRENs, arXiv preprint arXiv:2509.04506, 2025.
25. Al-Saadi, J., *et al.*, Advancing the Science of Both Coastal Ocean Biophysics and Atmospheric Pollution Chemistry: A White Paper Report to the NASA Earth Science Division by the GEO-CAPE Team (September 2015), National Aeronautics and Space Administration, 2025.

26. Sharma, D.C., *Space: The India Story*, Bloomsbury Publishing, New Delhi, India, 2025.
27. Anil, N., *et al.*, SHAKTI Dual Lockstep Microprocessor: Ensuring Functional Integrity for Robust Computing, in: *2024 IEEE International Conference on Contemporary Computing and Communications (InC4)*, IEEE, 2024.

4

AI and Neuromorphic Computing for Autonomous Deep Space Operations

P. Ashok[1]*, S. Lakshmi Sridevi[2], K. Murali Krishna[3] and Venkatesh Ramamurthy[4]

[1]Symbiosis Institute of Digital and Telecom Management (SIDTM), Symbiosis International (Deemed University) (SIU), Lavale, Pune, Maharashtra, India
[2]Department of CSE, Chennai Institute of Technology, Chennai, Tamil Nadu, India
[3]Department of ECE, Rajalakshmi Engineering College, Chennai, Tamil Nadu, India
[4]NETS International, Dubai, UAE

Abstract

Deep space exploration missions are, in essence, critically limited by enormously long communication latencies, bandwidth restrictions, and reliance on ground control. These challenges demand a paradigm shift away from pre-programmed robotic systems and toward fully autonomous spacecraft that have reached a stage of real-time decision making, adaptation, and scientific discovery. This chapter offers a technical overview of the capabilities of Artificial Intelligence (AI) and, in more detail, brain-inspired neuromorphic computing to make this next generation of autonomous deep space operations possible. It looks at how high level AI algorithms are being incorporated for purposes like autonomous navigation, onboard scientific data analysis, fault detection, and resource management. Special emphasis is given to neuromorphic computing, which is considered to be a paradigm change. It is event-driven, and a low-power processing architecture is extremely well suited to the stringent size, weight, and power (SWaP) constraints of space platforms. The chapter will describe the fundamentals of neuromorphic systems (Spiking Neural Networks [SNNs]) and review the state of the art of neuromorphic hardware, such as Intel's Loihi and IBM's TrueNorth. In addition, the key issues of radiation hardening for the emerging computing architectures under the demanding deep space environment are presented. The potential use cases are discussed, covering different phases of the mission, such as interplanetary cruise,

**Corresponding author*: p.ashok@sidtm.edu.in

N. Sathish, Prasanth Aruchamy, Rajesh Kumar Dhanaraj, Nithya Rekha and Karthik Palani (eds.) Space Electronics for Deep Space Exploration, (83–108) © 2026 Scrivener Publishing LLC

planetary landing, and surface exploration. Finally, by uniting the state of the art around research, hardware, and algorithmic challenges, this work provides a technology roadmap for implementing AI and neuromorphic systems, providing scalable solutions to implement AI-enabled systems for highly ambitious and scientifically interesting deep space missions.

Keywords: Autonomous systems, deep space exploration, neuromorphic computing, spiking neural networks, radiation hardening, onboard data processing, autonomous navigation, low-power computing

4.1 Introduction

Space travel has taken mankind to new heights of exploration and scientific and engineering knowledge for decades. Moon exploration, Mars exploration, and exploration of the outer planets have led to amazing discoveries that have changed how we fully understand the universe. However, the operational paradigm for these missions has not really differed much: spacecraft follow highly planned trajectories and sequences with critical decisions made by human operators on Earth. This "ground-in-the-loop" approach is slowly but surely becoming a constraint when mankind sets its sights on more distant and complex targets, such as plumes and icy mineral regolith of the Jovian and Saturnian midsts, or the distant Kuiper Belt. The basic shortcoming is based on the speed of light. Round-trip communication times to Mars can be between 8 and 48 minutes, or for a mission to Neptune, almost more than 8 hours. This communication lag makes real-time remote control impossible, especially during such important events as planetary entry, descent, and landing (EDL), or when unexpected scientific phenomena or system anomalies are encountered. The data downlink capability is also a significant constraint, and scientists are often forced to be very selective about the data that is returned, possibly missing transient scientific events. To overcome these barriers, a new generation of spacecraft is needed; one that has a high level of autonomy. But this autonomy must go further than mere scripted sequences and build to true cognitive functions: the definition's sense of being able to sense the world around you, to reason about it and make decisions, and to learn from experience without the intervention of a human. Speech and natural language processing, AI, and machine learning (ML) are the fundamental foundational technologies of this vision. Conventional AI, however, which is commonly run on high-power GPUs, is challenging from a size, weight, and power (SWaP) constrained environment of a deep space probe. This has led to the development of research into a new computing paradigm: neuromorphic computing. Like the biological brain, which integrates efficient computation and a modular architecture, neuromorphic systems have the potential to bring new verifying computational abilities for AI processes while offering up to triple the lower energy consumption. This chapter gives an extensive overview of the convergence of AI and neuromorphic computing in the specific context of autonomous deep space operations in the context of general space electronics.

4.2 Literature Review

Neuromorphic computing in conjunction with AI is quickly changing the way autonomous operations are carried out in deep space. Recent techniques in neuromorphic hardware and brain-inspired algorithms have allowed the creation of energy-efficient, robust, and adaptive systems to fulfill the unique needs of space missions (such as autonomous navigation, real-time data processing, and decision-making under stringent resource constraints), specifically [1–8]. AI-enabled autonomy plays an ever-increasing role in deep space mission activities because of the delay in communications and the harshness of space environments, which require intelligence and resilience on board [9–17]. Neuromorphic computing, with its low-power, event-driven architectures, can be an exciting solution to the problem of the limitations of traditional von Neumann systems, especially for edge computing and real-time control in space robotics and satellite operations [1, 3, 5–8, 18–22]. This paper introduces and summarizes results from recent papers that represent the state-of-the-art in AI and neuromorphic computing in the realm of autonomous deep space operations: hardware innovations, algorithmic developments, mission applications, and open issues. Recent research activities have shown significant advances in neuromorphic hardware, such as spiking neural networks (SNN), memristor-based circuits, and photonic neuromorphic systems, all of which have brought immense energy-saving capability and real-time processing capability for space applications [1, 3, 5–8, 18–26]. Recently emerged system technologies like MoS2 optoelectronic synapses (2D MoS2) [23], spintronic devices [25], and neuromorphic circuits (3D) [26] lead to miniaturized, durable, and adaptable in-Place computing systems. Techniques in the field of AI, deep learning, reinforcement learning, bio-inspired algorithms, and fungus learning, called federated learning, are being implemented in mission planning, navigation, robotic manipulation, and fault-tolerant control for deep space operations [2, 7, 10–12, 14–17, 27–31]. Hybrid deep learning and neuromorphic architectures are developing as organic routes, which blend deep learning and neuromorphic architectures, are an interesting principal way forward for high-performance and low power consumption [3–5, 7, 8, 22]. Neuromorphic and AI-driven systems are being applied in various types of space applications such as autonomous satellite operation, distributed satellite systems, robotic grasping, onboard data analysis, and mission autonomy [5–7, 10–12, 14–17, 19, 30, 32–34]. Notable examples are the use of neuromorphic processors for on-board satellite communication [6], AI-based guidance and navigation to uncooperative space object inspection [12], and energy-efficient mapless navigation for planetary rovers [7]. Even though there has been tremendous progress, there is still much to do to scale neuromorphic systems, ensure robustness in harsh space environments, standardize AI integration, and address legal, ethical, and explainability concerns [2, 5, 8, 10, 11, 21, 32, 35–40]. Research is still ongoing for the reliability, adaptability, and interpretability of AI and neuromorphic systems in deep space missions. The interconnection between AI and neuromorphic computing is transforming autonomous deep space operations, which allows systems to be energy-efficient and properly handle real-time, adaptive decision making [1, 3, 5–8]. As the literature review reveals, SNN-based chip and memristor circuit hardware offer significant power reduction and fault tolerance performance, which are important for space missions, below the critical resource limits [5–7, 19, 20, 22, 23, 25, 26, 32]. AI algorithms, especially those that exploit the capabilities of deep learning and reinforcement learning, however, are increasingly adapted for neuromorphic platforms, leading to hybrid platforms [3–5, 7, 8, 22] that exploit the capabilities of deep learning and reinforcement learning. Several challenges, however, still exist. The strength of neuromorphic and AI systems in the intense space environment, the hardware scalability, and the demand for explainable and trustworthy AI are still open research fields [2, 5, 7, 10, 11, 21, 32, 35, 36, 38–40]. Additionally, legal and ethical issues, as well as standardization issues of safe and reliable deployment, need to be addressed [10, 31, 35, 36, 39, 40]. The adopted technologies are leading to more integrated, adaptive, and resilient systems, but the state of the art is still far from giving full potential for the service of deep space operations through AI and neuromorphic computing. The literature proves that the use of AI and neuromorphic computing is propelling the capabilities of autonomous deep space operations to advance, providing both energy efficiency and strength. While the hardware, algorithms, and applications in space have made significant strides, there is still research needed to overcome challenges of scaling, reliability, and explainability. Future research should work on scaling of local neuromorphic hardware for space deployment, integration of explainable AI, and building vigorous and standardized systems for deployment in extreme environments. In conclusion, AI and neuromorphic computing offer a great opportunity to transform autonomous deep space operations; however, there is a lot of research required to overcome scalability, robustness, and explainability difficulties to deploy in mission-critical scenarios.

4.3 Deep Space Exploration: Architectures, Applications, and Implementation

The operational paradigm for deep space exploration is fundamentally dictated by the physics of communication across interplanetary distances. This model, characterized by long latencies and limited bandwidth, imposes severe constraints on mission complexity, reactivity, and scientific return. The "light-time" delay is the most significant factor.

4.3.1 The Imperative for Onboard Autonomy in Deep Space

For a mission at Mars, a signal can take up to 24 minutes to travel one way, meaning a 48-minute wait for a command to be acknowledged. For a probe orbiting Jupiter, this round-trip time is nearly 2 hours, and for Neptune, it exceeds 8 hours [8, 10]. This delay makes direct, real-time control of a spacecraft from Earth an impossibility for anything other than slow, predictable operations. Critical mission events, such as EDL, or a flyby of a dynamic object like a comet, must be executed autonomously. During the "seven minutes of terror" for a Mars landing, the entire sequence from atmospheric entry to touchdown is completed before ground control even receives the signal that the sequence has begun. Any anomaly during this phase must be handled by the spacecraft itself. Beyond critical events, the communication lag also hampers scientific discovery. A rover on Mars might encounter a unique geological formation, but by the time its images are sent to Earth, analyzed by scientists, and commands are sent back to investigate further, the rover may have moved on, or the lighting conditions may have changed. A transient event, like a plume erupting from the surface of Europa, could be missed entirely if the spacecraft is not equipped to recognize its significance and act upon it immediately. The second important limitation is data bandwidth. The bit rate of deep space is very low, even by terrestrial standards, subject to the inverse square law. The Deep Space Network (DSN) is a global network of powerful antennas, and it is a shared and over-subscribed facility. Missions are therefore forced to share communication time, and can only transmit a small percentage of the data they have gathered. As a result, there is a bottleneck of data to be transmitted through the data link, resulting in the mission planners making hard choices on how to prioritize their data downlink [2, 34]. High-resolution data, hyperspectral data, and other rich data types tend to be down sampled or selectively transmitted, and in doing so, potentially lose valuable scientific data. AI-based onboard autonomy is the direct solution to these challenges [9, 13, 16]. Also, it allows for a closed loop between sense, think, and act to be closed on the spacecraft itself, making the ground less significant. An autonomous system will be able to make time-critical decisions, for example, to make a safe landing site in real time or re-tasking instruments in an effort to come back with a rapidly evolving scientific event. It can also do onboard data triage as well, triaging and identifying the most scientifically valuable information in the full high-fidelity dataset collected by instruments and prioritizing it for downlink. Some call this "spacecraft-to-science conversion," but more importantly, this process operationalizes the spacecraft from an uncomplicated data acquisition system into a scientifically active system that increases the scientific yield of a probe to many orders of magnitude. These resolutions are increasingly important for many space missions where navigation in unstructured and unpredictable environments (e.g., diving into the subsurface oceans of icy moons or complex asteroid belt dynamics) is required. In these environments, pre-programmed sequences fall short and the spacecraft must be capable of adapting itself to an unknown and changing environment [41, 43].

4.3.2 Foundations of AI for Space Applications

AI for space applications has come a long way from the simple expert systems to today's sophisticated ml models [14, 17, 28]. These technologies form the "brains" of an autonomous system as they enable an autonomous system to interpret data gathered from sensors and make intelligent decisions. The Barde's AI technique can be divided into AI techniques that can be split into broad groups highly relevant to deep space operations. First, ML, which comes under AI, consists of algorithms that learn patterns from data without being explicitly programmed. Supervised Learning is one of the main examples of it, where in such a case, a model is trained on a labeled dataset. For use in space, this lends itself very well to tasks such as terrain classification, where a model is trained with the data analyzed from images that are labeled "safe terrain," "rock," "crater," etc. Convolutional neural networks (CNNs) are a type of deep neural network that is particularly efficacious in solving problems that have grid-like data, such as images. They have become the standard for computer vision missions in space for optical navigation, hazards, and scientific features. A typical CNN has layers of convolutions (as feature detectors), pooling layers (as down-samplers), and fully connected layers (for the task of classification). This hierarchy provides an opportunity for the network to learn more simple features, such as edges in lower layers and more complex features, such as crater rims or shapes within rocks, in higher layers [3, 4]. Unsupervised Learning is used when labeled data is unavailable [28, 49]. The algorithms seek to find hidden structures in the data. Clustering algorithms, for example, can be used to group similar instrument readings, which can help in identifying different geological compositions from spectrometer data or in detecting anomalous spacecraft behavior that deviates from the norm. Autoencoders are another powerful unsupervised technique, where a neural network is trained to reconstruct its input. By learning to compress and decompress data, it learns the latent features of the nominal data distribution. Inputs that result in a high reconstruction error can be flagged as anomalies. Second, sequence models are designed to handle data where order matters, such as time-series telemetry or commands. Recurrent Neural Networks (RNNs) contain feedback loops, allowing them to maintain an internal state or "memory" of past inputs [15, 32, 32, 45]. This makes them suitable for analyzing telemetry streams. However, standard RNNs suffer from the vanishing gradient problem, making it difficult for them to learn long-range dependencies. Long Short-Term Memory (LSTM) networks are a special kind of RNN that solves this problem using a set of "gates" (input, output, forget gates) that regulate the flow of information into and out of the cell's memory. This architecture allows them to effectively learn from data with time lags of thousands of steps, making them ideal for predictive health monitoring of spacecraft subsystems over long periods. Third, Reinforcement Learning (RL) is a paradigm that is focused on goal-directed learning from interaction. An "agent" (the control system of the spacecraft) learns to take a sequence of decisions (a "policy") in an environment to maximize a cumulative reward. This is extremely well suited to problems of autonomous control. For example, an RL agent can be trained in a high-fidelity simulation to drive a rover on a planetary surface, in which the "reward" would be the speed to reach a goal location without harming itself and with low energy consumption. This strategy can be used to find complex and non-intuitive strategies that score better than human-designed heuristics. Similarly, RL can be used to optimize complicated multi-burn orbital maneuvers or manage the power and thermal resources of the spacecraft. While these types of AI models have great capabilities, implementing them in conventional hardware poses a major challenge to their implementation in space. Training these models, particularly the deep neural networks, is computationally intensive but is usually performed on the ground itself. The thing that is important about the mission is inference - the process of using the trained model to make predictions. However, the invasion of a Gigantic CNN on

Table 4.1 Key AI techniques and their space applications.

AI technique	Description	Example space application	Key benefit for autonomy
CNNs	Deep networks for image analysis, learning hierarchical features.	Terrain classification, hazard avoidance (TRN), crater detection, and scientific feature finding.	Enables robust real-time visual perception.
LSTMs	RNNs are capable of learning long-term dependencies in sequential data.	Spacecraft telemetry analysis for predictive health monitoring (FDIR), solar wind prediction.	Enables proactive fault management.
Autoencoders	Unsupervised networks that learn efficient data coding to find anomalies.	Anomaly detection in telemetry, identification of novel scientific data.	Detects unforeseen issues and discoveries.
Reinforcement Learning (RL)	Learning optimal decision-making policies through trial-and-error with a reward signal.	Autonomous rover path planning, optimal trajectory control, and power resource management.	Enables adaptive and optimized control.

a normal non-McSHT CPU will be expensive and consume lots of energy. While Graphics Processing Units (GPUs) improve the speed of these tasks tremendously, their commercial versions are tremendously power hungry and sensitive to radiation. This has resulted in a desire for a new type of processing hardware that was inherently efficient and better suited to the structure of neural algorithms, which is a direct route to exploring neuromorphic computing, as in Table 4.1.

4.3.3 Neuromorphic Computing: A Paradigm Shift for Space Electronics

Neuromorphic computing is an essential step away from the conventional von Neumann architecture, which has been the basis of computing for nearly 70 years. Neuromorphic systems take inspiration from the structure and function of the biological brain, where memory and processing units are not separated, which results in the well-known "von Neumann bottleneck." The brain is an example of a masterpiece in terms of massively parallel computing with power constraints and fault tolerance. Neuromorphic engineering attempts to mimic these principles on silicon. The main principles of the neuromorphic paradigm are event-based processing, memory and processing close to each other, and massive parallelism.

I. **Event-Based Processing**: In contrast to the synchronous clock operations of a CPU, nervous systems are not synchronous but, instead, are events that can be triggered. Computation occurs only in response to a significant event, which is known as a "spike." This is similar to the way that neurons communicate in the brain. A neuron gets the incoming signals integrated, and only when the membrane potential of a neuron crosses a certain threshold does it fire a spike to other neurons. It is this nature of events that leads to extreme energy efficiency. There is no new information or activity. The system uses very little power, while a CPU that constantly runs instructions based on a clock is power-intensive even when idle. This capability is of unprecedented importance for deep space missions in which energy resources are provided by solar power array units or radioisotope thermoelectric generators (RTG), with power being a limited resource.

II. **Co-Location of Memory and Processing**: This causes the von Neumann bottleneck, as there is continuous transferring of data between the CPU and main memory *via* a comparatively low-speed bus. This movement involves a huge amount of total energy in modern systems. In neuromorphic chips, the memory (synaptic weights) and the processing elements (neurons) are distributed and reside in proximity to one another. Spike arrives on a neuron at the level of the local structure of the neuron, accesses the locally stored synaptic weights, does a computation, and its state is updated. This cripples data turnover in massive ways to save time and energy.

III. **Spiking Neural Network (SNN):** The Spiking Neural Network algorithmic model represents the algorithmic model of neuromorphic computers. SNNs are the third generation of neural networks after perceptrons and connectionist models. Unlike traditional Artificial Neural Networks (ANNs), whose activations are continuous values processed by every cycle, in SNNs, communication is done using discrete and sparse spikes over time. This temporal aspect enables the processing of information coded in the timing, rate, or phase of the spike cycles, making this kind of modification of neurons suitable for processing data from dynamic sensors (such as event-based cameras) and for tasks that involve temporal patterns, such as speech recognition, and for the analysis of time-series telemetry data. A typical model neuron used is of the type Leaky Integrate-and-Fire (LIF), which is computationally simple but captures the essential dynamics of spiking activities. By emulating the low-density, asynchronous, and massively parallel nature of the biological brains, neuromorphic computing offers a promising hardware solution to the SWaP limitations of planetary exploration, with the potential to enable orders of magnitude lower AI processing power cells as compared to traditional GPUs/CPUs.

4.3.4 Hardware Implementations of Neuromorphic Processors for Space

The theoretical promise of neuromorphic computing has been realized in several silicon implementations, creating a new class of processors with potential for space deployment. These chips vary in their architecture, scale, and learning capabilities, but they all share the core principles of event-driven and parallel processing.

I. **IBM TrueNorth:** One of the earliest large-scale neuromorphic processors, TrueNorth was developed as part of the DARPA SyNAPSE program. It features a massively parallel architecture with 1 million digital neurons and 256 million configurable synapses, organized into a grid of 4,096 "neurosynaptic cores." Each core is an independent, asynchronous unit with local memory and processing. TrueNorth is deterministic and highly energy-efficient, consuming only tens of milliwatts during complex real-time pattern recognition tasks. Its primary limitation is that it is designed for inference only; the synaptic weights must be pre-trained offline using a specialized toolchain. Its deterministic nature, however, is an advantage for verification and validation in critical systems.

II. **Intel Loihi & Loihi 2:** Intel's Loihi family represents a significant step towards more flexible and capable neuromorphic systems. The first-generation Loihi chip featured 128 neuro-cores, containing 131,000 neurons and 130 million synapses. Crucially, unlike TrueNorth, Loihi incorporates on-chip learning capabilities, using programmable microcode to implement synaptic plasticity rules like Spike-Timing-Dependent Plasticity (STDP). This allows the network to adapt and learn from new data in real-time. The successor, Loihi 2, manufactured on a pre-production Intel 4 process, offers up to 8 times the neuron density, improved performance, and a more generalized and programmable neuron model. The ability to perform online learning is a game-changer for deep space missions, enabling a spacecraft to adapt to its specific, unique environment, such as learning the particular characteristics of Martian dust to improve navigation. The software framework for Loihi, called Lava, is an open-source project aimed at creating a common programming model for the neuromorphic research community.

III. **SpiNNaker and SpiNNaker2:** The SpiNNaker (Spiking Neural Network Architecture) platform, developed at the University of Manchester, takes a different approach. Instead of custom neurosynaptic cores, it uses a large number of simple, low-power ARM processor cores (18 cores per chip on SpiNNaker1), connected *via* a bespoke communication fabric optimized for routing billions of spike packets per second. This makes SpiNNaker highly flexible and programmable, allowing it to simulate a wide variety of neuron and synapse models in real-time. It is designed to be a tool for computational neuroscience as well as a neuromorphic computer. The second generation, SpiNNaker2, developed at TU Dresden, leverages more powerful ARM cores and advanced power management to achieve significantly higher efficiency and computational power.

IV. **Other Commercial and Research Platforms:** Beyond these large-scale projects, several other platforms exist. BrainChip's Akida processor is a commercial event-based neural processor designed for edge AI, combining SNN and standard AI concepts. Startups like SynSense and GrAI Matter Labs are also developing neuromorphic chips for low-power edge applications. In the research domain, novel device technologies are being explored to move beyond standard CMOS. Memristor-based crossbar arrays are particularly promising. These devices may be used to bank the synaptic weights in the form of analog resistance values, so that extremely dense and low-power implementations of the multiply-accumulate operations at the heart of neural computation may be realized. When considering these processors for space, there are other considerations besides rude performance and energy. Temperature range of operation, long-term reliability, and most importantly, radiation tolerance are paramount. Most existing neuromorphic chips are COTS devices, which are not suited for the space environment. Therefore, strong research work is needed to characterize their response to radiation, as well as to develop strategies for hardening them, as in Table 4.2.

4.3.5 Applications of AI and Neuromorphic Systems in Deep Space Missions

The synergy between AI and neuromorphic computing will take deep space missions to the next level in almost every respect. Applications range from the basic capability of the spacecraft to navigate and survive

Table 4.2 Comparison of leading neuromorphic hardware platforms.

Feature	IBM TrueNorth	Intel Loihi 2	SpiNNaker2
Architecture	Digital, custom neurosynaptic cores	Digital, custom neuro-cores with programmable microcode	Many-core system using general-purpose ARM processors
Neurons	1 million	Up to 1 million	Simulates millions (depends on model complexity)
Synapses	256 million	Up to 120 million	Simulates billions (depends on model complexity)
On-Chip Learning	No (inference only)	Yes (programmable synaptic plasticity rules, STDP)	Yes (software-defined on ARM cores)
Power Consumption	Extremely low (~70 mW for typical tasks)	Very low (tens of mW to a few W)	Low (system-dependent, ~1W per chip)
Key Advantage for Space	Deterministic behavior, proven low power.	On-chip learning for adaptability, high flexibility.	High programmability for research and diverse models.
Primary Challenge	Inflexible, requires offline training.	COTS component requires rad-hardening.	Higher power than custom ASICs for specific tasks.

to its advanced scientific exploration capability. These missions can be spread out over a variety of mission phases. Guidance, Navigation, and Control (GNC) is probably the most important function of an autonomous spacecraft. AI promotes the transformation from ground-based radiometric tracking systems (intermittent) into continuous onboard autonomous navigation.

I. **Optical Navigation (OpNav):** This is a major technology by which the spacecraft can image the neighboring bodies of the solar system (Planets, Moons, Asteroids) against a background of stars using cameras. When these bodies are known and their positions measured against the known background of stars, their position and velocity can be calculated and known by the spacecraft itself. CNNs are also very good tools for finding out the extremities of a planet, or to spot specific craters on the surface of a moon for pinpoint location detection. Neuromorphic systems, when combined with event-based cameras, can facilitate this, identifying at extremely low delay tracking features continuously from instantiation rather than frame-by-frame.

II. **EDL:** This is a very critical phase, and AI is indispensable here. The terrain and relative navigation (TRN) system included in the Mars 2020 mission is taken as the reference case. As the craft descends and captures pictures of the ground, these are cross-referenced with a map of known hazards (big rocks, steep slopes) stored on the onboard computer, and the craft will diverge to a safe landing spot. In fact, the whole thing has to take place in minutes, way too quickly for intervention from the ground. Future systems will operate with more advanced AI, transforming into potential RL-trained agents to make even cuter decisions, balancing between being safe and getting close to a higher-priority science target. Neuromorphic vision processing has a potentially profound impact on limiting the power and latency of the hazard detection aspect of this loop.

III. **Trajectory Optimization:** Reinforcement learning can be used for Gospel Coyote Pneumatic Commanded Guidance (GNC) for optimizing curves for entering an orbit or interplanetary transfer using low-thrust propulsion systems. The RL agent can learn a policy that controls the continuous thrust of an electric propulsion system, and may find solutions that conserve large amounts of propellant, which may be the dominant life-limiting resource of a mission. Deep space missions are fundamentally about science, and AI has the power to turn science and transform spacecraft into intelligent scientific observers.

IV. **Data Triage and Prioritization:** Systems with hyperspectral imagers or particle detectors can provide terabytes of data per day, which is much more than the downlink bandwidth can handle. An AI system developed by scientists and trained by them is able to analyze this data onboard in real time. It can be charged to search for some particular biosignatures, geological structures, or phenomenal transients. An onboard AI could analyze imagery of a mission to Europa, looking for plumes of water vapor gushing from the ice shell. On detecting, it could automatically retarget other instruments, such as a mass spectrometer, to have it analyze the composition of the plume before the opportunity is lost. This technology of putting together a science on the go is a significant improvement in mission efficiency.

V. **Autonomous Discovery:** Anomaly detection algorithms can be used to examine scientific data for any unexpected or unusual patterns that do not conform to models of known data. This may result in a serendipitous discovery that would have been missed by a human, or would be lost in some of the data down-sampling. A neuromorphic system would constantly monitor information streams of several instruments, combining them to look for coincidences, the indicator of an interesting new event.

VI. **Data Compression:** Models of the type AI autoencoders can be used for intelligent data compression. Instead of using standard, lossless, or lossy compression algorithms, the trained model based on domain-specific data (e.g., magnetospheric data) will yield much higher compression ratios for nominal data, while at the same time maintaining the high-fidelity information of scientifically interesting anomalies. Components on spacecraft that go on deep space missions for years, or even decades, are put under a lot of stress put on them. The ability to detect and report faults and isolate them, and to report them back to ground control - Fault Detection, Isolation and Recovery (FDIR) - is critical to mission survival.

VII. **Predictive Maintenance:** As in the prior art, FDIR systems have a rule-based check (e.g., if voltage > X, switch to redundant unit) that already exists. AI can offer a more intelligent and predictive way. About their design, ML, and in particular LSTM, models can be trained with historical and simulated telemetry (voltages, temperatures, currents, component states) to learn the "normal" signature of operation of the spacecraft. By constantly monitoring live telemetry, these models can identify small deviations that are the precursors to a fault and hence allow predictive maintenance to be implemented. For example, it can sense a partial increase in current drawn by a reaction wheel motor, which indicates the wear of the bearing and warns the system of a backup wheel before a catastrophic failure occurs.

VIII. **Root Cause Analysis:** The root cause analysis of the anomaly can be challenging due to the interaction of the subsystems. AI's causal inference models can be used to analyze the chain of events from the onset of an anomaly to identifying the most probable cause of the fault, to speed up the recovery.

IX. **Low-Power Monitoring:** Neuromorphic systems are exceptionally well-suited for implementing FDIR systems. An SNN could be tasked to monitor thousands of telemetry channels simultaneously. Due to the event-based nature, it would consume almost no power as long as all systems are nominal. Only when a telemetry value begins to drift (creating "events") would the system become active, analyze the pattern, and raise an alarm. For surface missions, autonomy dramatically increases the pace and scope of exploration.

X. **Autonomous Traverse:** An autonomous rover can perform long-distance traverses that would take weeks or months to command from Earth. This requires a tight perception-planning-action loop. AI is central to this, particularly for path planning and navigation. Using stereo cameras and/or LiDAR, the rover builds a 3D map of its immediate surroundings. An AI planner (often using techniques from RL or search algorithms like A*) then identifies a safe and efficient path towards a long-term scientific goal designated by scientists. This involves reasoning about terrain traversability, slope angles, and potential obstacles. Neuromorphic vision systems can accelerate the perception part of this loop.

XI. **Autonomous Science Targeting:** Using onboard image and spectral analysis, a rover can identify rocks of particular interest (e.g., based on color, texture, or composition as determined by a CNN) and decide to perform a close-up analysis with its arm-mounted instruments without waiting for ground commands. This allows the rover to investigate many more targets than would be possible with the ground-in-the-loop model.

XII. Complex Manipulation: Future systems might use RL and other AI techniques to learn complex manipulation tasks, such as autonomous sample acquisition, drilling, or even assembling simple structures on a planetary surface. Neuromorphic implementations could provide the low-latency feedback control needed for such delicate tasks as those in Table 4.3.

Table 4.3 Mapping of AI/neuromorphic applications to mission phases.

Mission phase	**Key challenge**	**AI/neuromorphic application**
Launch & Early Ops	System checkout and calibration	Automated telemetry analysis for system health verification.
Interplanetary Cruise	Long duration, low activity; navigation	Predictive health monitoring (FDIR); Autonomous optical navigation.
Orbital Insertion	Complex, time-critical maneuvers	RL-based trajectory optimization; real-time maneuver validation.
Orbital Science	Massive data volume, limited downlink	Onboard science data triage; autonomous event/feature detection.
EDL	Extreme time criticality, unknown terrain	Terrain-relative navigation; real-time hazard detection and avoidance.
Surface Operations (Rover/Lander)	Interaction with the environment, navigation	Autonomous path planning and traverse; autonomous science target selection.

4.3.6 Radiation Hardening of AI and Neuromorphic Hardware

The deep space environment is awash with radiation from galactic cosmic rays (GCRs), solar particle events (SPEs), and trapped radiation belts around planets like Jupiter. This radiation poses a severe threat to electronic components. For any AI or neuromorphic processor to be viable for deep space, it must be hardened against this environment. Radiation effects are broadly categorized as Total Ionizing Dose (TID) and Single Event Effects (SEE). TID is the cumulative damage caused by long-term exposure to radiation, which creates trapped charges in the oxide layers of transistors. This can lead to shifts in transistor threshold voltages, increased leakage currents, and eventual device failure. SEEs are caused by a single high-energy particle striking the semiconductor. These include: A bit-flip in a memory cell or register (a 0 becomes a 1 or vice-versa) called Single Event Upset (SEU), A transient voltage pulse in a combinational logic circuit called Single Event Transient (SET), A potentially destructive high-current state created in the parasitic SCR structure of CMOS devices, requiring a power cycle to clear called Single Event Latchup (SEL). Traditional radiation hardening involves several approaches. Radiation-Hardened-by-Process (RHBP) uses special fabrication processes, like silicon-on-insulator (SOI), to make chips inherently more resistant. Radiation-Hardened-by-Design (RHBD) involves circuit-level and system-level techniques applied to standard CMOS processes. Examples include: Triple Modular Redundancy (TMR), where three identical logic circuits perform the same operation and a voter circuit outputs the majority result, masking any single fault. Error Detection and Correction (EDAC) Codes are used to protect memories by adding redundant bits that can detect and correct errors. Dual Interlocked Cell latches are storage elements designed to be immune to single-node charge collection, making them SEU-tolerant. These techniques, which have been established in the past, can be applied to digital AI accelerators and neuromorphic chips. For instance, the SRAM memory cells used to store synaptic weights or the registers that are used for storing the neuron states can be protected by EDAC. The digital logic of neurosynaptic cores can be implemented by TMR. However, there are special issues and opportunities with neuromorphic architectures. The massive parallelism and redundancy that are inherent in a brain-inspired design may convey a natural level of fault tolerance. If a few neurons or synapses are knocked out by the strike of radiation, it could be that the distributed nature of the computation in the SNN could enable the network to keep functioning but with a slightly degraded performance—a property known as "graceful degradation." This is a major possible advantage compared to conventional CPUs, where if the program counter were faulty even slightly, it could be devastating. Research is currently being conducted to quantify this inbuilt resilience, and also to create more understanding training methods for SNNs, in the form of specific methods that promote inbuilt resilience by injecting noise into the training. For analog or mixed-signal neuromorphic designs, the challenge is different. TID effects can cause significant drifts in the behavior of analog circuits. SEUs can inject a transient charge that looks like a spurious spike, introducing noise into the computation. Hardening these circuits requires careful analog design techniques (e.g., enclosed layout transistors, guard rings) and system-level calibration mechanisms. A hybrid approach is likely the most promising path forward: using established RHBD techniques like TMR and EDAC to protect the most critical parts of the chip (e.g., configuration registers, routing logic), while leveraging and enhancing the natural fault tolerance of the distributed SNN architecture to handle transient errors in the wider network of neurons and synapses.

4.4 Technical Challenges/Limitations

Despite the immense potential, the deployment of AI and neuromorphic systems in deep space faces significant technical challenges that must be addressed before they can be widely adopted for mission-critical operations. Firstly, the Verification and Validation (V&V) of non-deterministic, learning-based systems is a major hurdle. Mission planners and operators must have absolute confidence that the AI will behave safely and predictably, especially in unforeseen situations that were not part of its training data. How can one prove that a deep neural network, with millions of parameters, will not make a catastrophic error when presented with a novel input? This requires the development of new V&V techniques beyond traditional software testing. Research into formal methods for neural networks, which can provide mathematical guarantees on output bounds, is promising but currently limited to smaller networks. Extensive, high-fidelity simulation environments, or "digital twins" of the spacecraft and its environment, are needed to test the AI against countless edge cases before flight. Secondly, the development and training of Spiking Neural Networks for neuromorphic hardware is still a nascent field compared to conventional deep learning. The toolchains, software frameworks, and algorithmic understanding are less mature. SNNs cannot be easily trained with standard backpropagation due to the non-differentiable nature of the spiking mechanism. Current methods include converting pre-trained ANNs to SNNs, which often leads to a loss in performance, or using surrogate gradient methods to approximate the derivative of the spike function. These techniques are an active research area and require specialized expertise, which puts a higher barrier to entry for application developers. Thirdly, radiation-hardening aspects of these novel architectures are the main concern. While there is still the potential for graceful degradation, it must be demonstrated with a high degree of rigor by conducting numerous radiation test campaigns on a candidate hardware. Detailed information on their failure mechanisms is needed to accurately model their performance degradation when they are exposed to a realistic space radiation environment. Developing good RHBD techniques for architectures mixing digital, analog, and memory elements at a fine-grained scale is a difficult electronic design problem that requires a significant investment.

Fourthly, there is a practical bottleneck in the form of a lack of space-qualified neuromorphic hardware. Current leading-edge chips, such as Loihi 2, are a commercial-off-the-shelf (COTS) product. The entire qualification process of space, from testing radiation, thermal vacuum cycle, vibration testing, to creating a reliable supply chain, lasts long and costs a lot. This makes it difficult for them to find their way into near-term mission designs, which commonly consist of hardware with long flight histories. Bridging this "valley of death" between a promising COTS technology and a flight-qualified component is a major programmatic challenge. Finally, there is the challenge of the data. Large quality quality-labeled datasets are the most important resource for training high-performance AI models. For space applications, such data is often dirty, costly, or lacking for future environments that are not or have not been visited. While simulated data is very powerful, models trained on it are required to be robust to the "sim-to-real" gap. The models have to eventually operate correctly with actual sensor data, which may have different noise characteristics and features. This requires the development of powerful domain adaptation capabilities, as well as the capability for models to perform some level of on-mission refinement or the ability to learn to adapt to the environment in which they actually need to operate.

4.5 Future Enhancements

The future of autonomous deep space systems will be shaped by progressive advances in a number of key specific areas to move beyond what is currently possible for truly cognitive space platforms. The main direction is the construction of an integrating heterogeneous computing platform. The structure of future space processors will most likely not be one CPU or one neuromorphic chip, but a complicated System-on-a-Chip (SoC). This SoC would include a sequencing and high-level control general-purpose processor with radiation hardened using a high-performance AI/neuromorphic accelerator for pattern recognition and perception workloads, and might include reconfigurable logic (FPGA) to accommodate flexible data flow pipelines. This approach, which would often be termed "right-processor-for-the-task," would enable the system to have the best power efficiency for the computational element performing every specific function. Another big boost will come from the development of on-chip learning to maturity. Spacecraft with the capability to adapt and improve their performance during a multi-year mission will continuously apply and evolve these concepts of adaptive capabilities, which is called life-long learning. For example, the rover could learn the detailed soil mechanics of where it is, in order to drive better, or the orbiter could adjust its scientific models for atmospheric phenomena based on new observations. However, one of the major issues that must be tackled in this sphere is called "catastrophic forgetting," where a network trained on a novel task forgets previously learned tasks. Approaches such as elastic weights consolidation and experience replay are being researched as ways of surmounting this. Construction and incorporation of neuromorphic sensors will be an important addendum. Some of the sensors, such as event-based cameras (Dynamic Vision Sensors) and silicon cochleas, are themselves inspired by their human origins and generate sparse event-based data streams. They perfectly fit with neuromorphic processors to form a seamless ultra-low power perception pipeline running all the way from "sensation" to "action." This close integration will allow unprecedented reactivity for other types of tasks such as high-speed asteroid flybys, space debris tracking and avoidance, or dexterous robotic manipulation. Additionally, AI algorithms will be among the key drivers to be developed further, especially in terms of explainable AI (XAI), and ultimately help to build trust and perform V&V. If an AI system can give some human understandable reason for its choices (e.g. "I am diverting to this landing spot because my vision system detected a rock of size X with 98% confidence at the original target"), the behavior of the system can be more easily verified by mission controllers and any issues diagnosed. Methods such as LIME (Local Interpretable Model-agnostic Explanations) and SHAP (Shapley Additive exPlanations) go some way in this direction. Finally, the technology of underlying semiconductors will continue to push the boundaries. The use of 3D chip stacking will provide for the tight integration of different dies (e.g., a logic die stacked with a memory die), thus reducing even further the von Neumann bottleneck. Novel non-volatile memory devices such as memristors and phase-change memory are being discussed as potential candidates for dense, analog synaptic weight storage, which can create neuromorphic systems with billions of neurons and trillions of synapses, on the scale of a small mammalian brain. Passive systems would become available, which in themselves would produce levels of autonomy and scientific discovery unthinkable today.

4.6 Conclusion

A leap from the remotely operated probes to true explorers is the next big leap in our exploration of the cosmos. Given the near impenetrability of these environments to telecommunication by virtue of their communication delay, low bandwidth, and the requirements for real-time decision making and information exchange, this transition is not only desirable but obligatory for future scientific activities. The benefits of this leap come through AI, which provides the software and algorithmic backbone for autonomous navigation and landing, onboard scientific discovery, as well as predictive health management, among other things. However, there have been SWaP constraints of spacecraft, which, until now, have limited the complexity and scale of the AI that could be practically deployed. A brain-inspired, neuromorphic, and massively parallel architecture based on event-driven computing, the solution to this fundamental limitation in hardware is currently at hand. Upon providing a promise of AI-level computation for a part of the power cost of conventional systems, Neuromorphic electronics are quickly becoming a cornerstone technology for the next generation of space systems. The survey has traced the field of these converging technologies, including the fundamental principles of Spiking Neural Networks and the neuromorphic hardware design, to their implementation throughout all phases of a deep space mission. It has also indicated the critical way forward, highlighting the urgent issues in radiation hardening, algorithm development, and above all, the verification and validation to be addressed for the assurance of reliability of such systems. As these technologies continue to mature, as the headaches are carefully choreographed away, they will enable a new generation of deep space missions, of missions that are more desirable, more efficient, and more scientifically ambitious to bring about the transformation of our spacecraft from robotic servants to cognitive partners in the grand quest for knowledge.

AI disclosure statement: The author confirms that no artificial intelligence (AI)-generated content or AI-assisted writing tools were used in the preparation of this chapter (including summaries and the abstract). All text, analysis, and conclusions are the result of the author's original work.

Bibliography

1. Sandamirskaya, Y., Kaboli, M., Conradt, J., Celikel, T., Neuromorphic computing hardware and neural architectures for robotics. *Sci. Rob.*, 2022.
2. Ojo, O.L., *et al.*, Artificial Intelligence in Space Exploration: Improving Data Analysis and Decision-Making for Planetary Missions, in: *2024 IEEE 5th International Conference on Electro-Computing Technologies for Humanity (NIGERCON)*, 2024.
3. Schuman, C.D., *et al.*, Opportunities for neuromorphic computing algorithms and applications. *Nat. Comput. Sci.*, 2022.
4. Kumbhar, G., Synaptic AI: Bridging Neural Dynamics and Deep Learning for Next-Generation Computation. *Int. Sci. J. Eng. Manage.*, 2025.
5. Izzo, D., *et al.*, Neuromorphic Computing and Sensing in Space, arXiv, 2022.
6. Lagunas, E., *et al.*, Performance Evaluation of Neuromorphic Hardware for Onboard Satellite Communication Applications. *IEEE Wireless Commun.*, 2024.
7. Tang, G., Kumar, N., Michmizos, K., Reinforcement co-Learning of Deep and Spiking Neural Networks for Energy-Efficient Mapless Navigation with Neuromorphic Hardware, in: *2020 IEEE/RSJ International Conference on Intelligent Robots and Systems (IROS)*, 2020.
8. Roy, K., Jaiswal, A.R., Panda, P., Towards spike-based machine intelligence with neuromorphic computing. *Nature*, 2019.
9. Patil, S. and Patil, D.D., Autonomous Intelligence: AI's Role in Transforming Vehicles, Drones, and Space Missions, in: *2025 International Conference on Sustainable Energy Technologies and Computational Intelligence (SETCOM)*, 2025.
10. Oche, P.A., Ewa, G.A., Ibekwe, N., Applications and Challenges of Artificial Intelligence in Space Missions. *IEEE Access*, 2021.
11. Aguzzi, J., *et al.*, Developing technological synergies between deep-sea and space research. *Elem. Sci. Anth.*, 2022.
12. Brandonisio, A., *et al.*, Closed-loop AI-aided Image-based GNC for Autonomous Inspection of Uncooperative Space Objects. *Aerosp. Sci. Technol.*, 2024.
13. Thangavel, K., *et al.*, Artificial Intelligence for Trusted Autonomous Satellite Operations. *Prog. Aerosp. Sci.*, 2024.
14. Chien, S.A., *et al.*, The Future of AI in Space. *IEEE Intell. Syst.*, 2006.

15. Zhang, H., *et al.*, Leader Federated Learning Optimization Using Deep Reinforcement Learning for Distributed Satellite Edge Intelligence. *IEEE Trans. Serv. Comput.*, 2024.
16. Ben-Itzhak, S., The Use of Artificial Intelligence in Outer Space Capabilities. *SAIS Rev. Int. Aff.*, 2024.
17. Bernard, D., *et al.*, Autonomy and software technology on NASA's Deep Space One. *IEEE Intell. Syst.*, 1999.
18. DeWolf, T., Jaworski, P., Eliasmith, C., Nengo and Low-Power AI Hardware for Robust, Embedded Neurorobotics. *Front. Neurorobot.*, 2020.
19. Wang, X., *et al.*, Neuromorphic circuits based on memristors: endowing robots with a human-like brain. *J. Semicond.*, 2024.
20. Yang, J.Q., *et al.*, Neuromorphic Engineering: From Biological to Spike-Based Hardware Nervous Systems. *Adv. Mater.*, 2020.
21. Bersuker, G., Mason, M., Jones, K., Neuromorphic Computing: The Potential for High-Performance Processing in Space, 2018.
22. Croon, G.D., *et al.*, Insect-inspired AI for autonomous robots. *Sci. Rob.*, 2022.
23. Sahu, M.C., *et al.*, Multifunctional 2D MoS2 Optoelectronic Artificial Synapse with Integrated Arithmetic and Reconfigurable Logic Operations for In-Memory Neuromorphic Computing Applications. *Adv. Mater. Technol.*, 2022.
24. Chawla, K., An Analysis of Neuromorphic Computing in Modern Technology. *Int. J. Sci. Res. (IJSR)*, 2023.
25. Zhou, J. and Chen, J., Prospect of Spintronics in Neuromorphic Computing. *Adv. Electron. Mater.*, 2021.
26. Liu, X., *et al.*, Bio-Inspired 3D Artificial Neuromorphic Circuits. *Adv. Funct. Mater.*, 2022.
27. Jahanshahi, H. and Zhu, Z., Review of Machine Learning in Robotic Grasping Control in Space Application. *Acta Astronaut.*, 2024.
28. Russo, A. and Lax, G., Using Artificial Intelligence for Space Challenges: A Survey. *Appl. Sci.*, 2022.
29. Hilton, B., *et al.*, Intelligent Mission Planning for Autonomous Distributed Satellite Systems. *Acta Astronaut.*, 2024.
30. Siddique, M.A.B., The Integration of Neuromorphic Computing in Autonomous Robotic Systems, 2024.
31. Zhang, J. and Lyu, L., An Actor--Critic-Based Hyper-Heuristic Autonomous Task Planning Algorithm for Supporting Spacecraft Adaptive Space Scientific Exploration. *Aerospace*, 2025.
32. Lunghi, P., *et al.*, Energy efficiency analysis of Spiking Neural Networks for space applications, 2025.
33. Macdonald, A.J., *et al.*, Enabling Autonomy with a Deep Learning Framework for Planetary Exploration, in: *ASCEND 2022*, 2022.
34. Ghiglino, P., *et al.*, AI/ML Inference Engine Software for High-Reliability applications on Space Qualifiable Hardware, in: *2023 European Data Handling & Data Processing Conference (EDHPC)*, 2023.

35. Shastri, B.J., Tait, A.N., Ferreira de Lima, T., Pernice, W.H.P., Bhaskaran, H., Wright, C.D., Prucnal, P.R., Photonics for artificial intelligence and neuromorphic computing. *Nat. Photonics*, 15, 2, 102–114, 2021.
36. Martin, A.S. and Freeland, S., The Advent of Artificial Intelligence in Space Activities: New Legal Challenges. *Space Policy*, 55, 101408, 2021.
37. Marković, D. and Grollier, J., Quantum neuromorphic computing. *Appl. Phys. Lett.*, 117, 18, 180501, 2020.
38. Barkouki, T., *et al.*, XAI Design Goals and Evaluation Metrics for Space Exploration: A Survey of Human Spaceflight Domain Experts, in: *AIAA SCITECH 2023 Forum*, 2023.
39. Russi-Vigoya, M.N., *et al.*, Supporting Astronaut Autonomous Operations in Future Deep Space Missions, 2020.
40. Pasupuleti, M.K., Intelligent Silicon: Neuromorphic Systems and Semiconductor Innovations for the AI Revolution. *Int. J. Acad. Ind. Res. Innov. (IJAIRI)*, 2025.
41. Fratini, S., *et al.*, On-board autonomy operations for OPS-SAT experiment. *Appl. Intell.*, 2021.
42. Iaroshenko, O., Sornborger, A., Chavez Arana, D., Binary operations on neuromorphic hardware with application to linear algebraic operations and stochastic equations. *Neuromorph. Comput. Eng.*, 1, 3, 034002, 2021.
43. Kudithipudi, D., *et al.*, Neuromorphic computing at scale. *Nature*, 606, 204–221, 2025.
44. Barnes, C., *et al.*, Space Situational Awareness (SSA) and Quantum Neuromorphic Computing, in: *2022 IEEE International Conference on Imaging Systems and Techniques (IST)*, 2022.
45. Liang, D., *et al.*, Neural State Machines for Robust Learning and Control of Neuromorphic Agents. *IEEE J. Emerging Sel. Top. Circuits Syst.*, 9, 4, 702–713, 2019.
46. Tanaka, H., In-Materio Computing Devices Composed of Random Network Nanoparticles for Future Autonomous Robot Operation, in: *2022 29th International Workshop on Active-Matrix Flatpanel Displays and Devices (AM-FPD)*, 2022.
47. Marković, D., Mizrahi, A., Querlioz, D., Grollier, J., Physics for neuromorphic computing. *Nat. Rev. Phys.*, 2, 499–510, 2020.
48. Dunkel, J., *et al.*, Benchmarking Deep Learning Models on Myriad and Snapdragon Processors for Space Applications. *J. Aerosp. Inf. Syst.*, 20, 2, 54–66, 2023.
49. Ahmadvand, R., Sharif, S., Banad, Y., Enhancing Energy Efficiency and Reliability in Autonomous Systems Estimation using Neuromorphic Approach, arXiv preprint, 2023.

5

Quantum Computing and Secure Communication for Space Exploration

Harikrishnan M. P.[1]*, Arsha J. K.[2], Shyama M.[3] and Danilo Pelusi[4]

[1]*Department of Electronics and Communication Engineering, Al-Azhar College of Engineering & Technology, Thodupuzha, Kerala, India*
[2]*Department of Biomedical Engineering, Al-Azhar College of Engineering & Technology, Thodupuzha, Kerala, India*
[3]*Department of Computer Science and Engineering, Viswajyothi College of Engineering & Technology, Muvattupuzha, Kerala, India*
[4]*Department of Communication Sciences, University of Teramo, Teramo, Abruzzo, Italy*

Abstract

The deeper into space, the more secure and high-speed missions targeting Mars and lunar habitats are needed. As our needs for secure and high-speed communication and also for powerful computing capabilities grow increasingly critical, it becomes evident that classical systems are not capable of meeting the extreme demands of space. This chapter explores how recent advances in quantum computing and quantum communication could come forward to meet these challenges head-on. Based on QKD and entanglement-based protocols of communication, it investigates how mission data, commands, and telemetry can be shielded from intercept or tamper also how some complex calculations and various data processing tasks that require quantum computing can be performed in deep space. Real world projects regarding Micius satellite of China act as case studies to demonstrate how theory is pitted against practice. In addition, the major technical hurdles are outlined, ranging from photon loss to synchronization problems. The path ahead for enabling the next generation of space communications infrastructure is also presented.

Keywords: Quantum computing, secure space communication, quantum key distribution, deep space networks, satellite quantum links, quantum entanglement, post-quantum security, space mission technologies

**Corresponding author*: srihari.harikrishnan@gmail.com

N. Sathish, Prasanth Aruchamy, Rajesh Kumar Dhanaraj, Nithya Rekha and Karthik Palani (eds.) Space Electronics for Deep Space Exploration, (109–130) © 2026 Scrivener Publishing LLC

5.1 Introduction

Space exploration is changing rapidly. In the early space age missions, the short rockets were used only once and spacecraft was simple. Now both governments and private companies are planning for a future with lunar bases, Mars habitats, asteroid mining, and even interstellar probes using advanced propulsion like solar sails, fusion drives, and nuclear thermal rockets. Each of these missions generates prodigious quantities of data. Scientific data including high-resolution multispectral and hyperspectral imagery spectrographic analyses of planetary atmospheres and *in situ* resource utilization experiments. Engineering data covering life-support telemetry propulsion system performance power system health radiation environment monitoring and operational data encompassing autonomous robotic activities navigation and guidance signals and crew communications. Handling all these data with current computers and communication systems is very hard. Delays, limited bandwidth, and the risk of interception or disruption can threaten mission's success. especially for deep-space missions beyond Mars where communication latency becomes a fundamental constraint on operational capabilities [21].

5.1.1 Demand for Quantum Computing and Communication

Quantum computers are special devices that think differently. They depend heavily on properties of superposition and entanglement to delve deeper into numerous possibilities simultaneously, unlike normal computers which operate step by step. This gives the spaceship a mind of its own the ability to shift through large quantities of data replan its own mission on the fly and make precise decisions in the moment [4, 26]. Quantum communication gives us safe ways to transmit the information with the help of the no-cloning theorem and QKD. This means messages are safer from hacking or data theft which is very essential for a nation's security safeguarding commercial secrets and keeping the team safe [19].

The major benefits include high-performing onboard processors where quantum algorithms will be able to solve problems like optimization, simulation, and pattern recognition much better and faster than their opponents' strong commands and perfect control through the QKD helps to overcome eavesdropping and spoofing of spaceship commands and improved navigation and sensing through quantum sensors which enables exact positioning and timing without GPS or large ground-based stations [6, 28].

5.1.2 Scope of Chapter

This chapter provides the core ideas behind the basics of quantum computing and communication as they are needed for space exploration. Section 5.2 explains the quantum computing fundamentals, the unique features of qubits, the strong connections of entanglement, and the specific instructions known as quantum gates. Section 5.3 details Quantum Communication Systems from Generating Ultra Strong-networks with quantum keys and shared entangled states for proof it all works from free space with satellite-based experiments [2, 29]. Section 5.4 outlines precise Applications for space missions. Section 5.5 discusses current systems and future scopes. The chapter concludes with a synthesis of the transformative capabilities of quantum technologies in space.

5.1.3 Benefits and Transformational Potential

Imagine that a spaceship doesn't work based on the instructions from the Earth. It uses its own memory to locate asteroid fields and change its mission by reconfiguring its location, speed, and alignment accordingly in space. This is not an ordinary system. It is a giant step forward that is replacing the already existing system that is power hungry and having complex hardware systems with ultra-high speed self-reliant quantum communication-based systems. Ultimately, it allows a quantum Internet that could connect from Earth to Mars, securely helping the future settlements of humanity [25].

5.2 Quantum Computing Fundamentals

5.2.1 Basic Idea of Quantum Computing

In its memory a quantum computer shows a major change in how process the data. Unlike traditional machines that use binary number systems (0 or 1), quantum machines work on quantum bits, or qubits. A qubit is entirely different than the binary number system; it is existing in a state of superposition that is in effect being at 0 and 1 simultaneously. This core concept, in combination with the synchronized state of qubits and their guided wave reactions, permits conducting a massively parallel search along the solution space. This is the best way which helps to travel through the large distances of space where profound isolation is the main concern. A Mars rover which exists in a state of solitude even a simple instruction from the

Earth needs 20 minutes to reach it reducing any chances of direct control. By dealing with very large amounts of information with its quantum-based system a space ship achieves a total control of its own. It can easily move through new landscapes dynamically reconfigure its operations position velocity as it makes new areas and react speedily to dangers evolving from a mere ordinary equipment directed from earth into a thinking super machine that can work according to its own instructions and will be able to take decisions on the frontier of solar system [10].

5.2.2 Qubits and Superposition

A quantum bit or qubit is the major component of this emerging technology. The major difference is that an ordinary bit can have values 0 or 1, but a qubit exists in a state where it can be a linear combination of both 0 and 1. This property is called a superposition. For example, a qubit state can be explained as follows $|\psi\rangle = \alpha|0\rangle + \beta|1\rangle$, where α and β are complex probability amplitude constants and they satisfy the condition $|\alpha|^2 + |\beta|^2 = 1$. In a space mission scenario, a small group of rectangular array qubits could in effect generate thousands of possible mission channels or paths simultaneously hence it speeds up the search for better ideal paths allocation of resources or the detection of abnormalities in space.

5.2.3 Quantum Entanglement

Entanglement is a special quantum process in that two or more qubits will be correlated so that the state of one immediately affects the state of another one here distance would not be an issue. For example, two entangled qubits can be expressed as $|\Phi^+\rangle = (|00\rangle + |11\rangle)/\sqrt{2}$. In space applications, entanglement helps for ultra-secure high-speed communication between spaceships and Earth or between two space ships traveling distances apart [14]. This will provide a well-connected non destructible link that the time the system calculates one qubit its colleague quickly will take another matching state for that distance will not be an issue. This will produce an inherently clear and well-balanced stable network that will form the basis for this wonderful system and works well for the connected quantum encryption.

5.2.4 Quantum Gates and Circuits

In quantum computers, qubits are the main performers. To help them perform well the system use a set of instructions or commands known as quantum gates. These are not physical elements but specific instructions that will realign the qubit's state. For example, The Pauli-X Gate function works like a basic inverter changing a qubit from a 0 state to a 1 state. The Hadamard Gate produces a balanced condition, changing a qubit into a condition where it can represent both 0 and 1 simultaneously. The CNOT Gate can connect two qubits, that is any change in one qubit can directly affect the second one making the state of one directly affect the other. This correlation helps them work in perfect coordination. By assembling all these gates into a well-structured format, the proposed system can form a quantum circuit. System can create a path way for the qubits system can start from simple gates and with the help of these gates can build very large complex circuits for the qubits. These large complex circuits developed from qubits can solve many complex problems already faced by the scientific community and it will increase the computation speed, enabling computations that ordinary systems cannot perform [18].

5.2.5 Quantum Algorithms Relevant to Space Applications

The progress in quantum computer development continues to help for the emergence of new ideas for space exploration. The Quantum Approximate Optimization Algorithm (QAOA) is one of the finest methods for complex challenges in planning of trajectory and resource management [4]. The Quantum Fourier Transform (QFT) is a type of time domain to frequency domain transformation method which converts complex time domain space signals into frequency domain and it deals with enormous amounts of data hence speeds up the signal processing works, thus helping in the rapid analysis of spaceship sensor data. Grover's algorithm increases the search operations through the help of high-volume astronomical data banks and mission databases. The Variational Quantum Eigen solver (VQE) which works with molecular structures and chemical processes is very much needed for life-saving systems and space environment analysis. These set of algorithms could help the spaceship to easily understand the optimal flight channels, increase fuel efficiency, and process the sensor data in real time and it will make the remotely operated machines into a truly self-working machine.

5.2.6 Quantum Error Correction and Fault Tolerance

The quantum-based systems are extremely sensitive this may cause a lot of issues for space applications. The following are the main obstacles that can challenge the quantum-based system (1) The orbital environment having high radiation, (2) high temperature variations, and (3) electromagnetic interference. These three can affect the delicate state of qubits causing problems in the system thus the stability of the system will be in danger. To alleviate this system uses quantum error correction methods and one of the methods that can use is the surface code bundle data across the several physical qubits to form a single, largest, logical qubit [17]. This will help the system to find and solve the errors without changing the quantum state. Building long-lasting quantum computers that can withstand these problems is critical for future missions. Scientists are actively developing specialized shielding, ultra-cold reducing systems, and refined control systems to create robust strong space-ready quantum processors.

5.2.7 Physical Implementations of Qubits

Various emerging cutting-edge technologies are now competing with each other to generate the stable qubits necessary for space missions, each has its own distinctive strengths and weaknesses. Superconductive qubits will be able to process information quickly and can be fabricated at large but they are in need of complex ultra-cold refrigeration systems to work. Trapped ions have the following characteristics: they are well known for their excellent accuracy and stable quantum states this method brings raw speed for phenomenal accuracy and stability. Photonic Qubits using light particles can work at room temperature and will be able to transmit quantum data over large distances, making them the ideal choice for building a quantum Internet [9]. Neutral atoms and spin systems are another set of arising platforms that offer the potential for packing many strong qubits into a complex space [15]. The selection of the correct technology involves excluding vital spaceship constraints the platform's size, its weight, power requirements, and most importantly its capability to stop the severities of space.

5.2.8 Hybrid Classical-Quantum Architectures

Due to the problems in running a full quantum computer in space, a more practical solution has come to light The Hybrid quantum-classical architecture. In this model, a precise technologically superior quantum processor handles specific computationally difficult tasks where it excels. Besides a powerful classical computer function as the mission's major nervous system handling logic control achieving error-free and handling traditional data processing. This balanced approach drastically reduces the size, power, and cooling demands of the hardware while still supporting a quantum advantage for the mission's most crucial calculations [4].

5.2.9 Key Takeaways from Quantum Computing Fundamentals

Qubits can access exponentially bigger state spaces than ordinary bits, driven by the distinct properties of superposition and entanglement. These quantum effects help in new ways of information processing and communication. Through the use of applications like quantum gates and algorithms, they open crucial functions such as trajectory optimization and secure space communication. Securing dependability in space robust error correction and hybrid computing systems is important for long-duration missions.

5.3 Quantum Systems for Space Missions

5.3.1 Introduction to Quantum Transmission

Quantum transmission system utilizes quantum mechanics concepts especially entanglement and the no-cloning theorem to transmit information secretly across space. Unlike the ordinary communication systems that transmit and receive digital signals through electromagnetic waves quantum-based systems transmit qubits as photons and they conserve their quantum states or energy levels over long distances. For space missions this technology offers two crucial advantages. First, it ensures intrinsic security because any interception attempts made will upset the quantum state immediately showing eavesdropping [23]. Second it helps long-distance viability since photons travel through space with very low data loss when compared to Earth's fiber optics. These features make the quantum systems the best one for securing spaceship controls, inter-satellite connections, and information ion transfer from space telescopes or planetary probes.

5.3.2 Quantum Key Distribution

The most well-known benefit of quantum transmission systems is Quantum Key Distribution (QKD), is a technique that helps any two members in the system to create a shared secret key with a security code that works according to the laws of physics. Among the working concepts, the BB84 is a quantum protocol that encodes information through precise directions of light particles, ensures critical key exchange [3]. Meanwhile, the E91 technique uses intertwined light particles to distribute keys secretly avoiding the need to depend on the security of the communication system. In a space context QKD can be implemented between a spaceship and an Earth-based ground station or between two satellites in low Earth orbit or deep space, or a relay satellite acting as a quantum node for multiple spaceships [12].

Once the secret key is transferred traditional symmetric data encryption (such as AES) can be used to send large volumes of data securely.

5.3.3 Satellite-Based Quantum Communication Experiments

Several pioneering demonstrations have already proven the feasibility of quantum communication from space: In 2017, China's Micius satellite achieved a landmark by demonstrating quantum entanglement distribution over a distance of approximately 1,200 km, along with successful satellite-to-ground QKD [27]. Meanwhile, the Canadian Quantum Encryption and Science Satellite (QEYSSat) was planned to experiment with QKD using ground stations established in North America [2]. Additionally, the European Space Agency started projects under the "Quantum Internet Alliance" aimed at providing quantum links between ground stations and satellites [22]. From these experiments it can observe that quantum communication can achieve distances far exceeding the ordinary terrestrial fiber optic links their performance was very much limited due to processes like absorption and attenuation also due to the unavailability of quantum repeaters on Earth.

5.3.4 Deep Space Network Integration

NASA's Deep Space Network (DSN) and some of the space systems for research in Europe, China, and India currently depend on large parabolic antennas and high-power radio frequency (RF) amplifiers and optical repeaters for data transmission and reception Integrating quantum channels into these networks would provide ultra-secure communication and telemetry channels for high end operations inter-satellite quantum connectors will form a secure backbone for planetary networks and the basics for a Solar-based System Quantum Internet helping distributed quantum computing resources across large distances [19].

5.3.5 Quantum Signal Regenerators and Interplanetary Networks

One of the major technical challenges in quantum communication systems is the absence of amplifiers; the main reason for this is due to the no-cloning theorem. The amplifier operation in the quantum communication system is performed by the signal regenerators they are special types of nodes that perform specific operations like entanglement swapping and purification these operations help the quantum links to reach over long distances [13]. In the near future, quantum repeaters will be playing a major role in the quantum-based communication systems. Repeaters will be placed on satellites Earth Mars and in deep space could create a network of entangled links. Space-based quantum memory systems help to store qubits for a finite time while establishing an end-to-end network connection [9]. A fully functioning interplanetary quantum network could help for secure communication and distributed quantum computing across the solar system.

5.3.6 Photonic Qubits and Space Channels

Photons are the ideal choice for carrying quantum data across the vast area of space. Their typical uses are compelling great speed and minimum loss. Their velocity is exactly the same as that of light and they suffer very less signal loss in the vacuum space when comparing this with ordinary optical fiber networks that are installed in terrestrial networks. Versatile Qubit Encoding Techniques such as polarization and time-bin encoding provide reliable and suitable methods for etching and sending quantum bits. Improving existing technology, the global network of telescopes and optical ground stations can be rearranged and reconfigured to establish these new quantum links, by doing so it can save considerable infrastructure costs and time. Of course, there will be obstacles especially in order to hold on to the fragile quantum state of photons and synchronizing clocks between the fast-moving satellites and ground stations. However, the latest research going on in adaptive optics and specific timing protocols is steadily changing these challenges into manageable engineering problems.

5.3.7 Quantum Cryptographic Applications beyond QKD

The prospect of a quantum communication system is not just about making secret keys, it has many major potential applications far beyond that it can develop strong security protocol functions for space applications Quantum Digital Signatures (QDS) is a new concept they send important commands to space ships to establish network connection and also to confirm that the data or information receiving is from a fair source. Quantum Secret Sharing (QSS) it is the process of splitting highly susceptible data like a launch code or a deep-space maneuver parameter among many members. Here no single member can access the secret code alone need collaboration and alienate insider threats. Quantum Position clarification. It is the operation of verifying a spaceship's precise location in space before allowing a critical operation to stop hackers copying their position. These applications prepare the security focus from simply encrypting data to a comprehensive framework that ensures the authenticity origin and integrity of information [15].

5.3.8 Benefits to Space Missions

Integrating quantum communication technologies gives the following advantages they include the following and the most important one is Data Integrity, security and data privacy and Data Originality. It develops an inflexible audit trail system making it impossible for anyone to tamper with data or for a data sender to refuse sending an authenticated instruction. One of the main advantages of a quantum-based communication system is the Decentralization of Security that helps direct secure spaceship-to-spaceship links reducing the reliance on powerless ground station networks and building a more stable space-based Internet system [12]. Future Security Concerns QKD is working basically on the laws of physics it shows great security that will be safer even from the enormous computational power of future quantum computers [18]. Setting a Quantum Internet provides the major backbone for connecting the quantum computers and sensors across long distances supporting a distributed network of unprecedented processing power [24].

5.3.9 Challenges and Ongoing Research

Despite its great start, turning the concept of a quantum space network into its full potential involves overcoming many major technological challenges. Extreme Pointing Acquisition and Tracking (PAT). The main challenge is that of making and maintaining a stable narrow laser link connection between a fast-moving satellite and an unstable ground station or between two satellites millions of distances apart. In Hostile Space Environment Quantum signals are very delicate and the chances of degradation and loss of signals by solar radiation background star light cosmic rays and severe Doppler shifts. The Need for strong Miniaturized Hardware Developing photon sources, super-cooled detectors and quantum memories that are tough robust durable and enough for launch and space operations and also being small lightweight and energy-efficient [8]. Globally researchers are taking care of these problems with advances in adaptive optics to correct atmospheric turbulence efficient entangled photon sources and the development of integrated photonic chips.

5.3.10 Key Takeaways from Quantum Communication for Space Missions

The unique properties of quantum physics provide a great level of internal security and long-range capability that the already existing radio frequency communication systems cannot offer. Feasibility is no longer theoretical; successful satellite-based QKD-based systems have paved the way [7, 27]. The next biggest step is to expand this entire concept into a deeper space. The future quantum-based communication system architecture of an inter-planetary communication system will largely depend on a combination of photonic qubits for data transmission and quantum repeaters to speed up their range [9]. While the challenges in precision concept and hardware miniaturization are real, they are very dynamic and they are high areas of research and where progress is being made very rapidly.

5.4 Technical Challenges and Research Directions in Quantum Space Communication

5.4.1 Photon Loss and Signal Attenuation

One of the major issues that is affecting the quantum space communication is the core issue of loss of the photons. Even in the vacuum of space, the delicate quantum signal can never be strained by the following factors one of the factors is beam divergence. Over millions of kilometers, a laser beam naturally flares out into a wide cone. This makes capturing a large number of photons with a telescope extremely very difficult. Atmospheric effects for any downlink signals to Earth, the atmosphere acts as a rough and strong barrier. Clouds can fully block the signals while turbulence can scatter and warp the precious photons. Optical Component Imperfections, No hardware is fully perfect. Every mirror lens and detector in the system can generate small losses through scattering, absorption, or reflection, which ultimately ruin an already damaged signal. Mitigation methodologies, Researchers are actively developing several methods to counter these losses. By the deployment of highly collimated optical systems, it can develop the thinnest possible laser beams minimizing the split over large distances. The usage of advanced adaptive optics at ground stations, which uses special types of deformable mirrors can actively compensate for the atmospheric turbulence in real-time. Developing ultra-sensitive ultra-low-noise photon detectors, like the superconducting nanowire single-photon detectors (SNSPDs) can maximize the chance of detecting every single arriving photon. Designing refined quantum error correction codes particularly made to identify and recompense for the loss or corrupted qubits can maintain the virtue of the quantum information [5].

5.4.2 Decoherence in Quantum States

Decoherence is the process in which a weak quantum state loses its own unique properties by the interaction with its environment, in effect it causes demolishing its own desired entanglement. In the harsh backdrop of space, this can be caused by the cosmic radiation and stray magnetic fields, thermal fluctuations from both solar radiation and the spaceship's own device, and inadequate shielding or cooling systems on board satellites. Research Directions Quantum Memory Systems, Devices that can store and release qubits to continue their coherence during transmission [9]. Cryogenic Payloads Advanced cooling systems that needed to keep the sub-Kelvin temperatures in the desired levels for sensitive quantum components to work. Topological Quantum States Examining the status of naturally occurring strong types of qubits (like Majorana modes or anyons) that are traditionally resistant to environmental noise, making them the ideal one for long-term space missions [21].

5.4.3 Synchronization and Timing

A quantum communication system is in need of impeccable, precise timing controls to correlate the quantum states across large interplanetary distances. The Major Challenges The relative motion of the spaceship introduces Doppler shifts and subtle time dilation effects. Deep-space distances create communication delays making the real-time interaction impossible. A lack of highly stable clocks on the spaceships when compared to more of strong ground-based standards. Emerging Solutions, The deployment of next-generation high end optical lattice clocks and ultra-stable oscillators on a spaceship. Full duplex time-transfer protocols specifically adapted for quantum channels. AI-assistive predictive synchronization, which uses a set of algorithms to predict and re compensate for movement and environmental changes [17].

5.4.4 Quantum Signal Regenerators and Entanglement Connecting

To travel the quantum signal into interplanetary distances, quantum signal regenerators are very critical and are very much needed for keeping the strength of entanglement. However, current systems are weak and need significant amounts of power and they are sensitive to fidelity loss during the crucial entanglement swapping process. To overcome this miniaturization of quantum repeaters is needed for practical space applications. Integrating quantum error correction codes to create more efficient, low-power systems for satellites. Using satellite constellations as distributed repeater networks it can enhance the stability by removing the single points of failure [9, 13].

5.4.5 Hardware Limitations

Quantum devices for space applications must be designed to survive extreme climate conditions Drastic temperature variations from−150°C to +120°C, the hard vacuum and consistent radiation exposure of space, and strict restrictions on payload mass, volume, and power consumption. Key Developments Needed Radiation-hardened quantum chips that are capable of surviving long-term missions [17]. Integrated photonic circuits that can substitute bulky, traditional optical benches. Modular quantum payloads that can permit plug-and-play replacement or upgrading of components in orbit [1].

5.4.6 Standards and Interoperability

A global quantum space communication system-based infrastructure is needed for unified protocols for transmission and data encryption. Current systems of QKD protocols and photon encoding methods differ between different space agencies. Laser-based communication systems across national air spaces encounter regulatory and legal hurdles. Progress of system international organizations like ITU-T, ETSI, and ISO are actively involved in developing quantum communication standards. Collaborative efforts made by the agencies like NATO, ESA, and NASA aims to ensure that future quantum space networks are secure and can work together very easily.

5.4.7 Security Threats beyond Earth

Quantum cryptography, a theoretically safe concept, still faces many practical susceptibilities in space communication systems. Side-channel attacks will manipulate the timing or detector leaks, while Trojan horse attacks may inject light to remove data. Satellite capture creates new physical challenges. Counter measures that can be applied include solidified detectors, random measurements, quantum authentication, and error-free designs [8].

5.4.8 Policy, Ethics, and Legal Framework for a Quantum-Space Era

The creation of a global and interplanetary quantum network extends far beyond technical and technological challenges creating a complex set of international policy, ethical, and legal standards that at present lack a comprehensive uniform governance framework. The existing framework of space law primarily based on the Outer Space Treaty of 1967 establishes important rules and regulations such as the peaceful use of outer space and the prohibition of national claims of sovereignty. However, it is not sufficient in addressing the complex challenges happening nowadays and in the near future due to the advancement in technologies. Key issues that urgently require legal clarification is in association with the ownership and liability of quantum satellites. Although these satellites may be operated by multinational corporations, they fall under the jurisdiction of the specific launching country, which can lead to potential conflicts involving oversight and accountability. Moreover, data privacy and security evolve into a more complex issue with the advent of quantum networks. While QKD theoretically guarantees perfect security the storage and standard communication of encrypted information—along with the metadata associated with quantum channel usage it—present many new opportunities for strict surveillance and intelligence gathering and sharing.

The most important problem is the alarming threat of weaponization, where quantum sensing concepts should be used for hyper-precise submarine detections or it will be used for missile guidance, due to all these global securities will be challenged. Similarly, the ability to upset or unsettle on a quantum link could become a possible squint weapon in near future conflicts. To alleviate these issues and to promote an environment of joint development, the international community must venture into a path of diplomatic and legal innovation. This involves the extension and renovation of existing space agreements to clearly incorporate quantum information technologies by well defining allowed and legally banned activities. Solutions also depends on the shared ownership and on the operational management of critical infrastructure, such as satellite clusters and deep-space repeaters through international bodies similar to a quantum-era CERN or ITU. Establishing a set of open clear protocols Norms and laws that keeps Up the peaceful collaborations and setting an unbiased access for all countries, regardless of their present Technological or Economic conditions is not merely an ethical obligation but a practical necessary to stop a new "quantum divide" and also to ensure the continuity of the future interplanetary Internet. This can be achieved only by continuous multi-stakeholder diplomatic dialogues involving scientists, engineers, policymakers, and ethicists to create a perfect strong legal framework that can transform as swiftly as the technology itself.

5.4.9 The Integral Role of Artificial Intelligence in Quantum Network Management

The highly intricate and vibrant nature of quantum networks, specifically those expanding into the complicated and diverse environment of space, necessitates a fundamental change from contemporary manual management systems to fully Autonomous, Intelligent Modern control systems. Artificial Intelligence (AI) and Machine learning are not merely helping tools but are now becoming a major part to the stable and reliable working of these sophisticated systems. One of the most important applications is in the real-time monitoring and management of quantum channels, where AI algorithms will play a major role in continuously monitoring and characterizing the channel properties, such as propagation, attenuation, phase constant, noise, loss of data or information. The system can use the data from the AI algorithms and with the help of this data, the Satellite systems can easily adjust its own system parameters that includes photon timing, wavelength, and encoding bases with the help of AI algorithms can optimize the accuracy and response of the entanglement distribution and QKD sessions across large and volatile distances. This adaptive capacity is critical for continuing a stable quantum link between a satellite in motion and a ground station or between two satellites from different orbits.

One of the major applications of AI in Quantum communication other than optimization, is that it is very much critical and will be useful in handling network resilience and security. Advanced Machine learning algorithms can be useful in being capable of predicting fault detection, identifying slight variations in system response that will trigger a full component failure, thereby initializing a proactive service protocol that is needed for orbital regions where physical repair is not often possible. In the case of a bug or a suspected security lapse, such as a spying attempt ever happens to the system then the AI-based system will be automatically start acting against such kind of activities by taking countermeasure steps that includes rerouting of quantum data through the redundant channels avoiding shaky nodes and starting the key refreshment operations without human efforts. These automated capabilities guarantee that the network can self-operate and become fully automated and maintain service protocols. In addition to that for key balancing AI can control the complex scheduling of end-to-end key generation and distribution of keys across a multi-node network, and ensure that keys are updated at a rate that meets with the security norms and regulations of different users and applications. The correlation of AI with quantum communication thus creates strong faster dynamic flexible systems that are capable of handling the complex and volatile space eco system that are smart and form the major backbone for the future ready quantum-based systems.

5.4.10 A Comprehensive Research and Development Roadmap

The total realization and implementation of a fully developed quantum Internet will be a major change in the way it uses the Internet. It will be a bigger paradigm shift and can last for over many years to come and it is in need of a well-connected and systematically coordinated and controlled global research effort. A clear future plan is very much needed to organize and arrange the research objectives of academicians, industry experts, and government officials, to verify and ensure that the basic technologies evolve much before most of the desired ideas are attempted. The near-term future that is from 2025 to 2030 is a very important phase, it is mostly focusing on the efforts to consolidate and commercialize the Earth-based quantum network applications while expanding its core capabilities in space. This tenure is the ideal time for the deployment of stalwart satellite applications for QKD, providing constant, stable, secure communication channels for both government and non-governmental institutions. Simultaneously, research will be focusing more on constructing ultra-competent time and data synchronization between large continents by the use of entangled photons, a crucial forerunner for producing advanced quantum protocols. A dream goal for this period is the deployment of a strong and stable Quantum-based communication network between the Earth and the Moon, Exploiting the Lunar Gateway or lunar surface areas to test the limitations of quantum communications in absolute space.

The medium-term period, which runs from 2030 to 2040, marks the change in communication system from point-to-point connection to an incoherent multi-mode node quantum communication network. The major technological shift happened here is the improvement and in-orbit deployment of quantum repeaters, initially it was located at stable locations such as Earth-Sun Lagrange points. These repeaters can form the major backbone of a space-based quantum Internet, it helps in connecting any two points in and around the globe and for that a direct line-of-sight between the two points is not needed and helps in expanding the safer quantum-based communications to the regions in space that are far beyond the Moon. This period will also be described by the challenges faced by the quantum-based communication system in the area of standardization, creating a set of functional and operational standards for hardware, software, communication protocols, software interfaces and hardware interfaces, to monitor and safeguard those systems developed by different countries and organizations can function properly in tune with the system. The roadmap for this tenure logically supports sending humans to Mars by helping with strong, stable, flexible, and highly capable communication systems for very important telemetry and astronaut communication, much far greater than what current systems can offer. This long-term vision for reaching beyond 2040, is the installment of a true interplanetary quantum Internet. This involves developing and deploying a network of quantum memory-based repeaters and routers throughout the entire solar system, that are capable of automating the routing of quantum states between Earth, Mars, and interplanetary spacecraft. This system not only provides unsurpassable safety and security but also enables deep-space sensor networks and fundamentally strong new scientific devices, intertwining a quantum fabric across the solar system [19, 24].

5.4.11 Addressing Foundational Technical Challenges

Despite the promising early start, the route to a global quantum Internet is congested with intense technical challenges that require great research and development across multiple fields of physics, maths and engineering. The most crucial disadvantage is the loss of photon over large astronomical distances, which currently coerces a practical value of approximately 1,200 km for the direct optical fiber-based entanglement distribution due to the exponential signal loss found in optical fibers [5]. In space, while the vacuum avoids the fiber loss issue, diffraction and atmospheric turbulence remain major problems to counter. The support of two important technologies are required to overcome these issues and they are Adaptive optics systems that can actively correct the issue of wavefront distortions caused by the atmosphere, and with the help of high-performing quantum memories [9]. These memories are capable of storing the weaker quantum state of an incoming photon with high accuracy and can very easily retrieve the data up on any simple request, these memories also acting as buffers at the core of quantum repeaters to help in entanglement swapping and purification across intercontinental and interplanetary distances.

The second most important challenge is decoherence, which occurs when a quantum system loses its unique quantum properties due to the interactions with its surrounding environment. The existing quantum memories and processors, such as those based on trapped ions and superconducting qubits, can continuously maintain coherence for minutes or, in some unique cases, hours in highly protected practical settings. However, this duration is not enough for high end large distance space communication, where data storage and transit times has to be much larger. Research in the emerging concepts like topological qubits, which are intrinsically protected from the different noise effects due to their unique properties, provides an encouraging direction for attaining a fault-tolerant quantum data processing [21]. In addition to that, it is necessary to support better protection against cosmic radiation and temperature fluctuations for the space quantum communication system architecture. Synchronization also presents a crucial issue.

5.5 Future Directions in Quantum Computing for Space Exploration

Future space missions will increasingly depend upon polished quantum processors capable of operating in harsh and rough conditions in a reliable manner. Current quantum hardware are not able to perform well in the demanding harsh and rough conditions due to their limitations in physical size, speed, processor memory, and ability to withstand space environments like high radiation and the strength of the materials also posing a greater challenge. Research must be carried out in producing highly radiation resistant, strong, quantum chips that can withstand cosmic rays and other energetic materials that are located in space. Topological qubits give us another promising path, due to its inherent strength and very low error rates it is the ideal one for a very large number of missions [21]. Hybrid quantum processors which can be integrated with the traditional spacecraft systems with ordinary quantum processors will enhance the optimization operation by the circuit calculations, allocation of resources, and artificial intelligence, and machine learning applications of higher order [4]. These updated systems will help the spacecraft to perform complex functions locally without depending on ground-based systems.

5.5.1 Future Directions in Quantum Communication for Space Exploration

The emergence of quantum communication technologies will pave the way for setting up a quantum interplanetary Internet connecting Earth, the Moon, Mars, and future space habitats. This network would be making use of modern concepts like Entanglement connections and transport across millions of kilometer distances, with the help of quantum routers that are arranged at crucial Lagrange points. Near-term performances will likely include the hybrid architectures integrating with the quantum channels for key distribution with ways for data transmission and reception, all managed by the brilliant AI enabled quantum communication systems that optimizes the performance. Configuration-based algorithms using a large number of small satellites supporting quantum payloads will provide some repetition but help to improve the allocation of multi-hop entanglement. These networks will be launched from Earth orbit to deep-space networks, ensuring strong quantum communication links throughout the entire solar system [19].

5.5.2 Integration with Other Emerging Technologies

Quantum communication technologies will combine strategically with other advanced systems to speed up the space mission capabilities. Artificial intelligence-based systems work with some of the quantum links for safeguarded software updates and authenticated automated operations. Quantum-resistant shared ledgers will equip safety and security in space mission event records, allocation of resources, and all the information related with the crew members will be safely maintained, during the long tenure missions. Quantum sensors will be playing a major role by enhancing the energy management through the optimization of solar array alignment, while the Quantum high-speed machines will improve the distribution of power across the spacecraft systems. These updates will be creating a total technological ecosystem where quantum refinements will play a major role in boosting the advantages of other emerging applications.

5.5.3 Policy and Governance for Quantum Space Infrastructure

The growth of quantum technologies in space applications needs serious, thoughtful international joint government initiatives. Global partnership will be playing an important role in strictly monitoring and maintaining quantum communication-based satellites and setting frameworks for their operations. Agreements should be made between global countries and super powers to address the use of this advanced technologies for military quantum applications while efforts must be made to facilitate sharing of infrastructure and knowledge for peaceful operations. There must be some legal frameworks and a strong supporting system must be maintained in order to guarantee seamless QKD norms, verifiable space cryptographic protocols, and safe and secure upgraded devices for quantum systems. These policies will help to sustain the desirable results and help in the process of installing quantum technologies in space.

5.6 Conclusion

Quantum technologies are a very big positive step forward and are ready to revolutionize the space mission capabilities. They provide some unbreakable defense to the spacecraft systems through this latest advanced Quantum theoretical concepts *via* quantum secure key exchange and quantum entanglement-based standards with these concepts it can connect the Earth, spacecraft, and space habitats. Quantum computing improves and enhances the space mission planning, real time error detection, and automated decision-making capabilities. Despite the issues like the loss of photons, decoherence, and unavailability of resources, the ongoing research activities in this area continues to produce more practical solutions. The future plans include interplanetary quantum networks, Artificial intelligence enabled automated self-driven missions, and mixed quantum systems that will support humanity to safely and effectively analyze and settle throughout our solar system and beyond that through the use of integrated quantum and innovative concepts.

AI disclosure statement: The author confirms that no artificial intelligence (AI)-generated content or AI-assisted writing tools were used in the preparation of this chapter (including summaries and the abstract). All text, analysis, and conclusions are the result of the author's original work.

Bibliography

1. Anderson, W.A. and Meckling, G.N., Polymeric bicyclo-(2,2,1)-2-heptene, U.S. Patent 2721189, 1955.
2. Bedington, R., Arrazola, J.M., Ling, A., Progress in satellite quantum key distribution. *npj Quantum Inf.*, 3, 1, 30, 2017.
3. Bennett, C.H. and Brassard, G., Quantum cryptography: Public key distribution and coin tossing, in: *Proc. IEEE Int. Conf. on Computers, Systems and Signal Processing*, Bangalore, India, pp. 175–179, 1984.
4. Bharti, K., *et al.*, Noisy intermediate-scale quantum (NISQ) algorithms for space mission optimization. *Quantum Sci. Technol.*, 6, 1, 013003, 2021.
5. Boaron, A., *et al.*, Secure quantum key distribution over 421 km of optical fiber. *Phys. Rev. Lett.*, 121, 19, 190502, 2018.
6. Chen, L., Quantum-enhanced navigation systems. *J. Spacecr. Rockets*, 60, 2, 445–462, 2023.
7. Chen, Y.A., *et al.*, Integrated space-to-ground quantum communication network over 4,600 kilometers. *Nature*, 589, 7841, 214–219, 2021.

8. Diamanti, E., Lo, H.K., Qi, B., Yuan, Z., Practical challenges in quantum key distribution. *npj Quantum Inf.*, 2, 16025, 2016.
9. Gao, W., *et al.*, Quantum repeaters and long-distance quantum communication. *Nat. Photonics*, 14, 7, 377–382, 2020.
10. Hanzo, L., Quantum computing and post-quantum cryptography for space networks. *IEEE Access*, 9, 153401–153420, 2021.
11. Kiktenko, E.O., *et al.*, Quantum-secured satellite communications. *Phys.-Usp.*, 61, 6, 612–628, 2018.
12. Kumar, R., Quantum technologies for secure satellite communication. *J. Space Commun.*, 27, 3, 345–362, 2021.
13. Nemoto, K., *et al.*, Quantum repeaters with entangled coherent states. *Phys. Rev. A*, 68, 042326, 2003.
14. Pan, J.W., *et al.*, Satellite-based entanglement distribution and quantum teleportation. *Rep. Prog. Phys.*, 80, 5, 056001, 2017.
15. Pirandola, S., *et al.*, Advances in quantum cryptography. *Adv. Opt. Photonics*, 12, 4, 1012–1236, 2020.
16. Ren, J.G., *et al.*, Ground-to-satellite quantum teleportation. *Nature*, 549, 70–73, 2017.
17. Sanguinetti, B., *et al.*, Quantum random number generation on a mobile phone. *Phys. Rev. X*, 4, 031056, 2014.
18. Scarani, V., *et al.*, The security of practical quantum key distribution. *Rev. Mod. Phys.*, 81, 1301–1350, 2009.
19. Schmidt, M., Interplanetary quantum network architectures. *IEEE Trans. Quantum Eng.*, 4, 2500123, 2023.
20. Seshadreesan, K.P., *et al.*, Quantum communications and interplanetary networking. *Acta Astronaut.*, 187, 399–408, 2021.
21. Tabor, D.P., Quantum materials for space applications. *Nat. Mater.*, 22, 4, 345–362, 2023.
22. Ursin, R., *et al.*, Space-QUEST: Experiments with quantum entanglement in space. *Europhys. News*, 40, 3, 26–29, 2009.
23. Van Meter, R., *Quantum Networking*, Wiley, 2021.
24. Wehner, S., Elkouss, D., Hanson, R., Quantum internet: A vision for the road ahead. *Science*, 362, 6412, eaam9288, 2018.
25. Wittek, P., *Quantum Machine Learning: What Quantum Computing Means to Data Mining*, Academic Press, 2016.
26. Xu, F., *et al.*, Quantum secure direct communication for space applications. *Opt. Express*, 27, 26, 37070–37086, 2019.
27. Yin, J., *et al.*, Satellite-based entanglement distribution over 1200 kilometers. *Science*, 356, 6343, 1140–1144, 2017.

6

Edge Computing and Distributed Processing for Spacecraft Systems

Princy K. T. M.[1]*, Prasanth Aruchamy[2], S. Lavanya[3] and Afizan Bin Azman[4]

[1]*Department of Computer Science, School of Computing, Amrita Vishwa Vidyapeetham, Mysuru Campus, Mysuru, Karnataka, India*
[2]*Department of Computer Science and Engineering, Vel Tech Rangarajan Dr. Sagunthala R&D Institute of Science and Technology, Tamil Nadu, India*
[3]*Department of Computer Science and Engineering, Muthayammal Engineering College, Namakkal, Tamil Nadu, India*
[4]*Faculty of Innovation & Technology, Taylors University, Selangor, Malaysia*

Abstract

Space exploration is not just a journey, it is humanity's high-stakes adventure. Countries around the globe and dedicated teams dive into studying planets beyond our own, chasing cosmic knowledge. Each day, fresh data emerge like stars in a night sky. This flood of information sparks innovation, propelling space research to new heights. Through careful analysis, exploration evolves rapidly, unearthing mysteries that expand our understanding of the universe. To analyze this type of enormous data, a lot of processing power and multiple technologies are required. Achieving quick results requires fast decision making technologies. Traditional, centralized methods often take more time to process such data. Spacecraft interact with different terrestrial systems and produce large amounts of data, which require intelligent technologies and machines with high processing power. To reduce latency and enable quick decision-making, space exploration scientists are investigating edge computing, a new technology for spacecraft systems. Edge computing and distributed processing systems offer several advantages. In this chapter, we focus on how edge computing and distributed processing can benefit spacecraft, discuss their various techniques, and analyze their advantages and challenges compared to traditional methods.

***Keywords*:** Space exploration, edge computing, distributed processing, spacecraft systems, real-time data analytics, low-latency decision making

**Corresponding author*: princyraj22@gmail.com

N. Sathish, Prasanth Aruchamy, Rajesh Kumar Dhanaraj, Nithya Rekha and Karthik Palani (eds.) Space Electronics for Deep Space Exploration, (131–146) © 2026 Scrivener Publishing LLC

6.1 Introduction

Spacecraft systems consist of various components, technologies, and tools to operate the spacecraft and complete its mission. The main components of the spacecraft systems are terrestrial devices and the spacecraft. Spacecraft in outer space interacts with the devices in the terrestrial systems and machines. A spacecraft is a specially designed vehicle for operating in outer space.

It is used for a variety of operations, such as communication, earth observation, military operations, navigation, and the transportation of humans and cargo. Sputnik 1 was the first artificial satellite. It was developed by the Soviet Union. Crewed spacecraft are the spacecraft with people, and uncrewed spacecraft are the spacecraft without people.

6.1.1 Types of Space Craft

Transport spacecraft are divided into categories based on their orbit and capabilities, including manned spacecraft, near-space vehicles (NSVs), lunar and deep-space missions, and low-Earth-orbit (LEO) satellites.

Figure 6.1 shows the classification of aircraft. Low Earth orbit satellites are unmanned aircraft that orbit Earth at low altitudes. Low-earth satellites are satellites that have applications in navigation, earth observation, and military operations. Application satellites are used to meet national economic needs. Remote-sensing (RS) satellites, reconnaissance satellites, communications satellites, navigation satellites, and on-orbit servicing satellites are examples of application satellites. Scientific experiment satellites are used for scientific

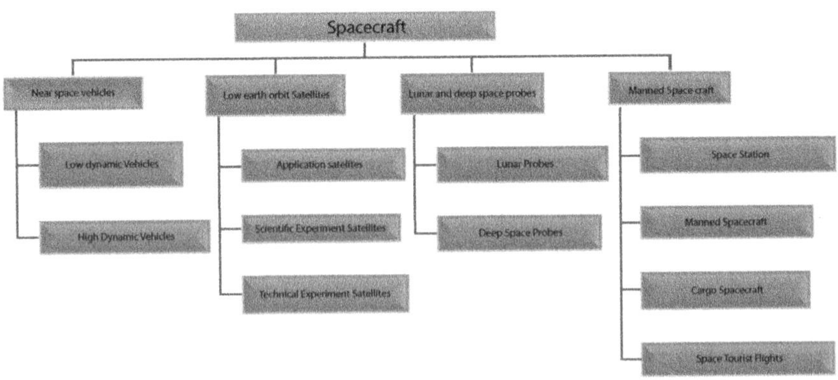

Figure 6.1 Types of space craft.

detection and research purposes. Astronomical satellites and satellites for physical testing are examples of scientific satellites. Mozi is the first quantum experimental space satellite launched by China. The technology experiment satellite (TES) is used to test new technologies or applications of future satellites before launch for real-time experiments. Lunar and deep-space probes are the unmanned spacecraft that fly in orbit and land on the Moon and other celestial bodies. Lunar spacecraft are used to explore the moon for different purposes. Chang'e 1 and Chang'e 2 are examples of lunar spacecraft that were used to explore the surface experiment and geomorphology of the moon. A manned spacecraft is a type of manned aircraft that orbits Earth or other bodies in space. Manned spaceships, cargo spacecraft, space stations, space shuttles, and space tourist flights are the different types of manned spacecraft. A manned spacecraft is a spacecraft that enables astronauts to live and work in space and to return safely to Earth with a vertical landing. A cargo spacecraft is a spaceship that carries supplies, such as food, water, air, fuel, and equipment, to a space station. Cargo spacecraft supports the other spacecraft. A space station is like a big house in space where astronauts can live and work for a long time. There are two types: single-piece (monolithic) and modular. A space shuttle is like a reusable space plane. It uses rockets to reach space, then glides back with wings to land on a runway, just like an airplane. There are two types, depending on how they are carried into space, one with a powerful engine and the other without. Space tourism means people travel to space as tourists to enjoy and experience space. Near-space vehicles (NSVs) are aircrafts that fly at very high altitudes, approaching space (20–100 km). Near-space vehicles are divided into low-speed vehicles and high-speed vehicles based on speed [1].

6.1.2 Components of the Space Craft

Spacecraft systems or spacecraft projects include science payload, spacecraft, launch system, tracking and data, and mission operations. The launch system is the vehicle that supports the spacecraft to launch it into orbit. The science payload is the set of instruments that perform the mission. The tracking and data subsystem collects and processes data from the spacecraft at the Earth station. The main components of a spacecraft include the structural system, power system, thermal control system, propulsion system, attitude and orbit control system, communication system, and payload system. The communication system is responsible for maintaining contact with the ground station. It collects processed data from the Earth station whenever a connection is established. But this process can take time, and as a result, communication may experience delays in both transmission and reception [2].

6.2 Edge Computing

Traditional cloud computing solutions are not the same as edge computing. In edge computing, the computation or data processing is near the sources. Calculations and data processing are performed near the source; as a result, these transmission delay problems and latency issues are reduced. These days, edge devices are spread across all fields, starting from agriculture to space missions. Edge computing is used in smart homes, educational institutions, agricultural fields, hospitality, the film industry, etc., with 17.7 billion devices connected to the internet, according to a 2024 report.

Edge devices are situated closer to the geographical area of the data-producing devices. By moving network, processing, and storage resources to the network's edge, edge computing technology offers intelligent help for real-time issues extremely close to their origins [3].

6.2.1 Architecture of Edge Computing

The Edge Computing Architecture is seen in Figure 6.2. Multiaccess computing is supported by edge computing. Then processing data, several devices can collaborate and operate independently at the same time. Cloud computing is expanded upon by edge computing. Communication between various Internet-based devices and applications [4]. Cloud computing systems are of higher quality thanks to edge computing.

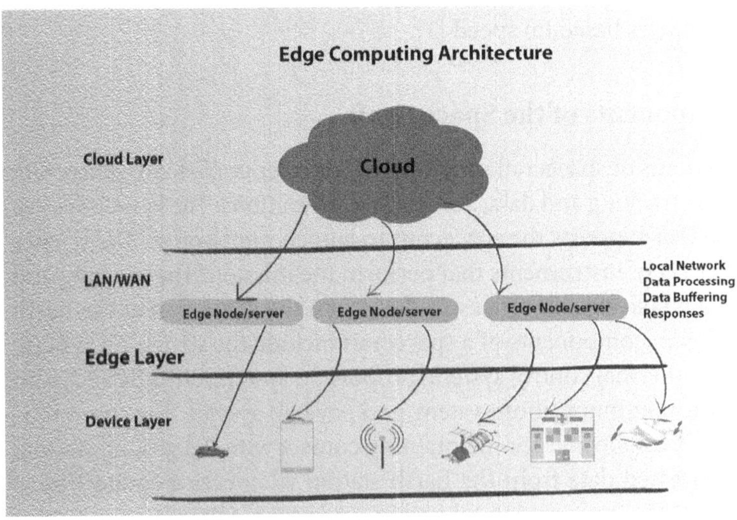

Figure 6.2 Architecture of edge computing.

The architecture of edge computing is often separated into three layers: Terminal layers, cloud computing layer and boundary layer. Every device is linked to the edge network at the terminal layer. Sensors, actuators, and other data acquisition devices are among the devices found in terminal layers. Devices may function as both data sources and consumers in terminal levels. It consists of the systems' base stations, gateways, routers, switches, and access points. The boundary layer, also known as the edge layer, is nearer the system. The server and the edge node are located in the edge layer. The edge layer manages the edge devices and their functions in addition to carrying out tasks including data processing and reduction [5].

The cloud layer is the third layer. The user and data sources are remote from the cloud layer. It is the most potent data sourcing hub. It has data storage devices and high-performance computing centers. It is useful for maintaining and analyzing data. The edge computing system's deployment and algorithm strategies can be modified by cloud computing layers [6].

6.3 Distributed Edge Computing

The distributed edge computing system's technical design and the rational operation of aircraft systems are presented in this part. Everyday, more and more sensors and gadgets are added to the aviation system. It generates a lot of data, but sending all of that data to a server that is far away from the data sources is not necessarily appropriate for devices. This issue can be resolved *via* distributed edge computing. The data from the sources are processed by edge devices in a distributed edge computing approach. However, under certain situations, inefficiencies in the edge devices or other problems with the devices make data processing at the node impossible. Cooperation between the devices and the cloud is necessary to solve this issue.

6.3.1 Components of Distributed Edge Computing

The edge computing system's intricate design is depicted in Figure 6.3. It features an offloading job, a compute node, and an edge device layer.

I. Edge computing model

To implement the computation of edge computing, typically, two models exist. They are the hierarchical model and (ii) the software-defined model. Figure 6.3 shows the hierarchical model of edge computing in a space spacecraft system. The model mainly consists of three parts: edge

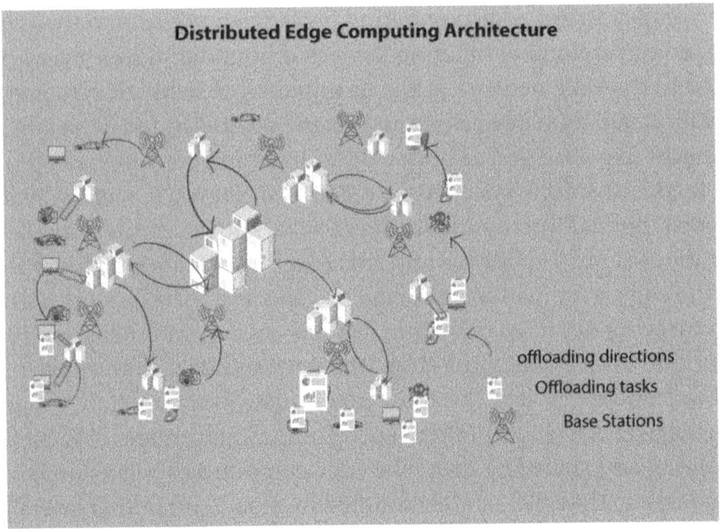

Figure 6.3 Architecture of distributed edge computing in space craft system.

devices, edge computation stations, and uploading and downloading link. The model is based on the distance of the edge computing devices from the base stations. Managing the large number of edge devices is difficult, so software-defined network model is proposed. It can integrate software-denied systems' capabilities to integrate multiple edge devices [7].

II. Edge devices

Edge devices are devices connected near the source that can produce data at the end of a system. Edge devices are of different types. Wang *et al* proposed isat8, isat is an edge computing node, which has a standardized architecture and an expandable operating system. Per the requirements, different applications can be loaded onto this node. It supports large-scale and distributed space network systems. Sensors and cameras are the primary edge devices. Edge nodes are devices connected very close to the sources. They handle the main computational system and can perform analysis, prediction, and processing. Edge node, An edge node can also store a few data items. When the edge fails to perform the computation services, it stops providing them for some reason, such as overload during peak hours. At that time, the edge server can terminate and hand off the work to the cloud.

III. Edge AI

Edge AI, or Edge Intelligence, is a new field. It helps devices make smart decisions quickly and locally. It has great potential. Neural networks and other AI algorithms apply to edge devices. Deep neural network performance depends on the large number of neuron and their connections. To process among different neurons, it needs high processing power and storage. But Edge devices have very limited processing power [8].

IV. Edge gateway

Edge gateways are devices that connect edge devices to the rest of the network and collect and send data to other network or cloud-based systems as required. The challenges of using an edge gateway include interoperability between devices, high energy efficiency, flexibility, fast allocation, and multi-tenancy issues. Robert Moribito and Riccardo Petrolo introduced the Internet of Things Lightweight Edge Gateway that makes use of a modular design, microservices, and lightweight virtualization to deliver adaptable, scalable solutions. LEGIoT supports a wide range of IoT framework protocols for the uplink and downlink communication. LEGIoT ensures resource management, energy efficiency, and interoperability. It features two modules: the northbound module, which manages communication with external internet and cloud services; and the southbound module, which handles connections with local sensors and devices. Edge networks are vulnerable to attack. An AI-assisted smart intelligence system with cybersecurity features can protect the network from edge attacks [9].

6.4 Edge Data Centers

These are the smaller data centers that are located near the edge devices. They offer processing, storage, and other resources for running algorithms. Edge data centers are typically managed by a single Organization. Edge data centers consume more energy. There are two possible locations for edge computing: the local edge and the network edge. Local edge data centers are very close to users, within 1 or 2 km, and they provide very low latency. But the network data centers are the ones that provide low to medium latency through different telecom companies and content delivery networks [10].

6.4.1 Edge Analytics

Edge analytics comprises software and algorithms that operate at the edge. It enables the control and processing of information to support decision making at the edge. It allows edge-level agents to gather and analyze information to process [11]. Edge devices are normally very tiny and cannot process large amounts of data. So, to process a wide variety of data that

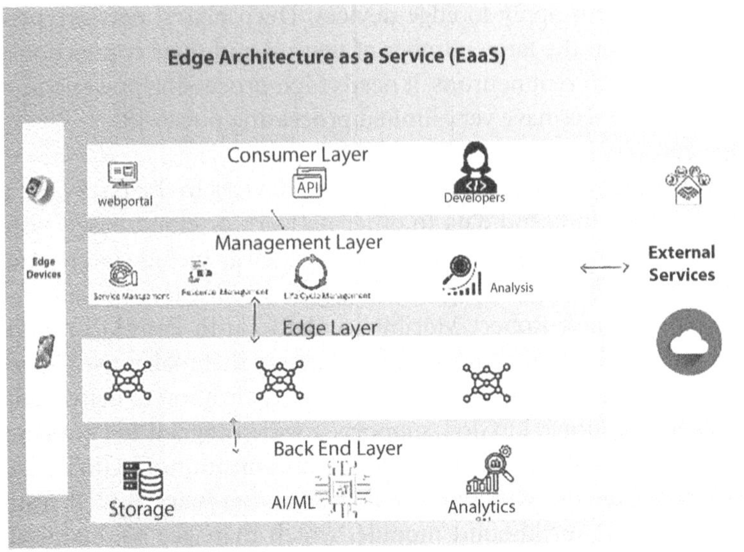

Figure 6.4 Edge architecture as a service.

comes to the edge, it needs to be reduced for efficient processing. Time series-based compression and aggregation and rule-based data filtering are the two primary techniques for lowering data traffic to edge devices. Microservices and edge analytics are offered by certain cloud services. AWS Greengrass provides local data analytics at the edge using Python and Lambda services. Edge analytics as a service provides real-time analytics for both cloud and edge devices. Edge analytics as a service architecture comprises devices, gateways, the cloud, and end users. Figure 6.4 shows the edge architecture-as-a-service models. EAaaS end users do not need to manually manage edge analytics. EAaas services have an IoT cloud platform. RESTful APIs and IoT time analytics services. Cloud Platform provides messaging queue telemetry transport (MQTT) communications for managing devices and exchanging data. In the EAaaS, RTAS enables users to develop, implement, remove, and track rule-based analytics [12].

6.5 Technologies of Distributed Edge Computing

6.5.1 Intelligent Edge-IoT Systems

Flexible infrastructure management is an emerging technology in distributed edge computing systems. Resource management, edge caching, data aggregation and analysis, and efficient frequency allocation and multi-access schema are used. Edge caching is the technique of storing popular content at the edge devices. Data aggregation processes data locally as and when it is generated. Smart frequency allocation methods are used to allocate the resource efficiently [13, 14].

6.5.2 Resource Management in Distributed Edge Computing

The main challenges of distributed edge computing systems are system management, resource allocation, and network issues. Since edge devices use variant devices, processing delays occur; to avoid these delays, a radio resource allocation schema can be used. When designing the resource allocation policy for a distributed edge computing system, reliability, performance, and implementation complexity are considered. Achieving reliability in a distributed edge computing network is a big challenge. Considering the algorithm's performance, it is challenging because reducing the number of days in the system may lead to other effects that increase the risk of losing important data. Implementation complexity is another challenge in creating and running the algorithm in distributed edge computing, which also requires knowledge of the hardware and software systems. There are various strategies and algorithms to reduce latency in a distributed edge computing network, including application placement. When placing the devices, scalability and locality are considered.

When additional devices are introduced to the system, scalability can assist in automatically updating the devices. wherein the optimal site is chosen to enhance performance. To avoid overloading, load balancing divides the burden among devices. Waiting times are shortened with the aid of caching. Software-defined networking (SDN) network is used to manage and optimize the resources. AI-powered decision making allows for smarter decisions. The main challenges of distributed edge computing systems are system management, resource allocation, and network issues. Since edge devices use heterogeneous devices, processing delays occur; to avoid these delays, a radio resource allocation schema can be used [15].

6.5.3 Edge Communication Protocols for the Spacecraft System

Edge communication protocols are the rules and regulations that govern the operation of edge devices. These protocols ensure smooth, reliable communication between edge devices, IoT devices, and the cloud platform.

I. MQTT protocols

IBM created the lightweight, open-source publish-subscribe protocol known as MQTT. It is best suited for low-bandwidth, high-latency systems that are highly resource-constrained, such as edge computing systems in distributed spacecraft. MQTT supports both decoupled and reliable messaging [16]. Major challenges of the spacecraft system are that the end devices are distributed across heterogeneous systems. For the smooth functioning of the system, working together can be challenging; MQTT protocols can help address these issues. It acts as a communication bridge between the Raspberry Pi devices and various web interfaces [17].

II. Data distribution service protocols in edge computing

The OMG created this middleware-level data management protocol. It offers an edge computing publish-subscribe model that is data-centric. Data distribution service (DDS) facilitates scalable, dependable, and low-latency communication. It lessens dependency on local systems and offers support for local processing. The three primary levels of DDS are the real-time publish-subscribe (RTPS) layer, the data-centric publish-subscribe (DCPS) layer, and the data local reconstruction layer (DLRL). Object-oriented data access is supported *via* the DDS's DLRL layer. Communication between publishers and subscribers is controlled by the DCP layer. Interoperability between several platforms is made possible *via* RTPS [18].

III. Constrained Application Protocol

The IETF created the constrained application protocol (CoAP) application-layer protocol to facilitate the seamless operation of edge and Internet of Things devices. CoAP functions similarly to HTTP, but it is tailored for the Internet of Things by minimizing protocol overhead and keeping only the most important functionalities. CoAP is more appropriate for contexts with limited resources because of its lighter approach, which sets it apart from HTTP. The request/response layer and the transmission layer make up the two primary layers of CoAP, which also facilitates multicast communication. While the transaction layer controls message flow between IoT devices, the request/response layer controls techniques like GET, POST, PUT, and DELETE. Confirmable (CON), non-confirmable (NON), acknowledgment (ACK), and reset (RST) are among the message types it supports. The version, type, option count, code, and message ID are all part of the CoAP message format [19].

6.5.4 SDN in Distributed Edge Computing

SDN is a network architecture that separates data forwarding and control planes. In the SDN network, control becomes programmable. Figure 6.5

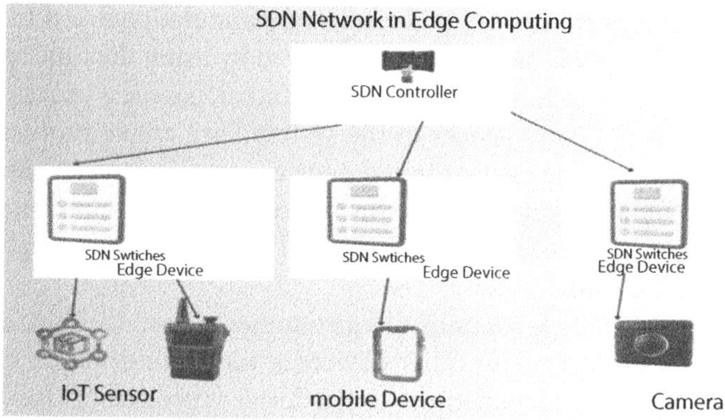

Figure 6.5 SDN network in edge computing.

shows the architecture of the SDN network in edge computing. The behavior of the distributed network and flow-based routing are managed by the central controller, which reconfigures the entire network in response to requirements. SDN controllers are centralized and reside in the cloud layer. So, it provides a global view of the network. SDN's focus is on improving network capacity and optimizing network performance. But a single failure in the SDN can affect the entire distributed system [20].

6.6 Advantages of a Distributed Edge Computing System in a Craft System

In this section, we are analyzing the advantages and of distributed edge computing in spacecraft systems.

I. Reduced transmission time and distributed recovery techniques

Fast transmission time is the main advantage of edge computing. It reduces bandwidth and latency by processing data very close to the sources. Edge computing solves the problem of the computer's bottleneck. Since most data processing occurs near the data source, it can reduce the network resource bottleneck [21]. In distributed recovery techniques, two or more copies of the same Work are available for processing. If one of the devices fails, other devices can activate and serve the distributed systems. In this, one node acts as the active node and the other node acts as the shadow node. The shadow node will not work all the time. The replication techniques are again divided into categories, such as passive, semi-active, and semi-passive [22]. System latency is the sum of computational and transmission latency. Computation latency is measured by using the time spent on the process in the data. Edge devices, like a camera, have less processing power, while other network devices in the system have ample processing power. The main challenge of distributed edge computing is determining which tasks to offload to other devices in the network before processing the data at the edge. To execute this task, several algorithms are available.

II. Bandwidth reduction

Distributed edge computing contains a large number of devices. The cloud, the upper layer of the system, or other devices at the ground station must process a large amount of data because every device is producing data. So, transmitting the data without compression is difficult. A large amount of data consumes large network resources. This leads to transmission delay and packet loss, but the edge devices are processing or pre-processing the data before sending it to the server or cloud resources. This approach can reduce bandwidth usage and improve the system's network performance [23].

6.6.1 Challenges of Edge Computing in the Distributed Spacecraft System

I. System integration

As the distributed system consists of a large number of devices, working with it is a difficult task. A distributed edge computing system comprises a heterogeneous set of network devices, edge devices, data processing components, and related technologies. Managing these resources is difficult. It is difficult to manage and process the heterogeneous application. In addition to storage and management issues, various devices are running different operating systems and processes. It creates difficulties in file naming, resource allocation, task scheduling, and other related tasks. Other challenges include naming and renaming multitasking and multi-source edge nodes. Since different types of devices are connected together, they use different proprietary firmware, and updating them is also difficult, leading to various security vulnerabilities.

II. Security and privacy

Edge computing utilizes multiple devices and various technologies, including peer-to-peer systems, wireless technologies, virtualization, and containerization. And it requires an integrated system with a well-versed team to manage the complex technologies and techniques. This system's heterogeneity creates many unseen security issues. The distributed edge devices and systems create security issues and data manipulation on local devices. Issues, such as data modification, tampering, and IP address spoofing, are common security threats. During message transfer between edge nodes and servers, various attacks can occur, including jamming, sniffing, and worm propagation. Therefore, privacy protection requires considering several privacy measures and methods. One of the main things you can do is use an authenticated gateway. Another way to mitigate this attack is to use the SDN. Several prevention methods are available, including network monitoring and prevention techniques. Notably, traffic isolation and a prevention mechanism are the most effective approaches. Migrating from one resource to another, as well as adding or changing devices, creates numerous issues. Migrating resources and modifying them are the other major concerns. Migrating devices from one mode to another creates numerous compatibility issues.

III. Edge devices security

In edge computing, users can access edge devices without restrictions. Users on the edge act as both creators and sharers of data. There is a chance that some users act maliciously and misinterpret the data. This creates the following security issues. It includes numerous touch statistics, and the attackers are using this communication channel to create threats.

IV. Side channel attacks

Since-edge devices are away from the central devices, attackers can easily collect sensitive information, such as passwords and other credentials, and attempt to attack the systems. Early side-channel attacks target general-purpose computers, cloud devices, and mobile devices, and now attackers are also targeting edge devices. Since edge devices are far away from the central controller, the possibility of attacking the edge through the side channel is high. Satellite-based edge devices are always on and make software installation on edge devices and small embedded devices easy, allowing them to easily integrate with all hardware. All these features make edge devices more vulnerable to side-channel attacks [24].

V. Denial of service attack on the spacecraft

End users or attackers can try to deny services as needed. Attackers try to create delays in computational resources, deny access to information, or cause delays in services. since the central or controlling devices are far from the sources. Dos targets the energy consumption rate, packet forwarding rate (PFR), packet sending rate (PSR), and packet receiving rate (PRR). Multiple node or edge devices may affect this, leading to the entire spacecraft system processing and a significant Loss Of time in both calculation and processing [25].

6.7 Conclusion

Distributed edge computing offers various advantages and use cases for spacecraft systems. This chapter highlighted the current distributed processing in the spacecraft system. There are different types of spacecraft systems. The processing and communication of data in these systems were different. In edge computing, data processing occurs on edge devices. There are different edge communication protocols and edge data processing technologies. Edge computing offers less transmission time and reduced latency. Edge computing helped reduce energy consumption in spacecraft systems. The integration of security and privacy across different devices is a challenge in distributed edge processing. Information injection and side-channel attacks are additional challenges in distributed edge computing for spacecraft systems. Moreover, the chatter highlighted the importance of edge computing, its advantages, and the differences between distributed edge computing in spacecraft systems.

AI disclosure statement: The author confirms that no artificial intelligence (AI)-generated content or AI-assisted writing tools were used in the preparation of this chapter (including summaries and the abstract). All text, analysis, and conclusions are the result of the author's original work.

References

1. Xie, Y., Lei, Y., Guo, J., Meng, B., *Spacecraft dynamics and control*, Springer, Singapore, 2022.
2. Brown, C.D., *Elements of spacecraft design*, American Institute of Aeronautics and Astronautics AIAA, 2002.
3. Cao, K., Liu, Y., Meng, G., Sun, Q., An overview on edge computing research. *IEEE Access*, 8, 85714–85728, 2020.
4. Hamdan, S., Ayyash, M., Almajali, S., Edge-computing architectures for internet of things applications: A survey. *Sensors*, 20, 22, 6441, 2020.
5. Taleb, T., Samdanis, K., Mada, B., Flinck, H., Dutta, S., Sabella, D., On multi-access edge computing: A survey of the emerging 5G network edge cloud architecture and orchestration. *IEEE Commun. Surv. Tutorials*, 19, 3, 1657–1681, 2017.
6. Bhandari, G.P. and Gupta, R., An overview of cloud and edge computing architecture and its current issues and challenges, in: *Research Anthology on Architectures, Frameworks, and Integration Strategies for Distributed and Cloud Computing*, pp. 2340–2370, Hershey, Pennsylvania, USA, 2021.

7. Dustdar, S. and Murturi, I., Towards distributed edge-based systems, in: *Proc. IEEE Second Int. Conf. on Cognitive Machine Intelligence (CogMI)*, pp. 1–9, 2020.
8. Liu, S., Ha, D.S., Shen, F., Yi, Y., Efficient neural networks for edge devices. *Comput. Electr. Eng.*, 92, 107121, 2021.
9. Morabito, R., Petrolo, R., Loscri, V., Mitton, N., LEGIOT: A lightweight edge gateway for the Internet of Things. *Future Gener. Comput. Syst.*, 81, 1–15, 2018.
10. Brocklehurst, F., International review of energy efficiency in data centres, IEA EBC Building Energy Codes Working Group, 2021.
11. Xiao, Y., Jia, Y., Liu, C., Cheng, X., Yu, J., Lv, W., Edge computing security: State of the art and challenges. *Proc. IEEE*, 107, 8, 1608–1631, 2019.
12. Xu, X., Huang, S., Feagan, L., Chen, Y., Qiu, Y., Wang, Y., EaaS: Edge analytics as a service, in: *Proc. IEEE Int. Conf. on Web Services (ICWS)*, pp. 349–356, 2017.
13. Alnoman, A., Sharma, S.K., Ejaz, W., Anpalagan, A., Emerging edge computing technologies for distributed IoT systems. *IEEE Network*, 33, 6, 140–147, 2019.
14. Chen, J., Li, K., Deng, Q., Li, K., Yu, P.S., Distributed deep learning model for intelligent video surveillance systems with edge computing. *IEEE Trans. Ind. Inf.*, 2019.
15. Mukherjee, A., De, D., Buyya, R., *Resource management in distributed systems*, Springer, Singapore, 2024.
16. Koziolek, H., Grüner, S., Rückert, J., A comparison of MQTT brokers for distributed IoT edge computing, in: *Proc. European Conf. on Software Architecture*, pp. 352–368, 2020.
17. Yeh, C.-S., Chen, S.-L., Li, I.-C., Implementation of MQTT protocol-based network architecture for smart factory. *Proc. Inst. Mech. Eng. B. J. Eng. Manuf.*, 235, 13, 2132–2142, 2021.
18. Jaroji, D., A Survey of DDS Implementation in Edge and Fog Computing. *INOVTEK Polbeng - Seri Informatika*, vol. 10, no. 2, pp. 1107–1118, 2025.
19. Donta, P.K. and Dustdar, S., Towards intelligent data protocols for the edge, in: *Proc. IEEE Int. Conf. on Edge Computing and Communications (EDGE)*, pp. 372–380, 2023.
20. Wu, D., Nie, X., Deng, H., Qin, Z., Software defined edge computing for distributed management and scalable control in IoT multinetworks, arXiv preprint arXiv:2104.02426, 2021.
21. El-Sayed, H., *et al.*, Edge of things: The big picture on the integration of edge, IoT and the cloud in a distributed computing environment. *IEEE Access*, 6, 1706–1717, 2017.
22. Fayyaz, M. and Vladimirova, T., Survey and future directions of fault-tolerant distributed computing on board spacecraft. *Adv. Space Res.*, 58, 11, 2352–2375, 2016.

23. Shi, W., Cao, J., Zhang, Q., Li, Y., Xu, L., Edge computing: Vision and challenges. *IEEE Internet Things J.*, 3, 5, 637–646, 2016.
24. Spreitzer, R., Moonsamy, V., Korak, T., Mangard, S., Systematic classification of side-channel attacks: A case study for mobile devices. *IEEE Commun. Surv. Tutorials*, 20, 1, 465–488, 2017.
25. Zaman, A., Khan, S.A., Mohammad, N., Ateya, A.A., Ahmad, S., ElAffendi, M.A., Distributed denial of service attack detection in software-defined networks using decision tree algorithms. *Future Internet*, 17, 4, 136, 2025.

7

Hybrid Communication Systems for Deep Space: The Role of Optical, RF, and Quantum Technologies

Indhumathi G.[1]*, Saranya G.[1], Naveen Kumar R.[2] and Raffaele Mascella[3]

[1]*Department of Electronics and Communication Engineering, Rajalakshmi Engineering College, Chennai, Tamil Nadu, India*
[2]*Department of Electronics and Communication Engineering, Sri Krishna College of Technology, Coimbatore, Tamil Nadu, India*
[3]*Department of Communication Sciences, University of Teramo, Italy*

Abstract

Deep space communicational brings about very difficult challenges like very long distances, time delays, limited bandwidth and very accurate data transfer, which are sometimes required to be ultra-reliable. This chapter introduces the idea that is starting to take shape about a hybrid communication system comprising of radiofrequency (RF), optical and quantum communication systems that will work together, thus assisting each other to finally put an end to the barriers traditional methods of space communications suffer from. RF systems are being integrated with high bandwidth optical links resulting in faster data rates and lower latency. At the same time, quantum communication does have some cool aspects like intractable encryption and entanglement-based connections that will have added to the security and access to singularity. This chapter enhances with a detailed and technical approach, the proposal of integrating such hybrid systems into the missions to deep space in the future. The technology advances and design considerations that will be needed to help enable this are also discussed. The workshop also considers practical case histories, currently running experimental missions and future architectures which may be implemented in the communication system that will bring seamless and resilient messages. Thus, it is long distances on Earth that is preparing to make possible the future of interplanetary and interstellar networks.

**Corresponding author*: indhumathi.g@rajalakshmi.edu.in

N. Sathish, Prasanth Aruchamy, Rajesh Kumar Dhanaraj, Nithya Rekha and Karthik Palani (eds.) Space Electronics for Deep Space Exploration, (147–174) © 2026 Scrivener Publishing LLC

***Keywords*:** Deep space communication, radio frequency systems, interplanetary networking, high-bandwidth data transfer

7.1 Introduction

Deep-space communication is the lifeblood of interplanetary exploration, navigation, and scientific discovery. The necessity of a reliable high-bandwidth secure communication system has been felt as humans venture beyond the influence of Earth toward Mars and the outer planets, and even interstellar destinations. During the course of its application, deep space communication has had to encounter many challenges such as distance, time delays, strength of signal, and restricted bandwidth availability. Although classical methods have a high level of reliability, their physical and technological limit applicability has started a new generation of hybrid communication systems mixed with RF, optical, and quantum communication systems to solve these challenges collectively.

For decades, RF communication is considered as the backbone of their respective space communication infrastructure. These systems are assuredly reliable and can maintain connecting capability for long ranges from spacecraft to Earth-based ground stations. RF, being an older technology, has been cost-effective and can be relied upon under varying atmospheric and environmental conditions. But still, the RF systems face critical limitations with respect to the available spectrum, bandwidth efficiency, and transmission speed as the complexity and data generated from the missions are increasing. The inverse-square law of propagation causes severe path losses over interplanetary distances, requiring larger antennas and higher transmission power for an acceptable signal-to-noise ratio to be maintained [1]. Hence, RF technology alone now cannot cater to the amplified data requirements of the incumbent and forthcoming deep-space missions.

The creation of optical communication systems can be imagined as a turning point in deep-space networking. Systems based on optics or lasers use very well-focused beams enabling high data rates with very low power consumption and high immunity to interference. Compared to radio wavelengths, optical wavelengths have greater directivity and higher data throughput. For instance, NASA's Lunar Laser Communication Demonstration (LLCD) and Deep Space Optical Communications (DSOC) projects have validated optical links able to transfer hundreds of megabits per second among astronomical distances [2, 5]. Nevertheless, despite their advantages, challenges abound in terms of very stringent pointing, acquisition, and tracking requirements, as well as vulnerability to atmospheric turbulence in ground-based systems [2].

In parallel, quantum communications are beginning to emerge as an innovative way to attain intrinsically secure deep-space communication. Quantum communication applies quantum entanglement and quantum key distribution (QKD) principles such that any eavesdropping attempt is detected instantaneously through the disruption of the quantum state. Long-distance entanglement distribution and quantum secure key exchange have been demonstrated with the Micius satellite from China [3]. Although quantum links in deep-space environments are still incipient, their ability to provide tamper-proof and unconditionally secure communication is highly valuable for future interplanetary networks.

Hybrid integration of RF, optical, and quantum offers a perfect remedy by exploiting the respective benefits of each one of them. In this kind of system, the RF link is used for robust control and telemetry, optical link is used for high capacity data transfer, and quantum channel provides end-to-end security. The collaboration among these channels greatly improves the strength, dependability, and flexibility of communication systems in ever-changing space environments. Also, the hybrid system enables the automatic switching and load balancing between the communication modes to the performance optimization according to the mission phase, weather, and resources [4]. Therefore, the hybridized design is rapidly being considered as the next-generation deep-space communication paradigm; thus, intelligent and secure interplanetary internet architectures are henceforth used in the system.

To conclude, the use of hybrid deep-space communication systems will set new limits on connectivity with the expansion of spacecraft distances from Earth and increase of mission requirements. The combination of RF, Optical, and Quantum technologies not only brings in increased bandwidth and efficiency but also new levels of security and fault tolerance. This partnership will be the basis of the future sustainable, scalable and intelligent communication infrastructures that will enable human beings to explore deep space in future decades [1–5].

7.2 Background and Evolution of Deep Space Communication

The improvement of deep space communications has kept pace with the increasing scientific aspirations of mankind, as well as with the improvement in technological sophistication and is parallel to the two developments. From early radio signals, it has progressively explored optical and quantum links to address problems associated with vast distance, high latency, and low bandwidth. A gradual shift from RF-only systems toward the current approach of using hybrid architecture for achieving high-speed, ultra-secure, and resilient deep space connectivity has begun.

7.2.1 Early RF-Based Communication Systems

Deep-space exploration is an era extended to its infancy, characterized by missions of radio frequency communication such as Pioneer, Mariner, and Voyager. The backbone of the spacecraft–Earth links was offered by RF systems operating mostly in S-band (2–4 GHz) and X-band (8–12 GHz) frequencies. In 1958, NASA established the Deep Space Network (DSN), which allowed continuous coverage of mission control and data reception using a set of large parabolic antennas. Such initial RF systems, as shown below in Figure 7.1, mark the primary stage in the evolution of deep-space communication technology.

The Voyager missions were launched in 1977 and have proved a point to the RF technology with its abiding power. Even though the data transmission extended 22 billion km, Voyager 1 continues to communicate through a 23-Watt transmitter, exemplifying the credibility of RF technology over the long haul. Nevertheless, connectivity limitations kept data rates very low, well below a few kilobits per second, which did not harmonize with the new complexities of planetary imaging and instruments [6, 7].

7.2.2 Advancements in RF Technology

Over the past several decades, RF systems benefited from a number of great improvements as a result of the optimization of antenna design,

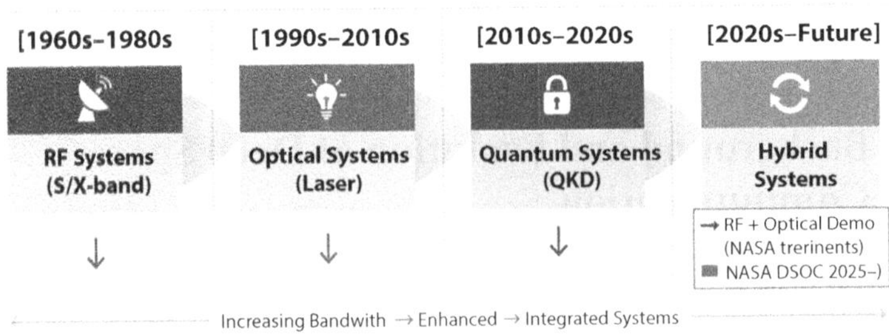

Figure 7.1 Evolution of deep space communication technologies.

implementation of highly efficient error correction coding, and implementation of adaptive modulation schemes. The switch from S-band to Ka-band (26–40 GHz) has significantly improved bandwidth and promoted higher data rates. Use of software-defined radios (SDRs) and deep-space transponder technology has also provided good versatility, in allowing frequencies and modulation parameters to be changed on a real-time basis [8].

Still, this novel approach for high-level research will get limited by inherent physical limits of RF systems, as they include the quadratic increase of free-space path loss, susceptibility to plasma scintillation in the solar environment, spectrum allocation constraints, and a rising volume of data need by scientific payloads. It has become clear that RF-only architecture will not be able to face the future challenges of deep-space mission support and stimulates the adoption of additional approaches [9].

7.2.3 Emergence of Optical Communication

The next milestone came with the development of optical (laser) communication, which offered bandwidths 10 to 100 times greater than RF links. By using narrow laser beams operating in the visible or near-infrared range, optical systems achieved high directivity, low power consumption, and reduced interference. The Lunar Laser Communication Demonstration (LLCD) in 2013 and the Deep Space Optical Communications (DSOC) system launched aboard NASA's *Psyche* mission in 2023 demonstrated record-breaking data rates, transmitting hundreds of megabits per second across interplanetary distances [10]. On the other hand, difficulties of a technical nature arise especially because of the necessity of very accurate beam pointing and tracking for a long-distance communication through the atmosphere, which is also affected by some disturbances. As a result, RF started being used as a backup or control channel in hybrid systems to ensure that the connection could still be maintained even if the optical links were interrupted [11].

7.2.4 Transition toward Quantum Computing

The decade of 2010 started the era of quantum communication with QKD and entanglement-based exchange as a basis to provide secure communication. Quantum channels exploit the peculiar characteristics of quantum states especially superposition and entanglement to monitor any form of eavesdropping attempt by signal alteration that any such intercept becomes immediately detectable [12]. The Micius satellite, launched by China in

2016, was the first to distribute entanglement and perform quantum-secured key exchange from orbit, thereby demonstrating the feasibility of quantum communication in space. For deep-space missions, quantum communication could deliver unique data security and synchronization between spacecraft and Earth. Currently, a rapid research direction is visible, which aims at integrating quantum subsystems within hybrid architectures to potentially combine the best of bandwidth, reliability, and encryption into an integrated system [13], while the implementations are still in the experimental stage.

7.2.5 Hybrid Architectures

RF, optical, and quantum technologies look to be the coming changes of the next space communications era of the evolution. Each of these subsystems, however, split but complementary plays a distinct role: the RF channel gives strength to command and telemetry, the optical channel sends the high-throughput data, while the quantum channel is used for the encryption and key management. An integrated hybrid scheme working is shown in Figure 7.2.

Hybrid architectures might gain from a software-defined control and AI-assisted adaptive modulation which dynamically allocate transmission resources depending on channel availability and environmental conditions as suggested by recent studies [10, 13]. This results in improved link resilience, fault-tolerant communication, and energy saving, which will be of utmost importance for interplanetary internet systems connecting Earth, Mars, and beyond.

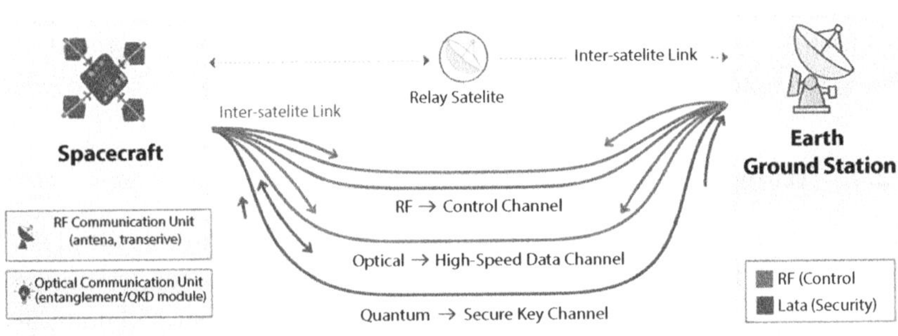

Figure 7.2 Conceptual block diagram of the hybrid deep space communication system.

7.3 Radiofrequency for Deep Space Communication

RF communication has been key in space exploration since the early missions. From telemetry data transfer by Pioneer and Voyager to real-time communication with Mars rovers, RF systems have faithfully provided the communication link between the spacecraft and Earth. RF systems use electromagnetic waves from the radio spectrum to carry information, typically in the S-band (2–4 GHz), X-band (8–12 GHz), and Ka-band (26–40 GHz). This particular band is chosen for consideration of factors such as link distance, required data rate, and spacecraft power availability.

7.3.1 Working Principle of RF Communication

The RF communication in lines with transmission of modulated radio waves between the spacecraft and a ground station. The transmitter aboard a spacecraft is generating an electromagnetic carrier signal, which is modulated to encode telemetry, command, or scientific data. The signal propagates in space and then it is received by large parabolic antennas of the deep space network (DSN) or other types of equivalent ground facilities.

The received signal power follows the inverse-square law, which means it decreases by the square of the distance. Hence, as it traverses millions of kilometres away, weak signal power from spacecrafts falls under the category of very weak—so much so that it even dips below a trillionth of watts at the receiving end. To counteract this attenuation, space agencies employ high-gain antennas, low-noise amplifiers, and advanced error-correction coding such as turbo and LDPC codes. These techniques improve link reliability even under extremely low signal-to-noise ratios (SNRs).

7.3.2 Frequency Bands and Data Rates

The most common frequency bands used in deep-space communication are given in Table 7.1.

Transitioning from S-band to Ka-band has enabled data-rate improvements of more than tenfold. For instance, *Mars Reconnaissance Orbiter* and *Juno* utilized Ka-band links to transmit high-resolution imagery back to Earth. However, higher frequencies require more precise antenna pointing and are susceptible to rain and atmospheric absorption at Earth-based stations.

Table 7.1 Frequency bands used in deep-space communication.

Frequency band	Range	Typical use	Advantages	Limitations
S-Band	2–4 GHz	Early missions, telemetry	Low atmospheric attenuation, robust	Limited bandwidth
X-Band	8–12 GHz	Modern planetary missions	Higher data rates, good balance between power and bandwidth	Moderate rain fade
Ka-Band	26–40 GHz	Recent and future missions	Very high data rates, efficient spectrum usage	Sensitive to weather and pointing errors

7.3.3 Components of a Deep Space RF System

A deep-space RF communication system typically includes the following major components:

1. **Spacecraft Antenna System:** High-gain parabolic or phased-array antennas are used for narrow-beam transmission and reception. Low-gain antennas provide omnidirectional coverage for near-Earth operations or emergency communication.
2. **Transmitter and Receiver Modules:** The transmitter modulates the data onto a carrier wave using schemes like Binary Phase Shift Keying (BPSK) or Quadrature Phase Shift Keying (QPSK). The input signal is demodulated at the receiver and the data is decoded.
3. **Amplifiers and Filters:** Electronic components that are used for amplification and filtering include the Travelling-Wave Tube Amplifier (TWTA) and Solid-State Power Amplifier (SSPA) for signal boosting, filters are helpful in blocking noise and interference.

4. **Error-Correction and Coding Unit:** The Forward Error Correction (FEC) code like Reed-Solomon or Turbo or LDPC can be used to increase the reliability of data against bit errors that may arise due to cosmic noise or the signal distortion.
5. **Ground Station Infrastructure:** The DSN makes constant contact with the spacecraft in the Solar System *via* its 34 and 70 m antennas that are located at strategic positions in California, Spain, and Australia.

7.3.4 Advantages of RF Communication

RF system is indeed the basis for space communication and soaring high with their own properties

- **Proven reliability:** Decades of operational heritage in interplanetary missions.
- **Wide coverage:** Omnidirectional capability for early-stage and emergency communication system.
- **Resilience:** Optical systems struggle more with dust, fog, and alignment errors compared to this one.
- **Mature technology:** The presence of reliable hardware, advanced coding methods, and worldwide infrastructure assistance.
- **Compatibility:** Optical and quantum systems can be perfectly integrated in a hybrid communication network without any compatibility issues.

7.3.5 Limitations and Challenges

However, a bevy of challenges out there that are beyond their scope to converge upon for RF systems in deep space missions over the abyss of our home planet:

- **Bandwidth saturation:** This is related to the radio spectrum, which is (most of the time, especially in X and Ka bands) overloaded.
- **Signal attenuation:** longer distances require more power in the quadratic ratio and hence, limit the data throughput for outer-planet missions.

- **Constraints of limited Antenna size:** They need large high-precision antennas which add weight and costs, besides the spacecraft and ground stations that consume more money.
- **Possessing Interference and noise:** Space radiation and solar corona plasma are the main factors that distort the signal and induce phase noise.
- **Lower data rates:** The comparison is still very limited to optical systems: there is rf links offering less bandwidth for high-volume scientific data.

7.3.6 Role of RF in Hybrid Communication Systems

RF systems are the essential components and parts of the hybrid architectures. They form a baseline for connecting spacecraft during disruptions in optical or quantum links caused by orbital disturbances or misalignment. Also involved are the RF channels that carry control and telemetry, alongside coordinating the transfer through high-speed optical links.

In the future interplanetary networks, the RF subsystems will also serve as a redundant and autonomous control layer in hybrid settings. Their missions constitute mission-critical commands and safety signals in such a way that they can never be lost. Therefore, in view of developmental advances in optical and quantum technologies, RF communication becomes solidly placed as an indispensable core of deep-space communication.

7.4 Optical Communication Systems for Deep Space

Optical communication, laser communication, is an incoming change for deep space missions. Unlike RF, which wide beam radio waves cover a limited amount of information (limited bandwidth) with its radio, optical systems transmit data using highly collimated laser beams operating in the visible or near-infrared spectrum. Extremely high throughput, narrow-beam directivity, lower power consumption, and immunity against electromagnetic interference constitute its outstanding features.

The increased necessity of sending heavyweight scientific data, high-resolution images, and real-time telemetering has been the main reason for the adoption of optical communication on lunar, Martian, and interplanetary missions. NASA's Lunar Laser Communication Demonstration (LLCD) and Deep Space Optical Communications (DSOC) experiment are examples of demonstrations which proved high-rate optical links over distances of millions of kilometers and that marked a significant change in deep-space communication [14–18]. As hybrid systems advance, optical communication will be a principal road in reality and will usher in high-bandwidth data channels along with the reliability of RF systems and the safety features of quantum communication.

7.4.1 Working Principle of Optical Communication

Optical deep-space communication relies on the processes of modulation, transmission, propagation, and detection of light waves. A laser source creates an outgoing signal that is the representation of coherent optical carrier. Then, the mission data is combined with this carrier through modulation methods like On-Off Keying (OOK), Pulse Position Modulation (PPM), or more advanced phase-based ones at higher speeds.

A telescope mounted on the spacecraft focuses the outgoing beam toward the receiving station. Because optical beams have extremely narrow divergences (micro-radian level), the transmitter must be aligned very precisely with the ground station. This alignment is provided by a Pointing, Acquisition, and Tracking (PAT) subsystem that continuously corrects the spacecraft orientation.

On the Earth side, the incoming beam is collected by a large optical telescope and transformed into electrical signals through a photodetector, for instance, avalanche photodiodes (APDs) and/or single-photon avalanche diode (SPAD). The incoming signal will then be demodulated and decoded to extract the received data. Optical links, as illustrated in Figure 7.3, require highly accurate alignment, very little jitter, and correct capture of a single photon for stable communication over long distances.

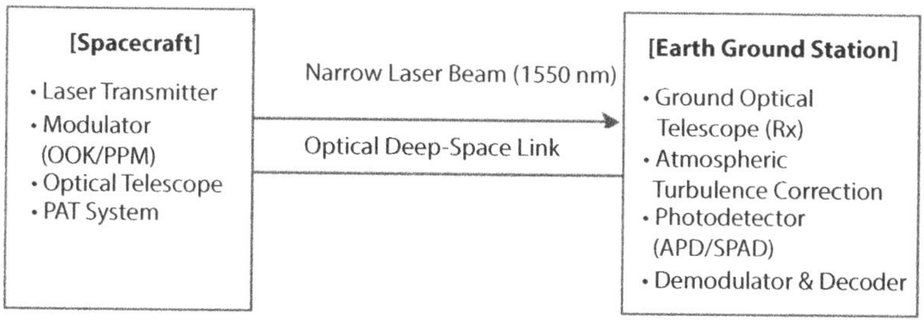

Figure 7.3 Optical communication link.

7.4.2 Components of an Optical Deep-Space Communication System

A full link constituted for deep-space missions is composed of several state-of-the-art subsystems that necessarily work in conjunction to ensure efficient, high-bandwidth, and long-distance data transfer. The heart of the system–at least an order of magnitude higher-rank importance than other subsystems–is the laser transmitter, which generates a coherent optical beam at a wavelength usually around 1064 or 1550 nm. This particular section of the spectrum has been chosen due to its low attenuation, high efficiency with respect to space-qualified semiconductor/solid-state laser sources, and relatively high point source stability. Before transmission, mission data were encoded onto the optical carrier with advanced modulation techniques like on-off keying (OOK), pulse position modulation (PPM), differential phase shift keying (DPSK), or advanced higher-order techniques. The picking of these modulation formats relies on the requirements of the mission, limitation in power, and data rates to keep the option of secure communication open even in the very low photon environment.

After the modulation, the telescope of the spaceship will expand and collimate the optical beam which is going to be very helpful in shaping the outgoing beam as well as in collecting very faint photons that are coming from the Earth. This is due to the very tight divergence of laser beams that a tiny error in pointing will completely ruin the link. To prevent this, the spacecrafts come with a pointing, acquisition, and tracking (PAT) subsystem that always improves the orientation of the spacecraft and the direction of the beam. This PAT system provides micro-radian alignment over interplanetary distances with the assistance of star trackers, gyroscopes, fine steering mirrors, and closed-loop feedback. At the other side of the communication chain, high-sensitivity photodetectors like APDs or SPADs turn the incoming photons into electric signals thus making it possible to recover data even when the optical power received is extremely low. The detectors are generally a part of the ground optical terminals that also incorporate huge aperture telescopes and adaptive optics that carry out atmospheric turbulence correction, and state-of-the-art signal processing units. These technologies work in concert with each other, and give optical communications systems the fantastic data throughput, long-distance reliability, and high efficiency in deep space environments.

7.4.3 Advantages of Optical Communication

It is hard to overstate the number of advantages that optical communication has to offer and it is no doubt that the technology would completely change such missions for deep space, especially as the great advancement in technology is very indicative of the scientific instruments that are progressively yielding enormous data volumes. The foremost benefit is the extremely high bandwidth associate with very slender laser beams, will make the data rates to this point excessive that they may vary between 10 and n100 instances of what could be done utilizing conventional RF systems. This very capability can letting to the sending of pictures at high-resolution picture of planets, spectroscopic measures and even of large-scale science data at a speed that is not just unprecedented but higher than at no point prior. Due to the very narrowness of optical beams, they have not only become the medium for communication but at the same time, they have been the reason for such significant savings in power that even under conditions of very low transmission energy spacecraft can still communicate at a very long distance. That situation is particularly favourable for deep-space missions where, despite the restriction on power availability, energy savings are greatly needed. One of the main drawbacks besides invisibility is the relatively low power usage of the optical communication hardware which also comes along with other drawbacks of weight and size being of lesser value. Compared to large, bulky parabolic antennas for high-gain RF transmission, laser transmitters, and optical telescopes are much smaller and lighter—they are even less so for compact spacecraft, CubeSats, and missions that want to keep a tight weight budget. They also gain an extra advantage due to the fact that their directivity, which means that both the dispersion of the signal and the increase of communication security are very much reduced. Laser beams also assure that there is always a cleaner, interference-free channel, even in an area where radio frequencies are being c and they are not so susceptible to electromagnetic interference and spectral congestion.

7.4.4 Limitations and Challenges of Optical Communication

Although it employs optical means for connecting, which could send data faster or more securely than traditional RF systems, it has some critical challenges to overcome before being feasible for deep-space transmission. The most important of these is the sensitivity of the systems to the atmospheric conditions. Optical signals, on their way to earth, have to traverse through different layers of the atmosphere; this would cause turbulence, scattering, absorption and distortion. It can significantly attenuate or completely block the optical path since cloud cover, fog, dust, and water vapor can make ground-based reception completely weather dependent. Adaptive optics and site diversity strategies can mitigate these effects but cannot be eliminated.

It is also extremely demanding in terms of pointing accuracy for effective communication. The narrow divergence of laser beams, usually in the order of micro-radians, means that less than a fraction of a degree might lead to devastating losses in acquiring the signal since it will miss the receiver entirely. Achieving this accuracy requires quite complex Pointing, Acquisition, and Tracking (PAT) subsystems, highly stable attitude control systems in the spacecraft, and very accurate fine-steering mirrors. Lastly, with increasing distance, deep-space optical links suffer from lack of photons. The smaller the number of photons that reach the detector, the more suitable ultra-sensitive detectors such as SPADs and superconducting nanowire single-photon detectors will be needed. These further complicate and raise the costs of the system. Parsing these challenges into why optical systems would more often accompany RF communication as backup or complementary channels within hybrid architectures.

7.4.5 Case Study: LLCD and DSOC

These two landmark demonstrations include: NASA's Lunar Laser Communication Demonstration (LLCD) which represents the first optical communication link at high rates between the Moon and Earth, and the Deep Space Optical Communications (DSOC) experiment, LLCD proved that optical systems could deliver high data throughput with relatively low power requirements, even under changing conditions of the atmosphere. LLCD received the best downlink achieved to date of 622 Mbps, and an uplink rate of 20 Mbps, all the while maintai. The distance from Earth to the Moon is 384,000 km; thus, LLCD established the technical foundation for future optical missions and proved that laser communication can still be successful in their collaboration with RF systems.

NASA's DSOC experiment is a continuation of what was achieved by LLCD; DSOC was launched aboard the Psyche spacecraft by NASA in 2023. This would now pave the path forward toward deep-space laser communication. While LLCD operated great distances approx. at the scale of the Moon's distance, on the other hand, DSOC constantly established high-rate optical communication over the wide span of 16 million kilometers. The mission finally reached a point where the systems used were quite advanced superconducting nanowire single-photon detectors, and high-precision pointing for very faint optical signals. It was proved that no optical links could operate dependable at such massive distances but the mission actually provided with a lot of useful information about deep-space channel characteristics, pointing systems, and long-range photon detection. Hence, these two missions do not only confirm but also demonstrate the extent of the impact optical communication will have and its necessity in the deep space exploration futures.

7.4.6 Role of Optical Links in Hybrid Communication Systems

Optical communications are undoubtedly featuring prominently among the future hybrid communication scenarios and they are a great contender in the case of deep-space missions. The gradual transition towards using optical links is akin to the utilization of a high-throughput channel which enables the rapid transfer of essential scientific datasets, high-resolution planetary images, and real-time telemetry that cannot be handled by RF channels alone. Moreover, due to their extremely high bandwidth and virtually no divergence in the beams, the data can be transmitted very selectively by the spacecraft, thus making optical systems one of the top choices for those missions with continuous or large-volume downlinks. The proposed system combines an optical link, RF links, and quantum subsystems that interconnect the diverse technologies and form a resilient and adaptable network. Whereas the RF link provides a strong and safe command and control that is not affected by very severe weather conditions, the optical links boost the overall capacity of the system in the areas where atmospheric and pointing conditions permit.

The varying nature of optical links in dynamic deep-space environments is a very attractive option when the channel conditions are highly time-varying. Real-time link assessment and the deployment of autonomous resources are combined in hybrid architectures to identify the most efficient mode of communication, wherein the spacecraft can fluidly switch between optical and RF channels. Subsequently, the optical links would take over the transmission of the data at a high rate for very short periods during good visibility, while the RF would provide constant, though by nature restricted, communication and, therefore, keep the mission operations going without accidental interruptions. Moreover, the optical systems could be co-integrated with quantum communications technologies which rely on the very same hardware components of narrow-beam optical carriers and photon sensitive detectors. With such voting behaviours, besides allowing data transmission at high speed, the optical channels could support the secure distribution of quantum keys, hence, enabling an extraordinary combination of bandwidths, reliability and security. It is by means of this comparison that the possibility of optical communication comes to occupy the rank of a sine qua-non in hybrid deep space networks, taking into account the scalability, flexibility and security that the communications infrastructure of the future interplanetary communications will require.

7.5 Quantum Communication Systems for Deep Space

Quantum communication seems to take the first few baby steps as the futuristic model for very safe, reliable information transfer by deep-space communication. Among classical communication systems, which encode data on ground amplitudes, phase, or frequency of electromagnetic waves, quantum communication is dependent on the fundamentals of quantum mechanics, especially quantum entanglement and quantum key distribution (QKD), in transferring data in a way that is, by nature, secure among these methods. QKD encodes encryption keys into quantum states of photons so that any attempt to intercept them disturb those states and is thereby detected. This feature carries security levels impossible with classical cryptographic methods, which future deep-space missions require, such as autonomous spacecraft, interplanetary data relays, and distributed scientific networks. China's Micius satellite demonstrated this through its message that with 1200 km between ground stations, there were demonstration of entangled photon distribution and QKD, making space-based long distance quantum communications feasible [19].

However, there are several engineering and scientific challenges to implementing quantum communication in deep space. Quantum states are extremely fragile and easily disrupted by photon loss, background radiation, or decoherence. The effective probability of a photon reaching its destination drops drastically after a few million kilometers, so it ultimately appears necessary to develop ultra-sensitive detectors such as superconducting nanowire single-photon detectors (SNSPDs). Timing synchronization, which is extremely precise down to the level of a few picoseconds, will also be provided because deep-space links must fulfill optical beam divergence, and advanced error correction techniques are fundamental in preserving quantum coherence. However, hybrid communication systems are promising avenues since they have been shown capable of integrating quantum links with both RF and optical channels. Such architectures ensure continuous telemetry, whereas high-rate scientific data will be carried by optical communication. The key distribution and authentication will be provided by quantum channels. Such layered approaches should augment the whole system's security and resilience, making quantum technology an invaluable asset for next-generation deep-space communication networks [20].

7.5.1 Working Principle of Quantum Communication

Quantum communication is mainly by using single photons, with their polarization, phase, or time-bin states representing the quantum information being transmitted. In most space-based applications, there were two main mechanisms: quantum entanglement distribution and QKD protocols. The entangled photons show that the measures remain correlated over any distance. If one photon's state is measured instantaneously, that will determine the state of its partner and can therefore create a secure coordination of nodes located at a large distance. Some QKD protocols, such as BB84 and E91, would use these quantum states for unconditional security in creating the keys for encryption. When deployed in space, these protocols would call for the sources of photons to have precise control, stable optical terminals, and high-efficiency detectors capable of working at very low levels of flux of an incoming photon. The Micius satellite successfully demonstrated both entanglement distribution and QKD, proving that quantum communication principles can withstand challenges posed by propagation in space.

7.5.2 Quantum Key Distribution for Deep Space Communication

QKD is significant because it is the most feasible usage of quantum mechanics when it comes to deep-space missions. In this QKD paradigm, the spacecraft and ground stations share keys for encrypting space communications, mathematically and physically secure from eavesdroppers. An attempt to read the quantum signal produces an error, which can be detected easily, thus ensuring that mission-critical data remain unscathed. For deep space, QKD will have tremendous advantages, as it will provide safety of navigation commands, scientific information, and will authenticate network traffic passing through the interplanetary network. Future missions will probably have quantum-repeaters and entanglement-swapping nodes at lunar, Martian, or deep-space relay stations, forming the end-to-end quantum-secure networks around the Solar System.

7.5.3 Challenges in Deep-Space Quantum Communication

Such great progress in research has opened doors to greater possibilities for quantum communication. However, quite a number of challenges still remain when attempting to extend this communication medium to deep-space applications.

- Photon loss: Arrival rates drop briefly over millions of kilometers, making it difficult to achieve quantum coherence.
- Decoherence: Cosmic radiation, solar particles, and background noise cause degradation of quantum states.

- Timing Synchronization: Timing accuracy at the picosecond level is required in QKD, and this becomes very daunting when measuring on interplanetary scales.
- Optical Pointing: Quantum links make use of extremely narrow beams and hence would require micro-radian-level pointing accuracy, similar to that of optical communication but more sensitive.
- Cryogenic Hardware: SNSPDs and several other quantum detectors usually require cryogenic operational temperatures, thus increasing the mass and also system complexity. Accordingly, the difficulties support the use of hybrid communication frameworks that interlink quantum systems with the current RF and optical channels such that they share redundancy and allow for operational flexibility.

7.5.4 Case Study: The Micius Satellite

China's Micius satellite mission is the biggest breakthrough in space-based quantum communication until now. Micius, which was launched in 2016, had a number of firsts, such as the distribution of entanglement and quantum key distribution over more than 1200 kilometers separating ground stations. The mission involved the use of entangled-photon sources, high-precision telescopes, and ultra-sensitive detectors to keep quantum links active despite the challenges of atmospheric turbulence and long-distance losses. Micius has successfully transmitted stable entangled photons through space in a vacuum. Additionally, it has given the world a glimpse of the future with its experiments proving the technical feasibility of deep-space quantum communication when coupled with advanced optics, precisely pointed, and high-quantum-efficiency detectors.

7.5.5 Role of Quantum Links in Hybrid Communication Systems

Quantum communication, working in a hybrid RF–Optical–Quantum framework, has become the most secure and the very foundation of the network. The RF band gives continuous and strong telemetry, the optical channel maintains high data transfer rates, and the quantum links provide secure key distribution, authentication, and cryptographic coordination which encompasses the whole network of deep-space communication. This allows the spacecraft to communicate securely even when they are passing through unpredicted areas or if there are several relay stations. It is also possible that quantum channels will bring about secure inter-satellite communications in the future interplanetary networks where quantum-secure routing, autonomous decision-making, and tamper-proof data exchange will be the characteristics of lunar, Martian, and deep-space missions. Therefore, as hybrid systems develop, quantum communication will play an important role in the ensuring of confidentiality, integrity, and authentication of deep-space data streams.

7.6 Hybrid Communication Architecture

The adoption of the hybrid communication architecture was inevitable as it is the most advanced deep-space networking technique combining RF, optical, and quantum communication technologies. These technologies will together support the entire range of concerns involved with communication to and between planets and stars. One of the high-power features of the RF system is that it can always be used for sending commands and telemetry regardless of the weather. Optical systems, on the other hand, are able to provide extremely high speeds for scientific payloads. Quantum links, however, are mainly used for the secretive and secure networking by using unbreakable key distribution methods. By using the integration of these technologies, high-reliability, large-bandwidth, and security will be won in keeping operational continuity under different environmental conditions. The most significant advantage of this synergy is for missions that would be performing at places where communication disruption or bandwidth limitation would cause significant delays or even data loss.

A hybrid deep-space communication system smoothly distributes the traffic among RF, optical, and quantum channels in a dynamic way depending on the real-time situation. RF paths play the role of primary traffic route while supporting a continuous link during spacecraft maneuvers, safe-mode operations, or adverse ground weather conditions that impair optical visibility. At the same time, high-capacity data transfer takes place through optical channels when the pointing accuracy and atmospheric conditions are favorable; on the other hand, quantum channels safeguard the network by carrying out periodic encryption key exchanges. This dynamic trafficking is made possible by developments in onboard communication controllers, software-defined radios, and AI-based decision algorithms that can handle channel quality monitoring, disruption prediction, and autonomous modality switching. Such an adaptive strategy provides both redundancy and efficiency at once making hybrid architectures attractive for long-duration missions where consistency, security, and resilience are the most important factors.

7.6.1 System-Level Integration of RF, Optical, and Quantum Channels

The effective merging of these three communication technologies requires not only excellent, but also well thought out onboard subsystems and ground infrastructures. Onboard, RF transponder, Optical terminals and quantum photon source are the three major components of the tripartite system that function together as per, a designated scheme. The sharing of timing systems synchronizes the data transfer and the use of integrated pointing mechanisms makes it possible for the optical and quantum beams to connect with the terminals placed on the earth's surface. The multi-modal ground stations fitted with the giant RF antennas, optical telescopes with adaptive optics, and cryogenic quantum detectors receive the signals *via* all these signals. The combination of the infrastructures will allow for hybrid routing, data fusion, secure key distribution, and link layer switching, all of which will occur seamlessly communication ecosystem during the operation that is independent of the changing mission phases and external constraints.

7.6.2 Adaptive Routing and Resource Allocation

Their knowledge of when and how to utilize the various communication modes is achieved with the help of flexible routing protocols and the management of the resources in a hybrid system. AI-powered algorithms analyze the quality metrics of the links and are based on signal to noise ratio, the atmosphere's transparency, the spacecraft's relativity, and the quantum detection rates. With the acquisition of these data inputs, the onboard system goes on to dynamically prioritize channels. For instance, it is to be expected that during a period of good visibility, the bulk data will be sent *via* optical links while the essential commands will be carried through RF links for stability confirmation.

Quantum channels will be employed from time to time for the exchange of keys that authenticate the entire chain of communication. Such an adaptive assignment achieves the best possible exploitation of the available bandwidth, less communication latencies, and sends mission data *via* the most efficient and secure methods of transmission.

7.6.3 Error Control, Synchronization, and Reliability

Any communication channel requires a full error correction and a synchronization plan as a minimum condition to be able to communicate reliably. In order to cope with noise that produces bit-errors in such systems, in RF and optical systems, error-correcting codes such as LDPC, turbo codes, and Reed-Solomon codes are used. Besides that, quantum communication is performed together with quantum error correction and entanglement purification aimed at the security functions integrity protection. Synchronization is realized through atomic clocks, precision timing systems, and photon time-tagging methods for the quantum links. Thus, hybrid architectures can be used to assure a continuous end-to-end communication safety even under very severe deep space scenarios by allowing each modality to keep its error-resilience and timing-accuracy.

7.7 Challenges in Implementing Hybrid Systems

The procedure for implanting the hybrid RF-Optical-Quantum communication system is full of technical, operational, and environmental troubles; nonetheless, the solutions it offers are very enticing for the future of the space missions beyond our planet. When a spacecraft has to combine three communication modalities which are fundamentally different, it means that the hardware for the coordination has to be enormous with a high system maturity level and the engineering effort needs to be extensive to provide a reliable system that will be able to keep extreme conditions of deep space. Among the challenges are hardware miniaturization and integration because optical telescopes, RF antennas, and quantum photon sources have to be non-interfering and at the same time coexisting. Moreover, the different thermal control, vibration, and pointing stabilities aspects of each subsystem make it too hard to put them together in a compact and lightweight configuration suitable for resource-challenged spacecraft.

On top of that, the environment poses serious challenges to the modules separately and is one of the main reasons why the manifestations have been slow-paced. For instance, besides the typical radiation, micrometeoroids, and extreme thermal and vacuum cycles that take place in deep space, a launch imposes different yet equally harsh conditions, such as severe acceleration, vibration, and shock, which the hardware must withstand together with the subsequent operation. Optical terminals should not lose a bit of their tight alignments even when thermal expansion or the maneuvers of the spacecraft happen; however, quantum superconducting detectors require the cryogenic temperature that is hard to maintain for long lapses in a mission. On top of that, mature RF communication subjected to plasma scintillation, solar interference, and antenna degradation is frustratingly hurting hybrid system performances. There are also weather-related limitations for ground-based optical and quantum reception that still require clear skies, light turbulence, and advanced adaptive optics. Environmental limitations like these give more reasons to have robust redundancy and fault tolerance built into hybrid architectures.

Also at stake is high-level concern about the compatibility and cooperation between the RF, optical, and quantum channels. To be able to function in real time, an onboard decision-making algorithm that is capable of hybrid systems interchangeably switching communication modes depending on prevailing conditions will only be effective with the help of highly advanced algorithms, software-defined communication modules, and high-speed digital processing units. The three modalities have to be synchronized as well, which is a more difficult task due to the fact that each modality has unalignable timing requirements as well as fundamentally different error statistics and in addition, even they are quite different by nature, but nevertheless, they need to be closely related. For example, quantum communication sets stringent requirements on timing-being-on-picosecond-level accuracy, ultra-low-noise photon detection, and secure authentication protocols-while these three parameters have to be correspondingly aligned with classical data channels. In this case, hybrid systems are very calculation-heavy which is why they need having light-weight, fault-tolerant onboard processors.

Besides that, the issues of scalability and the cost of the systems are quite serious ones. Both from a financial point of view and from a manufacturing angle, this will take a quantum leap to bring about the development of quantum communication hardware, lasers for deep space, adaptive optics, and telescopes for the ground to make them large enough. Subsequently, the hybrid communication system will require a network either terrestrial or interplanetary to function that is at least somewhat coordinated in terms of international standards, shared infrastructures, and a security framework. If there are no such standards, interoperability between agencies, spacecraft, and ground systems will become extremely difficult and, in the end, the most obvious consequence will be mission compatibility will be low. These challenges clearly show how important it is to do more research and collaborate through inter-agency networking and technological innovation, which will lead to the complete utilization of hybrid communication systems' potential to be used in missions going beyond the Earth's orbit.

7.8 Future Directions and Research Trends

Future hybrid deep-space communication systems are expected to evolve into highly autonomous, intelligent, and scalable networks that can sustain human life over the long term outside Earth. The most significant trend, among others, is the development of a Deep Space Internet that relies on Delay/Disruption-Tolerant Networking (DTN) protocols. These protocols are designed to cope with intermittent connectivity, long propagation delay and dynamic routing mechanism over planet distances. DTN is the technology for spacecraft, orbiters, landers, and relay satellites to communicate in a timely manner and is resistant to the conditions that are unfriendly in the context of traditional internet protocols.

As missions become more distant, e.g., going beyond Moon and Mars to asteroids and beyond, DTN will be at the heart of an interplanetary communication infrastructure, on which hybrid RF-Optical-Quantum systems can work in synergy.

The increasing integration of Artificial Intelligence (AI) and machine learning ideologies into communication architectures will be one of the emerging directions next to few others. AI is capable of optimizing resource allocation, predicting communication drops, improving pointing accuracy, and even deciding autonomously which channel to use i.e. RF, optical, or quantum depending on the environment and phases of the mission. A machine learning model trained with atmospheric parameters, spacecraft-to-spacecraft dynamics, and signal quality metrics would in fact enable real-time decision-making which in turn would increase link reliability and efficiency.

On top of that, the requirement for the intervention of the ground would be very much lowered if not eliminated altogether as AI-powered PAT systems, intelligent coding schemes, and adaptive modulation techniques make the spacecraft capable of operating and performing autonomously in deep space missions.

Research is also maturing toward space-based optical relay constellations, particularly in cislunar space, Mars orbit, and Lagrange points. Optical and quantum networks will stretch farther and reduce the point of contact space and improve communication windows because these relay stations keep line-of-sight with multiple spacecraft. Optical relay infrastructures will help create a multi-node communication mesh using lasers to transfer large volume data within the Solar System. Besides, there may be quantum repeaters and entanglement-providing nodes installed in space on the longer term to enable the linking of spacecraft, orbital platforms, and planetary bases in long-reaching quantum-secure networks.

In hardware, continued miniaturization of optical terminals, photonic integrated circuits, and quantum detectors will enable feasibility for hybrid communication systems in small spacecraft missions, including CubeSats and distributed swarms. Adoption of photonic chips, micro-lasers, and cryogenic-free single-photon detectors would minimize mass, reduce power consumption, and simplify the entire system architecture. Such innovations will democratize the deep-space exploration by allowing smaller, cheaper missions to use high-speed and secure communications capabilities initially available for flagship missions only.

International collaboration is expected to play a crucial role in defining the future of hybrid communication systems. Global standards, interoperable protocols, and shared infrastructure lead to compatibility among space agencies and commercial operators. Such agencies include NASA, ESA, JAXA, CNSA, as well as those from the private aerospace industry, and are already oriented toward interoperable communication frameworks for ensuring mission cross-support. The probability increased through cooperative efforts that in the long term hybrid deep-space networks will be available, mission costs reduced, and resilience increased by increasing numbers of terminals across the Solar System.

7.9 Ethical and Security Considerations

Deep-space communication systems, blending RF-Optical-Quantum-Almost-non-ethical architectures, are evolving, with ethical and security considerations becoming weightier. High-capacity optical links and unconditionally secure quantum channels imply additional responsibilities concerning data privacy, mission integrity, and international governance. One of the foremost ethical considerations has to do with the security and the authenticity of spacecraft command signals, the compromise of which could endanger mission safety, or, in the alternative, lead to unintended behavior of the spacecraft. Quantum communication is an area that is immune to eavesdropping and unauthorised access but the protocols of encryption, key distribution and authentication need to be protected to ensure resilience across inter-planetary distances.

Fair and responsible use of interplanetary communication resources is one more. Space frequencies, optical relay stations, and orbital infrastructure are coming under some pressure from different nations and private developments increasingly laying their claim to cislunar and deep space. Setting up common protocols and fair access guidelines will be crucial in reducing monopolization or interference that will impede science. Ethical frameworks need to ensure that communication assets and specifically those used for planetary exploration or for long-term colonization, are being managed in an open and collaborative fashion. In this light, development of international agreements similar to modern Outer Space Treaty will also be important in order to define the usage of fronts, data-sharing, and assign of responsibilities in case of communication argument or failure.

The issues of security sanction the smooth transfers of data as well as communication network integrity. Hybrid systems rely on civil and other terrestrial infrastructures that should be protected from cyber-attacks, signal spoofery and physical destruction. On the other hand, with the advent of autonomous spacecraft and AI communication routing, robust cybersecurity measures need to be established in order to prevent the unauthorized modification or exploitation of onboard decision-making systems. Above those, the deployment of quantum secure communication keeps on bringing up questions about long-term archival strategies to be adopted, as ephemeral quantum keys would necessarily need a solution for the preservation of mission-critical data. On top of that, the ethical processing, storing and forwarding of data through hybrid communication platforms will be the key to maintaining scientific trust and operational transparency.

Deep-space communications technologies also have to consider the environment. The placement of large numbers of optical or quantum relay satellites - specifically in cislunar or Martian orbits - could be the source of space garbage and long-term orbital congestion. Ethical stewardship of the orbital environments would need to consider the debris mitigation and end-of-mission disposal concepts in order to minimize the environmental impact. By taking good care of these ethical and security issues, space agencies and international organizations will ensure that the development of hybrid communication architectures is safe, just, and sustainable for the future of space exploration.

7.10 Conclusion

One of the foremost ethical considerations has to do with the security and the authenticity of spacecraft command signals, the compromise of which could endanger mission safety, or, in the alternative, lead to unintended behavior of the spacecraft. Quantum communication is an area that is immune to eavesdropping and unauthorised access but the protocols of encryption, key distribution and authentication need to be protected to ensure resilience across inter-planetary distances.

Fair and responsible use of interplanetary communication resources is one more. Space frequencies, optical relay stations, and orbital infrastructure are coming under some pressure from different nations and private developments increasingly laying their claim to cislunar and deep space. Setting up shared protocols and fair access guidelines will be vital in minimizing monopolization or interference that would possibly impede science. Ethical frameworks need to ensure that communication assets and specifically those used for planetary exploration or for long-term colonization, are being managed in an open and collaborative fashion. In this light, development of international agreements similar to modern Outer Space Treaty will also be important in order to define the usage of fronts, data-sharing, and assign of responsibilities in case of communication argument or failure.

The issues of security sanction the smooth transfers of data as well as communication network integrity. Hybrid systems rely on civil and other terrestrial infrastructures that should be protected from cyber-attacks, signal spoofery and physical destruction. On the other hand, with the advent of autonomous spacecraft and AI communication routing, robust cybersecurity measures need to be established in order to prevent the unauthorized modification or exploitation of onboard decision-making systems. Above those, the deployment of quantum secure communication keeps on bringing up questions about long-term archival strategies to be adopted, as ephemeral quantum keys would necessarily need a solution for the preservation of mission-critical data. On top of that, the ethical processing, storing and forwarding of data through hybrid communication platforms will be the key to maintaining scientific trust and operational transparency.

Deep-space communications technologies also have to consider the environment. The introduction of large numbers of optical or quantum relay satellites - in cislunar or Martian orbits in particular - could be part of space debris and long-term orbital congestion problems. Ethical stewardship of orbital environments would require consideration of strategies for debris mitigation and end-of-mission disposal to minimize environmental impact. As such, by proactively addressing these considerations with regards to both ethics and security, space agencies and international organizations will ensure that the evolution of hybrid communication architectures will be safe, just and sustainable for the future of space exploration.

AI disclosure statement: The author confirms that no artificial intelligence (AI)-generated content or AI-assisted writing tools were used in the preparation of this chapter (including summaries and the abstract). All text, analysis, and conclusions are the result of the author's original work.

References

1. Raible, D.E. and Romanofsky, R.R., On the Physical Realizability of Hybrid RF and Optical Communications Platforms for Deep Space Applications, AIAA 2014–4482, 2014, DOI: 10.2514/6.2014–4482.
2. Karmous, S., *et al.*, How Can Optical Communications Shape the Future of Deep Space Communications A Survey. *IEEE Commun. Surv. Tutorials*, 2024, DOI: 10.1109/COMST.2024.3403873.
3. Xu, G., *et al.*, Error Performance of Deep Space Optical Communication. *Opt. Express*, 27, 9, 13344–13358, 2019, DOI: 10.1364/OE.27.013344.
4. Investigations of Free Space and Deep Space Optical Communication Scenarios. *CEAS Space J.*, 2021, DOI: 10.1007/s12567–021-00388-y.
5. Operational Results from the Deep Space Optical Communications Project Ground Demonstration, in: *SPIE Conf. Proc.*, 2024, DOI: 10.1117/12.3041400.
6. National Aeronautics and Space Administration (NASA), Deep Space Network Overview, NASA Jet Propulsion Laboratory, 2023.
7. Berman, D., *et al.*, Voyager Mission Communications Performance and Longevity. *IEEE Aerosp. Electron. Syst. Mag.*, 35, 9, 26–35, 2020, DOI: 10.1109/MAES.2020.3020310.
8. Kaur, R. and Kumar, V., Advances in RF Communication Techniques for Deep Space Missions. *Acta Astronaut.*, 196, 315–326, 2022, DOI: 10.1016/j.actaastro.2022.05.003.
9. Hasan, M. and Ouyang, J., Challenges in RF-Based Deep Space Communication Systems. *IEEE Access*, 9, 135472–135484, 2021, DOI: 10.1109/ACCESS.2021.3119923.
10. Ivanov, H., Di Mira, A., Mejri, S., Maroquin, S., Heese, C., Schulz, K.J., Deep-space optical communications: challenges and technological advancement, in: *International Conference on Space Optics—ICSO 2024*, 2025, July, vol. 13699, SPIE, pp. 596–606.
11. Biswas, A., Srinivasan, M., Andrews, K., Velasco, A., Alerstam, E., Allmaras, J., ... & Rogalin, R., Overview of the deep space optical communications (DSOC) technology demonstration, in: *Free-Space Laser Communications XXXVII*, 2025, March, vol. 13355, SPIE, pp. 125–136.
12. Yin, J., *et al.*, Satellite-based Entanglement Distribution over 1200 Kilometers. *Science*, 356, 6343, 1140–1144, 2017, DOI: 10.1126/science.aan3211.
13. Bennet, C.H. and Brassard, G., Quantum Cryptography: Public Key Distribution and Coin Tossing. *Theor. Comput. Sci.*, 2014, DOI: 10.1016/j.tcs.2014.05.025.
14. Zhang, G., Xie, Z., Bian, D., Sun, Q., *et al.*, A Survey of Deep Space Communications. *J. Electron. (China)*, 28, 145–153, 2011.
15. De Cola, T., Paolini, E., Liva, G., Calzolari, G.P., Reliability options for data communications in the future deep-space missions. *Proc. IEEE*, 99, 11, 2056–2074, 2011.

16. Lazio, T.J.W. and Lichten, S.M., Radio Astronomy with NASA's Deep Space Network. *Galaxies*, 13, 6, 123, 2025.
17. Posner, E.C., Rauch, L.L., Madsen, B.D., Voyager mission telecommunication firsts. *IEEE Commun. Mag.*, 28, 9, 22–27, Sept. 1990, doi: 10.1109/35.57695.
18. Boroson, D., *et al.*, Overview and Results of the Lunar Laser Communication Demonstration, in: *SPIE*, 2014, DOI: 10.1117/12.2037231.
19. Ren, J.G., Xu, P., Yong, H.L., Zhang, L., Liao, S.K., Yin, J., ... & Pan, J.W., Ground-to-satellite quantum teleportation. *Nature*, 549, 7670, 70–73, 2017.
20. Liao, S.-K., Cai, W.-Q., Liu, W.-Y., *et al.*, Satellite-to-Ground Quantum Key Distribution. *Nature*, 549, 43–47, 2017, DOI: 10.1038/nature23655.

8

Energy Harvesting and Wireless Power Transfer Technologies for Space Systems

P. Arul[1*], N. Sathish[2], G. Ayappan[2] and Xiaozhi Gao[3]

[1]*Department of ECE, University College of Engineering Panruti, Anna University, Tamil Nadu, India*
[2]*Department of ECE, Sri Venkateswara College of Engineering , Sriperumbudur, Tamil Nadu, India*
[3]*School of Computing, Faculty of Science, Forestry and Technology, University of Eastern Finland, Joensuu, Kuopio, Finland*

Abstract

Energy Harvesting and Wireless Power Transfer Technologies for Space Systems chapter explores the innovative advancements in powering space missions through sustainable and efficient energy solutions. This chapter reviews in detail state-of-the-art energy harvesting systems including solar, thermal, vibrational, and radiation-based systems, which can operate in the harsh space environment, as deep space travel imposes strong constraints on energy supply and provision. It also investigates the technology of wireless power transfer (WPT), which is fundamental in long-distance energy provision without involving any physical connector. Such technologies are microwave and laser-based systems. Design challenges, system integration, reliability and the potential to support satellites, planetary rovers, and autonomous space platforms are highlighted. Using detailed analysis and case studies, this chapter aims at showing the revolutionary nature of energy autonomy in future space missions. This will aid to increase mission duration, lessen reliance on conventional energy storage, and give long-duration deep space exploration. The chapter also discusses futuristic trends, adaptive systems, and AI-driven energy management for scalable deployment.

***Keywords*:** Energy harvesting, wireless power transfer, deep space exploration, sustainable energy, AI-driven energy management

**Corresponding author*: arulmaran786@gmail.com

N. Sathish, Prasanth Aruchamy, Rajesh Kumar Dhanaraj, Nithya Rekha and Karthik Palani (eds.) Space Electronics for Deep Space Exploration, (175–198) © 2026 Scrivener Publishing LLC

8.1 Introduction to Energy Challenges in Space Systems

8.1.1 Overview of Energy Needs in Space Missions

Space missions, in the earth orbit or in deep space, are essentially resourceful to dependable and constant energy supplies. Energy is the lifeblood of all space operations, whether it is operating communication systems, sensors, scientific instruments, and propulsion units or keeping up life support systems in crewed missions. As our missions become longer and more complex like lunar habitats, Mars missions and deep space probes there is now a tremendous increase in energy needs.

The space missions that are being implemented nowadays use a broad spectrum of subsystems- power management units, on-board computing, active thermal controls, and scientific payloads that all need constant power access in harsh environmental factors. Power systems, in turn, are supposed to be strong, light, efficient, and able to maintain the operation, additionally, months or years without human interventions.

Vehicles in deep space also encounter the extra demands of diminishing solar radiation and communication lags with the Earth, which require very much autonomous and energy-hardened systems. Therefore, the provisioning of energy is a serious design concern in space systems engineering.

8.1.2 Limitations of Conventional Power Sources

Traditionally, space missions have relied on conventional power sources, such as:

a) Batteries

Rechargeable batteries, such as lithium-ion or nickel-hydrogen types, are commonly used to store energy onboard spacecraft. While they provide immediate and consistent power, they suffer from several drawbacks like mentioned below.

- Limited lifespan due to degradation after multiple charge-discharge cycles,
- Weight and volume constraints for long missions,
- Thermal sensitivity and performance degradation in harsh space conditions,
- Finite energy capacity, making them unsuitable for long-duration missions without replenishment.

b) Radioisotope Thermoelectric Generators

Radioisotope thermoelectric generators (RTGs) convert heat from radioactive decay (typically of Plutonium-238) into electricity. They have been successfully used in missions like Voyager, Cassini, and Curiosity. The limitations of RTGs include:

- High cost and limited availability of radioactive material
- Stringent safety regulations for launch and operation
- Low energy conversion efficiency (typically 6–7%)
- Political and environmental concerns regarding nuclear material in space

c) Solar Panels

In satellites and missions close to the earth, solar energy is popular. Nonetheless, its efficiency decreases along the way to the Sun and thus it cannot be used in deep space missions (i.e. farther than Jupiter) [1]. Moreover, solar panels are also susceptible to the dust (like those in the missions to Mars), damage by micrometeorites, and disturbance in orientation during maneuvers of the spacecraft.

8.1.3 Importance and Motivation for Autonomous Energy Systems and Advanced Power Technologies

The autonomous energy systems and advanced power technologies are important and have a need to be motivated. The further space missions get away and the more autonomous they have to be, the more energy systems must develop autonomy and intelligence. Independent energy systems are able to respond to the environment of the mission, distribute power optimally, and reduce the risks of the shortages or failures. Their advantages are:

- Less reliance on ground-based control,
- Increased resiliency and redundancy,
- Flexibility and adaptability of mission,
- Isolated or extreme environments palatability of robotic missions or human missions.

These systems will be important in applications in the future, such as lunar habitats, planetary rovers, and interplanetary probes [2]. The innovations required in order to attain reliable autonomy are to be made in energy harvesting and wireless power transfer (WPT). Energy harvesting is a set of technologies that converts the ambient environment such as sunshine, thermal gradient, vibrations, and electromagnetic fields into useful energy in-store devices, such as solar arrays with MPPT, thermoelectric generators, and piezoelectric materials.

The systems offer long-term, renewable energy, and low-maintenance energy whereby battery change is not possible. WPT allows transfer of energy without using physical connectors, which finds application in beaming power between spacecrafts, rovers, and space-based solar stations. WPT systems based on microwave and laser have scalable, modular architectures suitable to distributed missions and lunar infrastructures. Combined with technologies, they will enable the next-generation autonomous, sustainable, and resilient energy ecosystems, which are needed in deep-space exploration.

8.2 Energy Harvesting Mechanisms for Space Applications

In the realm of space exploration, energy autonomy is essential due to the limited feasibility of refueling or repairing spacecraft. As missions push further into deep space and aim for longer durations, energy harvesting technologies provide a promising pathway to support power needs sustainably. This section explores the primary energy harvesting mechanisms suitable for space, focusing on their operational principles, applications, and viability.

8.2.1 Solar Energy Harvesting: Photovoltaic Cells and Concentrators

The most proven and trusted source of power use in space missions is solar energy which drives satellites, rovers and orbital platforms. Photovoltaic (PV) cells directly transform solar radiation into electricity by means of a semiconductor material (silicon or sophisticated multi-junction compounds e.g. GaAs, InGaP) [3]. These space-qualified cells are highly reliable, and can have long life cycles (10–15 years or more), and efficiencies up to approximately 18% on single-junction or more than 30 percent on triple-junction designs.

Solar concentrators, such as parabolic mirrors and Fresnel lenses, concentrate sunlight on smaller PV areas or thermal receivers to maximize the level of energy captured and concentrate the sunlight (a factor of 2 to 10 times) on local areas. This method enhances power density and allows small hybrid systems to be used in small spacecraft.

Nevertheless, there are drawbacks in both technologies: the solar intensity decreases with the distance between the Sun ($\propto 1/r^2$) and dust deposition, radiation damages and orientation may hamper the performance of the panel. The use of concentrators further needs high accuracy in tracking and alignment and is therefore sensitive to vibration and deployment mistakes. The combination of PV cells and concentrators is the basis of solar energy collection in space as it balances the efficiency, space, and reliability of long-term missions.

8.2.2 Thermal Gradient Energy (Thermoelectric Generators)

Thermoelectric generators (TEGs) use the See beck effect -a phenomenon in which a temperature gradient through a substance generates voltage [4]. In space, temperature variations between the sun facing side and shadowed side of a space ship or waste heat produced by onboard electronics are used.

Radioisotope Thermoelectric Generators:
Use heat from radioactive decay (e.g., Plutonium-238) to power TEGs. Used in missions where sunlight is insufficient (e.g., Voyager, Cassini, Curiosity, Perseverance).

Solar-Thermal TEGs:
Employ focused solar heat and radiative cooling for electricity generation. Under active research for smaller missions or lunar night operations.

The advantages are no moving parts, high reliability, and operates continuously regardless of solar exposure. But it has limitations like low efficiency (~6–10%) and radioisotopes pose handling, cost, and political concerns.

8.2.3 Kinetic- and Vibration-Based Harvesting

During space missions, spacecraft are constantly exposed to small mechanical vibrations caused by deployment, attitude adjustments, and even micro-meteoroid impacts. Instead of letting this energy go to waste, it can be converted into usable electricity through piezoelectric, triboelectric, or electromagnetic transducers. Piezoelectric harvesters, made from materials like PZT or PVDF, generate voltage when stressed and are ideal for flexible or built-in systems such as robotic arms or panels. Triboelectric generators (TENGs), on the other hand, produce energy through the friction between different materials and can be easily incorporated into lightweight, thin-film structures.

These technologies are especially useful for powering small sensors or capturing energy during rocket launches and spacecraft maneuvers. However, their output is typically quite low ranging from microwatts to milliwatts and the availability of vibration energy can be inconsistent.

8.2.4 Radiation and Magnetic Field-Based Harvesting

Space presents exceptional radiation and magnetic conditions that can be harnessed to energy, but the technologies are only developing, and are still mostly experimental. The comparative Efficiency and Feasibility in Space for various mechanisms has been highlighted in Table 8.1.

In radiation harvesting, the beta-voltaic and gamma-voltaic devices utilize the radioactive materials or cosmic rays energy particles to generate electricity. They have the ability to provide extremely high life cycles yet extremely small output power. The Magneto dynamic Harvesting tries to harness the motion of spacecrafts by the magnetic field of the earth in order to generate current within coils similar to an effect of the Faraday law [5]. It is experimentally applied in electrodynamic tethers.

Table 8.1 Comparative efficiency and feasibility in space.

Mechanism	Typical efficiency	Power output	Application suitability	Challenges
Solar PV (multi-junction)	25–35%	High (Watts to kW)	LEO, Mars missions, satellites	Radiation degradation, angle of incidence
Solar concentrators	Enhances PV/TEG	Medium to High	Small satellites, deep space	Tracking systems required
Thermoelectric (RTGs, TEGs)	6–10%	Low to Medium (W to 100W)	Deep space, lunar night, rovers	Radioisotope handling, low efficiency
Piezoelectric/ Vibration	<10%	Very Low (μW–mW)	Sensors, structural health	Inconsistent mechanical energy
Radiation/ Magnetic Field	<5%	Very Low	Experimental, tether systems	Location-specific, low conversion rates

8.3 Wireless Power Transfer Technologies for Space

Due to the complexity of space missions and their duration, the need to ensure flexibility, contacts, and reliability of power transmission is increasing at an alarming rate. The technologies of WPT are becoming critical enabling factors in future space missions- they will remove the physical connector requirement and can distribute power throughout modular, mobile, or distributed space systems [6]. This section discusses the principles, most important WPT techniques, innovations, and challenges related to space systems.

8.3.1 Principles of WPT

- WPT is the transmission of electrical energy to a load without necessarily coming into physical contact with a source of power. The underlying concept is the electromagnetic coupling in which power is transmitted *via* electric or magnetic fields.

Core Mechanisms:

- Near-field transfer (short-range, non-radiative): It is based on magnetic or electric field coupling.
- Far-field transfer (long range, radiative): It involves electromagnetic radiation, e.g., microwaves or lasers.

A WPT system consists of the following basic components:

- Transmitter: Produces and channels energy with the help of oscillators, amplifiers, and antennas or coils.
- Transmission medium vacuum (in space), or air (near-Earth system).
- Receiver: Transduces the sent energy back into consumable electric energy (e.g., rectennas or photovoltaic cells).

In space, WPT can be used to remotely supply power to landers or rovers, satellites or in-space infrastructure, where any physical cables have become impractical, whether by movement, range, or unfriendly conditions.

8.3.2 Types of WPT Technologies

The various mechanisms used in WPT in space systems include several, namely, induction, capacitive, microwaves, and laser-based—each having an appropriate range and power demand. Inductive coupling is based on the magnetic fields between coils to provide high efficiency (70% to 95%) over short ranges, suitable to wireless charging between spacecraft modules or rovers, but necessitating close alignment and sensitive to microgravity. Capacitive coupling is based on electric field between plates to transmit low power energy in compact satellite modules; it is appropriate to microelectronics, but is affected by dielectric variation and spacing of electrodes. Hotter distances are served by microwave power transfer (MPT) which uses high-frequency beams (2.45 or 5.8 GHz) converted to DC through rectennas and allows beaming of power across meters up to hundreds of kilometers with 40% to 60% efficiency. It is in favor of solar power satellites and energy sharing between the satellites but has challenges, such as beam positioning, thermal controls, and interception. WPT uses laser-based transmission of focused optical beams to photovoltaic receivers with moderate efficiency (10–40%), and high accuracy. It can be helpful to operate the spacecraft, drones, or lunar instruments at night but needs to have a clear line of sight and must be handled with care to avoid errors in alignment or damage to the optical property.

8.3.3 Recent Innovations in Long-Distance WPT for Spacecraft and Rovers

Potential of long-distance WPT systems in space is confirmed in a few projects and demonstrations in recent years:

I. Space Solar Power System (SSPS) of JAXA

Japan aerospace exploration agency has been developing actively microwave-based transmission of solar power by orbital platforms to earth, which forms the ground work to space based solar power satellites [7].

II. NASA's BEAM Experiment

NASA has researched on laser beaming as a way of powering the lunar and Martian surfaces in order to provide power to the shadows such as the craters of the moon.

III. The Wireless Rover Charging of ESA

European Space Agency showed how the inductive charge of the rovers could be achieved using the landing stations without mechanical connections, which can be lost following dust or mis possessing directions.

IV. DARPA's POWER Project

DARPA began the creation of microwave beaming of satellites to ground receivers, which could also be used in spacecraft-to-spacecraft energy transmission as well.

V. LEO-Ground Microwave Transfer

In 2024, Caltech researchers were the first to beam power to a ground station orbitally, and this demonstration indicated that future beaming power relays in space were possible [8].

8.3.4 Efficiency and Safety Concerns

Despite the enormous potential of WPT in space applications, there are a number of efficiency and safety issues that need to be addressed first before massive implementation. The conversion inefficiencies in transmitters and receivers, beam divergence of far-field systems, and atmospheric attenuation of operations connected to the Earth are the factors that create efficiency losses [9]. High thermal loads created by the high-power transmitters are also problematic and managing excess heat is an issue because it is challenging to cool these loads in the vacuum of space.

Safety is also crucial, though the microwave and laser beams might disrupt onboard electronics or communication connections, and out of alignment beams might pose a danger to astronauts or destroy delicate instruments. This will need stringent eye-safety protocols, radiation shields, and automatic alignment controls [10]. Mobile or rotating receivers, like rovers, require precise tracking and beam steering. Besides, environmental risks, such as space dust, micrometeoroids and lunar regolith, may impair the work of receivers or block the photovoltaic surfaces, which influences the reliability of energy collection.

8.4 Design Considerations and Constraints in Space Environments

The design of space systems spacecrafts, satellites, landers, and rovers demand a careful knowledge of the hostile environment of space, which is unusual. Extreme temperatures, vacuum, radiation, microgravity, and mass limitations are very challenging factors to consider in engineering [11]. The section gives the design considerations and constraints that are critical to deploying energy systems and electronics in space.

8.4.1 Impact of Radiation, Vacuum, and Extreme Temperatures

Space conditions expose electronic circuits to radiation, vacuum, and huge temperature changes that are very dramatic to performance and life expectancy. Single-event effects are caused by radiation, ranging from Galactic Cosmic Rays (GCRs), Solar Particle Events (SPEs), trapped belt particles, and accumulated ionizing radiation damages semiconductors and materials. The mitigation of these is done by radiation-hardened components, error correction systems, and high-Z shielding, such as aluminum or tantalum.

Vacuum conditions present such problems as outgassing, limited cooling, and arcing of high-voltage systems. One can use low-outgassing materials, radiative cooling (e.g., deployable radiators), and smooth designs of electrodes to be stable.

The drastic variations in temperature (−150°C to +150°C) cause mechanical fatigue, inefficiency of the battery and sensor drift. Multilayer insulation, thermal coating, active heaters, and low thermal expansion materials are some of the measures of protection to maintain the integrity of the system [12].

8.4.2 Structural and Weight Constraints

The mass is a significant limitation in the design of spacecrafts, with the cost of launching a spacecraft being between $2,000 and $10,000 per kilogram. High payload weight sets more fuel needs and reduces the flexibility of mission and high-g launch forces require powerful but lightweight structures. To solve this, engineers make use of a high standard of composites like carbon fiber and titanium alloys, and the topology optimization to reduce mass without harmfulness in terms of integrity.

Spacecraft also need to meet tight volume constraints given by the requirements of the launch vehicle (e.g., CubeSat standards, Ariane or Falcon fairings) [13]. These modular and deployable systems, such as foldable solar arrays, antennas, and radiators, allow efficient stowage and deployment in space thus allowing the efficient use of payload space with structural integrity.

8.4.3 Materials Selection and Miniaturization

The choice of materials is very important so as to make sure that it is mechanical, thermal, and radiation tolerant in space. The ideal materials should be able to withstand thermal cycling, degradation by radiations, and also have low outgassing rates according to NASA and ASTM guidelines [14]. Typical include lightweight alloys of aluminum and alloys of beryllium in high stiffness modes, polymers such as Kaplon or Mylar in insulation and flexible electronics. Adhesives based on silicone are used since they have low outgassing and bonding characteristics.

Small satellites and CubeSat missions have led to miniaturization, which focuses on small and integrated designs. System-on-Chip (SoC) and multi-functional sensing, control, and power management multifunctional sensing. These trends provide an increase in system efficiency with robustness in line with the requirement of lightweight and scalable and low-cost space architectures [15].

8.4.4 Power Management Systems and Storage Integration

Good power control gives spacecrafts independence when there is a variation in energy supply. The power management units (PMUs) stabilize and control the power input of solar panels or WPT systems, charge batteries, and understand the voltage, thermal, and surge protection [16].

Lithium-ion batteries are used as the primary choice of energy storage due to their high-energy density, and peak loads are assisted by supercapacitors, and solid-state batteries are being proposed as safer and long-life options. There are also studies on flywheel systems to be used in the missions of long duration.

Some design priorities are that the battery should be efficient in extreme conditions, that radiation should withstand degradation, and redundancy will be incorporated to avoid a thermal or chemical failure [17]. The concept of maximum power point tracking (MPPT) is still necessary to sustain solar power production in different illumination levels as a way of maximizing the use of energy in essential spacecraft operations.

8.5 Integration with Spacecraft and Robotic Systems

Power integration in space missions is not merely about generating or storing energy—it is about designing a holistic, resilient, and intelligent power ecosystem. The integration of energy harvesting, WPT, and storage systems with spacecraft and robotic platforms ensures mission autonomy, durability, and adaptability in the unforgiving environment of space. This section covers the architecture, real-world applications, hybrid power strategies, and emerging intelligent power management technologies.

8.5.1 Onboard Power Architecture and Modular Integration

The concept of power integration in space missions is not about producing or storing energy, but about coming up with a holistic, resilient, and smart power ecosystem. Combining energy harvesting, wireless power transmission and storage systems with the spacecraft and robotic platforms will protect mission autonomy, life, and flexibility in the space environment, which is inhospitable. This part deals with architecture, real-life applications, hybrid power plans, and new smart power management technologies.

The spacecraft and robotic systems developed today have a layered and modular architecture of their power systems to accommodate dynamic mission profiles and subsystem autonomy [18].

The basic elements in power architecture:
The space power systems are made up of subsystems that are integrated to make sure that there is efficient generation, storage, and distribution of the power.

1. Power Generation Units:
   Solar panels (deployable or body-mounted).
   Radio thermoelectric generators (RTGs).
   Antenna energy harvesting or transducing.
2. Power Distribution Unit (PDUs):
   Assign, allocate and distribute power to different subsystems.
   Install DC/DC converters, isolators and fault-detection modules.
3. Power Storage:
   Primary: LiFePO4 or LiFePO3 battery.
   Secondary: Generative flywheels, Supercapacitors.
4. Power Management and Control Unit (PMCUs):
   Carry out maximum power point tracking (MPPT).
   Process redundancy, fault recovery and load prioritization.

8.5.2 Case Studies: Mars Rovers, CubeSats, and Lunar Landers

The various case studies for different systems in terms of power has been highlighted in Table 8.2.

Table 8.2 Case studies.

System	Primary power source	Supporting power/features	Key highlights/ mission examples
Mars Rovers (Curiosity, Perseverance)	Radioisotope Thermoelectric Generators (RTGs) providing ~110 W continuous power	Battery banks for peak loads; thermal power management for night survival	Intelligent power scheduling prioritizes science operations; sustained operation during dust storms
CubeSats	Deployable solar panels generating 2–20 W	Rechargeable batteries for eclipse periods (~35 min/orbit); energy-efficient electronics	LightSail 2 demonstrated solar sails for propulsion and energy augmentation
Lunar Landers	Solar panels and rechargeable batteries (short missions); hybrid systems with solar + laser WPT or fuel cells (long missions)	Thermal insulation for extreme temperature management	NASA's VIPER rover (2024) uses a solar-battery hybrid to explore the lunar south pole

8.5.3 Hybrid Power Solutions: Harvest, Storage, and WPT

Hybrid power solutions are also being used more extensively in spacecraft in order to improve redundancy, resilience, and adaptability through the combination of different sources of energy, like solar, RTG, and WPT systems. When operating in deep space, such combinations as Solar + RTG provide uninterrupted energy supply, and Solar + Battery + WPT are common in orbital relays and on the rovers sent to the moon. In the case of smaller platforms, piezoelectric harvesting combined with super capacitors can be used to promote local power production and building monitoring [19]. Such hybrid systems minimize reliance on a single power source, allow sharing of energy between spacecrafts swarms, as well as provide non-contact charging of mobile platforms. Nevertheless, there are smart power flow management, efficient source-switching and load balancing connections, and backup storage necessary to decelerate power imbalance at transitional times. Also, it is important to be compatible with energy-aware communication protocols, e.g., duty-cycled RF or optical links to ensure system efficiency and mission reliability in dynamic space environments.

8.5.4 Autonomous Power Management Algorithms

Independent power control is also necessary in the long or deep space missions that experience delays or blackouts in communications. Smart PMCUs are being integrated with AI and rule-based algorithms in real-time to make intelligent decisions regarding power production, storage, and use to guarantee the efficiency and operational life of spacecrafts. Its major features are its ability to schedule tasks energy aware, i.e., non-critical operations are delayed during low-power conditions; fault detection and recovery to isolate failures and reroute power; battery health optimization, i.e., the charge-discharge rates are controlled; and prediction-based control based on solar flux models and past data to predict energy availability. Control algorithms range between fuzzy logic and rule-based systems of early-stage designs to model predictive control (MPC) of forecast-driven management, and machine learning (ML) and reinforcement learning (RL) of adaptive and self-cultivating power strategies. The practical applications are NASA Mars Helicopter, dubbed Ingenuity, which plans flights according to the peak of sunlight, and the ESA Proba-V that plans its payload use in accordance with energy limitations [20]. The recent literature also examines the concept of decentralized energy sharing between a constellation of satellites to collaboratively and effectively use power in space.

8.6 Ground to Space and Space-to-Space Power Transfer

Long-duration and large-scale space missions, which include orbital mega-constellations, lunar bases, Mars missions, and other have highlighted the necessity of non-traditional energy logistics. One such frontier in this regard is beamed wireless power, which allows energy to be sent off Earth into orbit, and between spacecraft in space. It talks about the existing and developing technology of ground to space and space to space power transfer and specifically beamed energy systems, satellite power sharing, and the possibility of wireless refueling in space.

8.6.1 Beamed Power Beamed Power (Space-Based Solar Power) from the Earth to Orbit

In space-based solar power (SBSP), solar collectors are located in orbit to collect continuous solar energy and transmit this energy (either by microwave or laser) to receiving stations located elsewhere in space or on the surface of the Earth. Alternatively, it is possible to beam energy on Earth to orbit, and drive satellites during launch, eclipse or shadow [21].

Architecture of SBSP Systems
The Figure 8.1 shown below depicts the key parts of SBSP system.

- Solar power satellite (SPS): Solar energy is collected with the assistance of high-efficiency photovoltaic (PV) arrays.
- Microwave/laser transmitter: DC power is changed to microwaves (0.45GHz or 5.8GHz) or coherent lasers (1–2 mm).
- Beaming unit: It is a high-gain antenna or optical lens system that deflects the beam of energy.
- Receiving station (rectenna): The station on the ground or in the orbit is an array of rectifying antennas that receive the incoming energy and convert them into the usable electric power.

SBSP offers continuous solar power and less dependence on heavy onboard power systems; however, it has limitations in that it has beam problems, atmospheric losses, safety, and high costs of deployment of large-scale orbital structures.

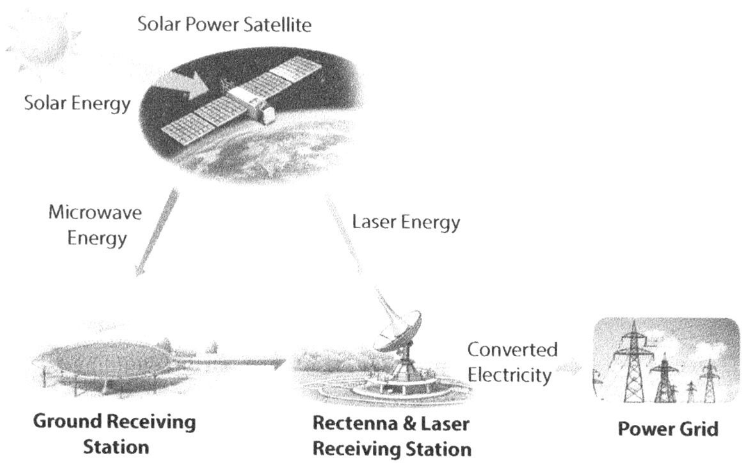

Figure 8.1 Architecture of space-based solar power systems.

8.6.2 Wireless Energy Sharing between Satellites and Platforms

Wireless energy sharing has been created as one of the fundamental facilitators of mission resilience, resource optimization, and redundancy as satellite constellations and modular space habitats grow in size. It has been applied to inter-satellite power routing of space constellations, such as Starlink or OneWeb, common modular space station power grids, and as a power source to lunar rovers when operating in the dark or in the shade of a space hub. These systems are based on microwave and laser-based WPT technologies, which are enabled by rectennas to convert RF-to-DC and autonomous beam to steer systems to provide greater accuracy in the dynamic orbital environments [22]. Examples are the POWER Program (2023) by DARPA to beam power to space by use of microwaves and the Solaris Initiative (2024) by ESA to transmit energy to the moon, as well as the Luna Net Framework by NASA bringing power and interplanetary communications together. Wireless energy transfer means that one does not need large solar arrays, can do distributed energy pooling, and can have better operational flexibility, but beam alignment, power loss mitigation, and buffering latency in severe space conditions are currently challenging.

8.6.3 Feasibility of In-Space Refueling *via* WPT

Traditionally, in-space refueling has been considered with reference to propellants. But an emergent idea is the system of refueling electrical energy charging battery-depleted satellites or spacecrafts wirelessly through WPT as opposed to docking them physically.

Space-based wireless energy transfer concept framework consists of a central power station or satellite that transmits the energy to other orbital and surface objects. This covers provision of power to spacecraft in eclipse or operating on a broken array of solar panels, small probes, and drones in cislunar orbit or even working in Martian orbit. It may also be useful *in situ* robotic platforms (autonomous servicing platforms), so that maintenance, exploration, and communication can be done continuously even in places with minimal sun exposure.

System Requirements

- Beam stabilization and high-precision tracking systems.
- Storage of energy: The energy received will have to be stored in onboard capacitors or batteries.
- Normative energy interfaces and safety measures.

Advantages

- Eliminates the necessity of large onboard power storage and eliminates weight.
- Prolongs the life of satellites that fail to operate.
- Supports deep space missions and rendezvous missions with low solar exposure.

Key Developments

- OSAM (On-Orbit Servicing, Assembly, and Manufacturing) Program of NASA: Researching on robotic servicing which may involve power replenishment.
- Wireless Laser Charging Tests at MIT (2022): They showed that it was possible to charge CubeSats at least up to 20 m in the lab.

The major limitations to removing the space based WPT include low energy density and low conversion efficiency of beam receivers as opposed to conventional wired systems [23]. Dissipation of heat in receiving units is another critical challenge because the surplus of thermal energy may have an impact on performance and life of the components. Also, some of the necessary prerequisites to keep the energy transmission safe and reliable is to take care of collision and interference in the growing crowded orbital space.

8.7 Problems, Dangers and Dependability Concerns in Space Systems Energy Harvesting and Wireless Power Transport

With technological improvements in WPT and energy collection, there are major engineering and operational issues with the use of these technologies in the space systems. These are technical constraints that comprise power losses, efficiency bottlenecks, loss of material under harsh space conditions and system failure and human safety risks [24]. This section presents major challenges, risks and reliability issues of power technologies in the space missions.

8.7.1 Power Losses and Conversion Efficiency

The inefficiency in the conversion and transmission of the energy is one of the main issues in the WPT and energy harvesting systems. Such losses can severely decrease the effectiveness of the whole system especially in the long-range space applications.

I. Conversion Losses

Photovoltaic Efficiency: Solar cells used in space (standard) have an efficiency around 28% to 34% i.e. more than half of the solar power is wasted as thermal energy.

Rectenna Efficiency: Microwave or RF-to-DC conversion efficiency is typically 60% to 85% depending on operating frequency, load matching, etc.

Laser power transfer laser-based WPT has a problem of beam divergence and conversion losses at the transmission and reception ends. The total efficiency of the system may decline to less than 20 per cent [25].

II. Transmission Losses

Beam Divergence: It is particularly important in long-range beaming (e.g., ground-to-orbit). Energy diffuses, unless higher levels of concentration are employed.

Atmospheric Interference: When a system is in the ground it gets interfered with by cloud cover, dust and humidity which contributes to the absorption and scattering of the signal.

Alignment Sensitivity: A mismatch of the transmitter and receiver causes a large amount of energy loss.

III. Thermal Losses

Those systems which have large power (e.g., laser transmitters) need to deal with thermal dissipation because heat radiation in a vacuum is not as efficient as conduction or convection.

8.7.2 System Degradation Over Time

Long-term reliability is critical in missions that require deep-space probes or Mars rovers, or lunar environment. Materials and components are degraded in the environment therefore performing poorly or fail with time. The space power systems are subjected to various challenges of degradation with time. Radiation caused by cosmic rays and solar particles decreases solar array efficiency through defects in a semiconductor junction. Exposing materials to thermal cycling between high and low temperatures (between −150°C and +150°C) leads to fatigue, cracking and dielectric degradation of components [26]. The process of mechanical wear impacts deployable equipment such as solar panels, and dust settling on planetary surfaces is an additional source of reduced efficiency. Also,

Li-ion or LiFePO 4 cells that have grown old in battery form cause a loss of capacity and increase of internal resistance and reduced cycle life because recharge or discharge cycles occur under different WPT conditions.

8.7.3 Safety Requirement and Recovery of Failure

Under deployment of high-power space energy systems like microwave or laser-based power transfer, a strong observance of safety regulation and failure recovery processes should be in place. Astronauts and delicate electronics can also be exposed to harmful levels of electromagnetic radiation, and there are standards such as IEEE C95.1 and ANSI Z136 that specify the acceptable exposure limits. In order to be reliable, WPT systems are designed with redundant power supply (e.g., hybrid solar-RTG), radiation-resistant electronics, and emergency shutdowns to WPT beams. Software-based load-shedding to keep priorities going, autonomous reconfiguration to avoid breaking power distribution paths, and smart diagnostics to find and isolate faults (rectenna degradation, power anomalies, etc.) are all part of recovery and failure management.

8.7.4 Communication and Control Issues in WPT

WPT systems require real-time, autonomous, and accurate feedback to manage in space which requires real-time communication. These systems need a constant tracking of beam and its alignment with the help of GPS, inertial sensors or vision-based tracking. The AI-centered feedback controls are utilized to optimize the beam focusing, power modulation, and diagnostic, and machine learning is utilized to predict the best power transfer windows, depending on orbital information [27]. Autonomous decision making is important because of delay in the communication involved in deep-space operations. Nevertheless, like the software-defined WPT networks, they are vulnerable to cybersecurity threats such as signal jamming, spoofing, and thus secure communication and blockchain-based control validation is implemented.

8.8 Future Directions and Emerging Research Trends

As space missions expand toward deep space exploration, lunar colonization, and interplanetary networks, the future of energy harvesting and WPT hinges on disruptive technologies and intelligent systems. Emerging trends such as artificial intelligence driven energy systems, quantum-enhanced materials, space-based solar power satellites, and robust architectures for deep space missions are set to revolutionize how power is generated, managed, and distributed in space.

8.8.1 Smart Adaptive Power Systems Using AI

Artificial intelligence is playing a transformative role in smart power management and adaptive control for space missions. Traditional energy systems are static and deterministic, often requiring human intervention [28]. AI-based systems promise real-time optimization, predictive diagnostics, and self-reconfiguration of energy pathways. The Key Capabilities are

- Load Forecasting and Scheduling: AI models such as neural networks or reinforcement learning agents predict upcoming energy demands (e.g., during eclipse, propulsion, science operations).
- Autonomous Load Shedding: In the event of a power shortfall, AI prioritizes critical subsystems, shutting down non-essential modules intelligently.
- Adaptive Beam Steering in WPT: AI helps maintain alignment of microwave or laser beams in dynamic orbital environments using feedback from sensors and inertial units.
- Fault Detection and Prognostics: AI algorithms monitor system health and predict component degradation, extending operational lifespan.

Artificial intelligence is increasingly integrated into space power systems to enhance autonomy and efficiency. For example, NASA's EDGE AI Initiative employs onboard machine learning for real-time decision-making, while CubeSats and Small Sats use AI-driven maximum power point tracking to optimize solar energy utilization. These advancements reduce human intervention in deep space missions, enable adaptive energy routing among interconnected spacecraft, and support resilient, decentralized systems capable of self-healing and maintaining continuous operation even after component failures.

8.8.2 Quantum and Nano-Materials in Energy Harvesting

The limitations of conventional materials in terms of efficiency, radiation tolerance, and weight are driving interest in quantum-enhanced materials and nanostructures for energy systems in space. Emerging energy technologies such as quantum dot photovoltaics, 2D materials, and plasmonic nano antennas are transforming spacecraft power generation. Quantum dot solar cells use nanoscale semiconductors with tunable bandgaps, achieving higher efficiencies and better low-light performance, ideal for Martian or icy environments. Two-dimensional materials like graphene and MoS_2 enable transparent, flexible, and radiation-resistant devices for lightweight power systems. Thermoelectric harvesters exploit temperature gradients for continuous energy generation, while plasmonic nanoantennas can convert light at specific frequencies for laser-based WPT. However, challenges such as radiation stability, integration complexity, and large-scale fabrication remain significant barriers to deployment.

8.8.3 Space-Based Solar Power Satellites

Space-based solar power systems offer a promising approach to harvesting continuous solar energy in orbit and wirelessly transmitting it to Earth or other space platforms. Large solar arrays in geosynchronous orbit capture sunlight, converting it into microwaves or laser beams that deliver energy to ground-based rectennas, spacecraft, or lunar rovers. Recent advancements include JAXA's 1 MW SSPS demonstrator planned for 2030, Caltech's 2023 CubeSat demonstration of orbital solar-to-laser transmission, and ESA's SOLARIS initiative exploring lunar microwave beaming [29]. While SSPS provides uninterrupted power and lightweight receiver infrastructure, challenges such as high deployment costs, beam safety, atmospheric losses, and international regulatory concerns remain key barriers to large-scale implementation.

8.8.4 Prospects for Long-Term Deep Space Exploration (e.g., Mars, Europa)

Sustaining long-duration missions to Mars, Europa, Titan, or the asteroid belt will require radically different power systems from those used in LEO or even lunar orbits. Future space energy architectures emphasize adaptable power solutions for diverse planetary environments. On Mars, reduced solar irradiance and frequent dust storms make hybrid nuclear-solar systems, such as NASA's Kilo power Reactor (1–10 kW for up to a decade), ideal for sustained missions. In distant icy moons like Europa, where sunlight is scarce, RTGs provide limited but reliable power, while emerging research explores subsurface probes and orbital laser-based WPT relays. Interplanetary microgrids, featuring distributed solar towers, reactors, and mobile receivers, aim to support long-term exploration and habitation.

These systems increasingly rely on AI-driven energy management for predictive distribution and maintenance, with self-repairing nanomaterials and modular architectures ensuring scalability and resilience [30].

The future of space power systems is being shaped by intelligent, quantum-enhanced, and scalable solutions. AI-based smart power grids, nano-enabled harvesting materials, and orbital power satellites form the backbone of future-proof architectures. As humanity prepares for Mars colonies, Europa ice drilling, and planetary defense missions, next-generation power technologies will serve as both the foundation and the enabler of these ambitions. While technical, financial, and regulatory hurdles remain, the research trajectory indicates a compelling shift toward autonomous, adaptive, and highly efficient energy systems in space.

AI disclosure statement: The author confirms that no artificial intelligence (AI)-generated content or AI-assisted writing tools were used in the preparation of this chapter (including summaries and the abstract). All text, analysis, and conclusions are the result of the author's original work.

References

1. France, R.M., Geisz, J.F., Song, T., *et al.*, Triple-junction solar cells with 39.5% terrestrial and 34.2% space efficiency enabled by thick quantum well superlattices. *Joule*, 6, 1121–1135, 2022.
2. Xu, X., *et al.*, A brief review of high efficiency III–V solar cells for space application. *Front. Phys.*, 8, 631925, 2020.
3. Ziani, Z., *et al.*, Enhancing multi-junction solar cell performance: advanced predictive modeling and cutting-edge CIGS integration techniques. *Energies*, 17, 4669, 2024.
4. Hadj Dida, A., Bourahla, M., Bekhti, M., Comparative analysis and simulation of highest efficiency multi-junction solar cells for space applications, in: *Renewable Energy for Smart and Sustainable Cities, Lecture Notes in Networks and Systems*, vol. 62, pp. 535–543, Springer, 2019.
5. Electromagnetic vibrational energy harvesters: a review. *Sensors*, 22, 5555, 2022.
6. Hu, Y., *et al.*, The application progress of wireless power transfer in space utilization field, in: *Proc. Int. Conf. on Wireless Power Transfer (ICWPT)*, Lecture Notes in Electrical Engineering, vol. 1158, Springer, pp. 520–532, 2024.
7. JAXA, Research on microwave wireless power transmission technology, 2014, https://www.kenkai.jaxa.jp/eng/research/ssps/ssps-mssps.html.

8. Ramos, R.M. and Yuan, J.-S., Wireless laser power transmission: recent progress and future trends. *Int. J. Eng. Res. Technol.*, 2018.
9. He, L., *et al.*, Efficiency analysis and optimization for laser wireless power transfer, in: *Laser Wireless Power Transfer*, Springer, 2022.
10. Clerckx, B., Kim, J., Choi, K.W., Kim, D.I., Foundations of wireless information and power transfer: theory, prototypes, and experiments. *Proc. IEEE*, 110, 8–30, 2022.
11. Anwar, A., Khan, F., Hussain, M., Outgassing effect on spacecraft structure materials. *Int. J. Adv. Struct. Eng.*, 6, 145–154, 2019.
12. Power bus management techniques for space missions in low Earth orbit. *Energies*, 14, 7932, 2021.
13. Singal, J., Schindler, R., Chang, C., *et al.*, A multi-chamber system for analyzing the outgassing, deposition, and associated optical degradation properties of materials in a vacuum, arXiv, 2009.
14. Fedchak, J.A., *et al.*, Outgassing rate comparison of seven geometrically similar vacuum chambers of different materials and heat treatments, arXiv, 2020.
15. Mamun, M.A., *et al.*, The effect of heat treatments and coatings on the outgassing rate of stainless steel chambers, arXiv, 2013.
16. Study on serial MPPT electrical power system stability for space application. *Adv. Mater. Res.*, 1016, 441–445, 2014.
17. Anwar, A., *et al.*, Power bus management techniques for space missions in low Earth orbit. *Energies*, 14, 2021.
18. Vishnuprakash, B., *et al.*, Advanced power management in CubeSats: optimization of solar harvesting and energy storage technologies. *Acceleron Aerosp. J.*, 3, 738–757, 2023.
19. Pinkowski, M.A., *et al.*, MSL MMRTG Power Modeling for Lifetime Performance Predictions, NASA JPL Technical Report 20230005751, 2022.
20. Edpuganti, A., A comprehensive review on CubeSat electrical power system architectures, ResearchGate, 2025.
21. Steitz, D., NASA study: clean space-based solar power beaming possible, SpaceNews, 2024.
22. SpaceDream (JAXA), Microwave power transmission experiment on ground for SSPS demonstration, 2014, https://spacedream.sakura.ne.jp.
23. NewAtlas, DARPA's wireless laser power relay sets new distance record, 2025, https://newatlas.com.
24. Phys.org, NASA is building a mission that will refuel and repair satellites in orbit, 2022, https://phys.org.
25. NASA, OSAM-1, Small Steps, Giant Leaps Podcast, Episode 91, 2022, https://www.nasa.gov.
26. Jowett, P., European Space Agency mulling feasibility of space-based solar power, PV Magazine, 2024.
27. Soleimani, A., High-power radio frequency wireless energy transfer system: comprehensive survey on design challenges. *IET Wireless Sens. Syst.*, 2024.

28. Wireless power transfer: systems, circuits, standards, and use cases, PMC, 2022.
29. Amarel, T., Wireless power transmission: separating fact from fiction, 2025, https://tayloramarel.com.
30. Advancements and challenges in antenna design and rectifying circuits for radio frequency energy harvesting, PMC, 2023.

9

Advanced Imaging and Sensor Technologies for Deep Space Navigation

S.M. Mehzabeen[1]*, R. Kousalya[1], B. Sarala[1] and Mariya Ouaissa[2]

[1]*Department of Electronics and Communication Engineering, Sri Venkateswara College of Engineering, Sriperumbudur, Tamilnadu, India*
[2]*Network and cyber security, Cadi Ayyad University, Marrakech, Morocco*

Abstract
Advancements in the field of imaging and sensor technologies extend its applications in the field of space missions and deep space exploration. Since space mission are moving out of the orbit of the earth to planet, asteroids, and to the deep space, the need to have accurate, autonomous, and reliable systems of navigation is growing. Deep space navigation requires imaging and sensor systems. They help spacecraft to track and react to their surroundings, especially when there is a limited communication with Earth .This chapter suggests recent sensor and imaging technologies that support autonomous navigation across huge interplanetary distances. It covers optical sensors like star trackers and multispectral cameras, as well as inertial and proximity sensors, such as light detection and ranging (LiDar) and MEMS-based IMUs. It also highlights important features like radiation-hardened hardware, AI-powered image processing, and sensor fusion methods to improve accuracy and autonomy. Real-world examples from Mars rovers and asteroid missions show practical usage of these technologies. This chapter also looks at future trends, including quantum sensing, neuromorphic vision systems, and smart on-board navigation, offering a glimpse into the next generation of space navigation systems.

Keywords: Deep space navigation, imaging sensors, autonomous spacecraft systems, sensor fusion, LiDaR, MEMS-based IMUs, quantum sensing

**Corresponding author*: mehzabeen@svce.ac.in

N. Sathish, Prasanth Aruchamy, Rajesh Kumar Dhanaraj, Nithya Rekha and Karthik Palani (eds.) Space Electronics for Deep Space Exploration, (199–228) © 2026 Scrivener Publishing LLC

9.1 Introduction

Space missions usually that goes beyond Earth heading to the Moon, Mars, distant asteroids, or even farther face the toughest challenges in navigation, determining position and moving safely through the vast area of space. On Earth, guidance depends on GPS satellites and real-time communication. In deep space, however, there is no GPS network, and communication with Earth can involve long delays. A single message sent to Mars, for instance, may take up to 20 minutes to reach its destination. This delay imposes limits on how far engineers on Earth can control the process of a spacecraft in real time, particularly in such crucial times as landing or evading barriers. Because of this, spacecraft will need to work on their own and make decisions without always getting help from Earth.

Importance of imaging and sensor technologies plays a vital role in making the spacecraft to work autonomous. Several spacecraft navigation and control systems like LiDar systems, star trackers and cameras plays an essential role in control and navigation. These systems behave like an assistive eye in identifying stars, surface maps to prevent dangers and also to know its original position. Inertial sensors like accelerators and gyroscopes can also be used for these operations. The onboard sensors are very advanced mechanisms of spacecraft known as the inner ear that tracks the movement of the spacecraft even in the absence of an external reference. A combination of these sensors enables a spacecraft to sense and act without delay for Earth's instructions. These technologies have been proven to be very important during recent missions. The special cameras and navigation software were used on perseverance rover on Mars for safe landing on precise location and identifying hazardous crater and also finds asteroid even when they are in a distant location for example about 300 million kilometers. Due to the advancement in the field of imaging and sensor this can be easily accomplished. These systems are not merely useful, and also, they are needed to investigate the places unreachable by humans and for quick decision making. In the future, further sophisticated tools are being produced. Quantum sensors could be one among the pioneer technology that determines the spacecraft location with a very high accuracy without the use of any external signals. Neuromorphic cameras which are more like human brains and eyes can enable real-time vision processing on a very low power consumption. Together, these technologies form the backbone of autonomous deep space navigation, helping spacecraft travel farther, explore smarter, and operate more safely as humanity reaches out into the Solar System and beyond.

9.1.1 Motivation for Deep-Space Navigation - Challenges Beyond Earth Orbit

When spacecraft travel far from Earth, they face long delays in communication, weak and sometimes broken connections, and very limited support from ground stations. Even within the inner Solar System, one-way light time to Mars regularly reaches approximately takes twenty-one to twenty-three minutes, which makes any closed loop, ground in the loop control impossible during time critical operations, such as entry, descent, and landing (EDL). Scientists and architects conduct thorough reviews through analog studies, which simulate space conditions on Earth, and currently recognize communication delays as significant challenges for space exploration missions [1].

Bandwidth is another major limitation for space communication. NASA's deep space network (DSN) relies mainly on these three main facilities namely Goldstone, Madrid, and Canberra. Each equipped with large 70-m antennas capable of tracking spacecraft billions of miles away. However, DSN time is in high demand and tightly scheduled, forcing missions to carefully limit the amount of data they send, especially during critical events when communication windows are short. This scarcity of communication resources requires strict planning to ensure essential data can be transmitted without overloading the network.

In order to push the bandwidth limits well beyond, space agencies are now testing deep space exploration using optical communications. In the year 2023 and 2024, NASA tested a psyche mission to break records by using laser signals to send data across interplanetary distances, proving the data rates as much as ten to a hundred times higher than traditional radio transmissions. This technology signals a major shift and also gives a promising result that greater data return achieved using this technique for future autonomous spacecraft exploring deep space [3].

The extreme distance in deep space makes navigation even more difficult because of constant exposure to radiation, extreme temperatures, and weak or uneven gravity especially near small bodies like asteroids. These harsh conditions interfere with the accuracy of inertial sensors, making it harder to predict a spacecraft's exact path, carry out close up maneuvers, or achieve precise landings. Recent missions to small, rocky destinations such as Bennu have revealed that low gravity, cluttered environments require frequent, real time navigation updates are necessary to keep spacecraft safe during exploration [4].

9.1.2 Challenges of Space Navigation

Navigating in space poses a set of challenges even more complex than those encountered in air or maritime navigation. Unlike travel on Earth, where reference points are relatively fixed, space navigation operates in a constantly shifting environment. Every celestial body is in continuous motion planets revolve around the Sun, moons orbit planets, and the Earth itself rotates and revolves around the Sun. Spacecraft must travel at extremely high velocities, often thousands of kilometers per hour, while also aiming for targets that are moving along their own orbital paths. This dynamic setting demands highly precise calculations to ensure a spacecraft and its destination align at the exact point in space and time.

Figure 9.1 shows the bar graph showing the impact levels in motion, distance, communication, and gravity, these are the major deep space navigation challenges. Out of the four different factors, distance stood in the top with an impact level of 10, and gravity stood with the least impact level of 7. Navigating deep space is especially challenging because the destinations are both extremely far and small compared to Earth [5]. For example, if Earth was the size of a softball, the Moon would only be a few meters away, and Mars would be over a kilometer distant. Hitting such tiny, fast-moving targets demands precise calculations and careful long term planning. Even slight mistakes in the flight path can grow into major errors over time, leading a spacecraft to end up with some far locations.

Communication restrictions make the exploration of deep space even tougher. Spacecraft have only limited power and rely on solar panels that

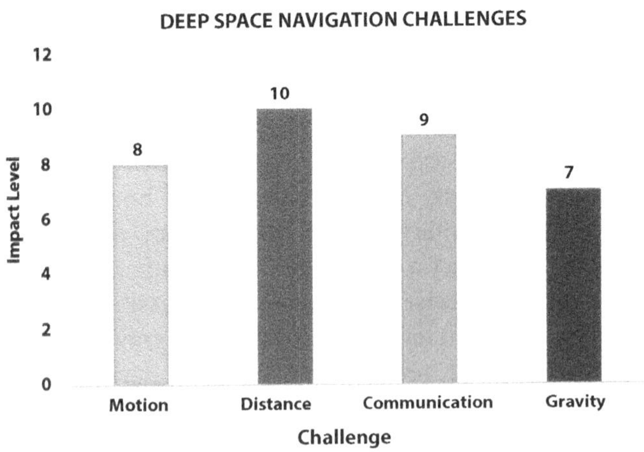

Figure 9.1 Bar graph depicts impact levels in motion.

become less effective the farther they travel from the Sun. Because of these constraints, the signals they send are extremely weak and must cross enormous distances often millions or billions of kilometers to reach Earth. This journey can take anywhere from minutes to hours, and the signals must be picked out from background cosmic noise. The long delays mean ground control cannot respond quickly, so spacecraft must make critical decisions and navigate largely on their own.

Navigating through deep space is not just about following a straight path. Every spacecraft feels the pull of gravity from the Sun, but as it travels, planets, moons, and even smaller bodies can approach it off course. Space mission teams build complex models to predict these gravitational impacts and also account for things like solar radiation. If any mishap happens the spacecraft can drift far from its target.

Space navigation stands out as especially demanding, since missions must track moving targets at immense distances, deal with lengthy communication delays, and factor in ever changing gravitational forces. There is a high demand for advanced prediction software, high-tech onboard systems, and careful timing to ensure a spacecraft safely reaches its goal, making navigation one of the hardest parts of deep space exploration.

9.1.3 Traditional versus Autonomous Navigation Technologies

Traditional space navigation has heavily relied on ground-based communication networks, such as the DSN, to track spacecraft positions, update trajectories, and issue maneuver commands. However, as the spacecrafts travel farther into the Solar System, the communication time lag may be up to several minutes to hours. From Earth station to process this time lag is really a challenging one and also it is impractical. To overcome these kinds of issues one of the key solutions that have been developed in order to minimize is autonomous navigation technologies [6]. This emerging autonomous navigation actually reduce the dependence of Earth communication delays to a very large extent.

The process of decision making is very slow due to the communication delays from the Earth station. The communicating commands need to cover great distances, and spacecrafts usually comply with a pre-loaded orders and is unable to change with the times. These systems are also intensive in resource usage, and having to be monitored by the DSN, which leads to increase in operational costs. The major challenges associated with deep space navigation, including motion dynamics, communication constraints, gravitational effects, and long-distance trajectory planning, are summarized in Table 9.1.

Table 9.1 Major challenges and their key aspects in deep space navigation.

Challenge	Description	Impact level (1–10)	Key considerations
Motion	Spacecraft, planets, and moons are all in constant motion, requiring precise trajectory calculations.	9	Requires continuous updates of position and velocity using onboard sensors.
Distances	Enormous distances make targets appear small and moving, requiring extreme accuracy.	10	Long-duration missions need highly accurate long-range planning and correction.
Communication	Weak radio signals, long transmission delays, and limited power far from Earth.	8	Needs high-gain antennas, data compression, and autonomous decision-making.
Gravity	Complex gravitational influences from the Sun, planets, and moons.	7	Must use gravity assists and account for perturbations in navigation models.

On the contrary, autonomous navigation systems makes use of on board sensors, such as cameras, gyroscopes, accelerators, and sophisticated AI tools to analyze information in real time and navigate the spacecraft without external instructions necessary to change its path and orientation. This onboard decision making can enable spacecraft to respond to hazardous conditions in real time. The unpredictable occurrences of it, aiding more accuracy in mapping and localization and reducing the level of error against the historical methods. It is necessary to have autonomous navigation. Earth engagement, which translates to a higher mission resiliency, decreased operation cost and the capacity to make contact with the ground that is either restricted or delayed.

9.1.4 Communication Limitations in Deep Space

Satellites visiting other planets or asteroids are faced with a number of communication delays. Signals at distances longer than Mars, two-way communication is constrained by signal latency delays, which experiences a delay of more than 30 minutes, and it is impossible to control in real time. Due to the cosmic noise and limited power constraints, the signal attenuation are more difficult to detect as the radio signals travel a long distance. With lower data rates, spacecraft cannot send high-resolution telemetry or imagery continuously to Earth for navigation purpose Ground-based tracking networks must support multiple missions simultaneously, leading to limited communication windows. These challenges underscore the need for onboard decision-making and real-time navigation capabilities that do not rely on immediate instructions from Earth.

9.1.5 Role of Autonomous Navigation

Autonomous navigation refers to the capability of a spacecraft to determine its own position, velocity, and attitude and to execute trajectory corrections without ground intervention. This greatly reduces the burden on Earth-based tracking systems and enhances mission flexibility. Spacecraft use onboard sensors like star trackers, sun sensors, and inertial measurement units to determine their position and orientation relative to celestial bodies. This eliminates the need to constantly transmit raw data to Earth for processing. Autonomous algorithms can compute necessary trajectory corrections and execute thruster burns locally. This ensures timely response to perturbations or hazards, which would be impossible if waiting for instructions from Earth. During time-critical operations, such as orbit insertion, landing, or flybys, autonomous systems can respond instantaneously based on sensor inputs something Earth-based operators cannot do due to signal delay. Spacecraft passing behind planets or moving out of line-of-sight can continue navigating accurately even without radio contact, resuming communication once the link is restored.

9.1.6 Enabling Technologies

Several advanced sensor and processing technologies support autonomous navigation and reduce communication dependence includes Instruments such as star trackers, horizon sensors, and LiDar units help determine location using images of celestial bodies or nearby landmarks. MEMS-based IMUs (Micro-Electro-Mechanical Systems Inertial Measurement Units) provide continuous motion data for dead-reckoning when external references are unavailable. Artificial intelligence and computer vision algorithms enable the spacecraft to recognize terrain features, match them with onboard maps, and autonomously localize [7]. Combining data from multiple sensors (inertial, optical, and radiometric) improves accuracy and reliability without depending on Earth-based updates.

9.1.7 Autonomous Navigation Workflow

Autonomous navigation plays a pivotal role in reducing reliance on Earth-based communication, especially for deep space missions where communication delays can extend to several minutes or even hours. The autonomous navigation workflow adopted in deep space missions, which minimizes reliance on continuous Earth-based communication, is illustrated in Figure 9.2. On board high-technology imaging and sensing equipment such as Star trackers, LiDar, Multi spectral imagers, and MEMS-based IMUs constantly collecting the real-time data. The data includes information such as concerning the spacecraft orientation, speed as well as the surrounding environment, which allows it to have accurate

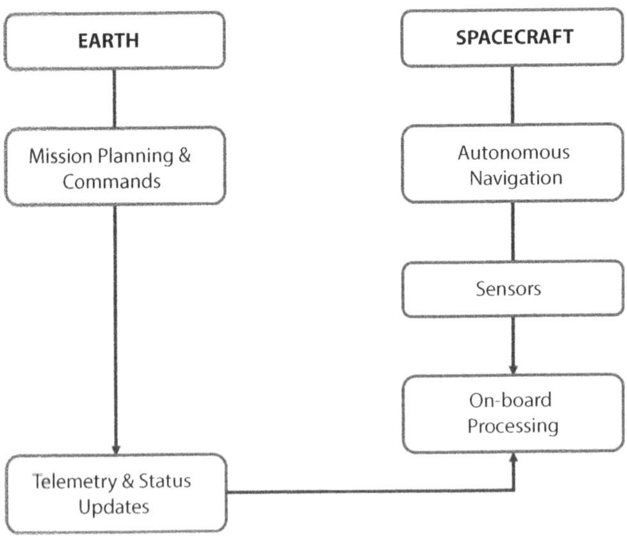

Figure 9.2 Autonomous navigation workflow reducing dependence on Earth-based communication in deep space missions.

decision making capability without the intervention of continuous information from ground stations [8]. The advanced artificial intelligence executed on board to process the data, that is being received by these sensors and image processing algorithms involve in activity such as object detection in the sky, object recognition in the sky, landmark tracking, and predictive trajectory modeling, which enables the spacecraft to observe the space and predict independently in the future or possible risks.

The processed data is subsequently inputted into the onboard decision engine of navigation which combines the data of the different optical, inertial and proximity sensors with sensor fusion techniques. This engine works out the present parameters of the spacecraft such as attitude, position, and velocity and decides the best path of action and course corrections [9].

This engine has autonomous guidance, navigation and control (GNC) logic incorporated that enables the spacecraft to make informed navigation decisions without the need to have real-timeground support [12]. After decisions are taken, the system directly starts control activities by directly triggering them controlling the thrusters of the spacecraft in changing trajectory and the reaction wheels or control moment gyros to fine control the attitude. These movements are performed in seconds or minutes, and by passed the long line of waiting time per two-way communications with Earth. The spacecraft rather than relaying large amounts of raw sensor data back to the earth only provides summarized status updates and health reports. This severely lowers the demand on intense radio systems and band bandwidth, which is frequently constrained during deep space missions. What it did was to transfer the majority of the navigation jobs to the space vehicle instead of the Earth autonomous systems save a lot of communication cost to Earth-based systems. Due to this fact, ground teams are not direct as much as they are monitoring and supervisory control, and this has significant benefits on the mission safety and the efficiency of operation. This autonomy becomes particularly important in the event of communication breakdown, e.g., when the space craft is goes behind a planet or acts in deep space areas where there are weak signals. Finally, this workflow allows robust, accurate, and ongoing navigation, which guarantees mission success even in the farthest and most challenging space conditions [11].

9.1.8 Why Imaging and Sensors are Central to Navigation?

Deep space navigation relies so much on the capability of a spacecraft to navigate identify accurately its location, direction, and speed in relation to heavenly bodies and its intended path. Cosmogonic navigation unlike on Earth, where satellites, GPS networks, and ground-based referencing is easily accessible, deep space does not have any established external infrastructure. In this context, onboard sensors, imaging systems are the major tool of spatial awareness and direction, and guidance, and they are mission success factors [10].

A spacecraft is able to determine its orientation (attitude) and can also estimate its position in space without depending on Earth-based tracking. Sensor systems can present supplementary information to precise positioning. MEMS technology are used to measure acceleration and angular velocity in order to do dead reckoning. When fused together by sensor fusion algorithms, these systems are able to provide accurate and real time even when there is a communication blackout or the device is not close to some external, navigation solutions can be provided.

There are also imaging and sensors which enable spacecraft to identify hazards, including dust rocks, or irregularities of the road, and automatically anticipate diversion of their courses. This capability is particularly crucial when the real time advice of operators of DSN cannot be obtained because of the lengthy delays of communication. Simply, the imaging and sensor technologies provide spacecraft with the capacity to see and know their surroundings, as their eyes and their inner ears in empty space. Without them, navigation would be dark, slow, and unnecessarily dependent upon tardy Earth-based instructions.

9.2 Traditional Navigation Methods

Space missions have relied heavily on Earth-based tracking networks, especially the DSN, to provide precise navigation data. Various sensors and imaging systems used for spacecraft navigation and environmental perception are presented in Figure 9.3. The DSN consists of large ground-based antennas located at three sites around the world, strategically placed to maintain continuous contact with spacecraft as the Earth rotates Traditional Earth-based tracking techniques employed for monitoring and guiding spacecraft are shown in Figure 9.4. Using radio communication, the DSN sends signals to the spacecraft and measures the returned signals to compute its distance (ranging) and velocity (*via* the Doppler effect).

Two-way ranging involves sending a radio signal from the ground, which is reflected back by the spacecraft. The round-trip time of the signal is measured to determine the spacecraft's distance with high accuracy. Similarly, Doppler tracking measures the frequency shift in the returned radio signal caused by the spacecraft's motion relative to Earth, which reveals its

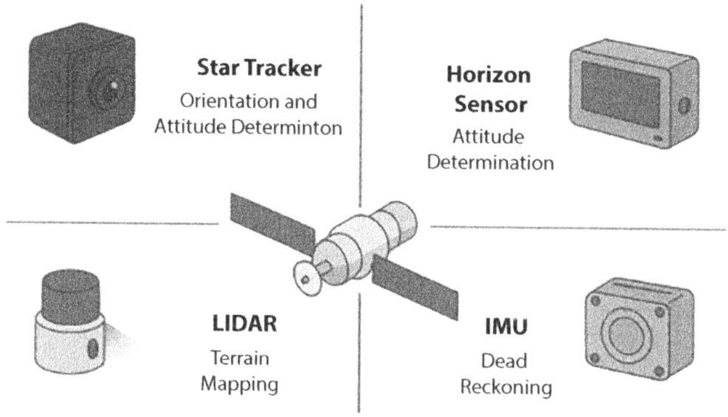

Figure 9.3 Sensors and imaging systems for navigation.

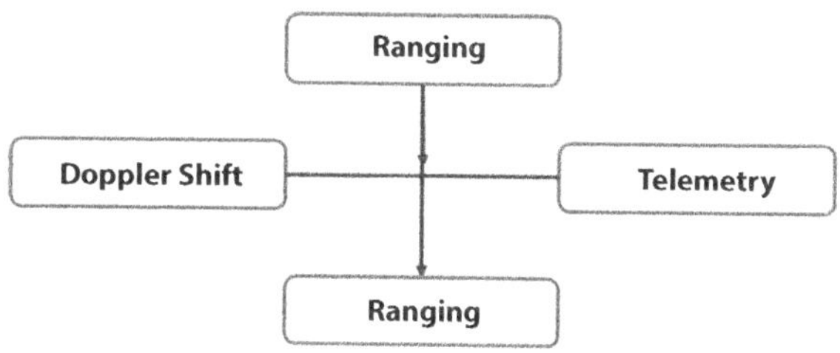

Figure 9.4 Earth-based tracking techniques.

velocity along the line of sight. These methods, when combined with Delta-Differential One-Way Ranging (Δ-DOR) techniques, allow ground controllers to triangulate the spacecraft's location in three-dimensional space.

9.2.1 Limitations in Deep Space

Although the old, radio-based methods of navigation have been used to achieve decades of success planetary missions they face major constraints since spacecraft travel deeper away Earth. Communication delay is one of the challenges. In the case of radio signals, it has a value of approximately 8 minutes to travel between the Earth and Mars at the nearest point and more than 40 minutes at their farthest. Deep-sea navigation cannot be supported in real time by the Earth because this causes a time lag space operation, where real-time decision making is more often important [20]. The other limitation is the attenuation or weakening of the signal strength. As spacecraft travel even farther, radio signals sent by them are incredibly weak when they reach the earth. This requires huge antennas, high power transmission, and extensive integration periods to extract useful

Table 9.2 Deep space navigation limitations.

Near-Earth missions	**Deep space missions**
Communication delay Seconds	**Minutes to hours**
Signal strength Strong	**Very weak** (due to vast distance)
Bandwidth availability High	**Limited**
Real-time Control Possible	**Impossible** (must be autonomous)

information on the noise of background [23]. In addition, missions to deep space tend to be bandwidth limited the data rate available is smaller the further the distance, which this implies less navigation data can be sent back to Earth. The major limitations associated with deep space navigation systems, including communication delays, positioning challenges, and environmental uncertainties, are summarized in Table 9.2.

9.2.2 Key Requirements for Deep Space Sensors

In order to counter the difficulty of navigating space, spacecrafts need to be highly loaded special sensors and imaging devices, which can identify their state without the presence of the Earth-based input. The deep space navigation sensors should address some important requirements:

I. Precision

Deep space navigation is very precise as even the most basic types of precision are very high even small malfunctions will multiply millions of kilometers and lead to significant changes in the trajectories measures of orientation and LIDAR systems and Star trackers are optical instruments precision of less than a sub-arcsecond and centimeter respectively.

II. Radiation resistance

The deep space environment exposes spacecraft to intense radiation from solar flares, cosmic rays, and charged particles trapped in planetary magnetic fields. Navigation sensors must be built using radiation-hardened electronics to ensure long-term functionality. Even temporary sensor malfunctions caused by radiation-induced bit flips can corrupt navigation computations, so robust shielding and error-correction designs are critical.

III. Autonomy

Due to long communication delays, deep space sensors must be capable of operating autonomously. This means they should not only collect raw data but also process it onboard to generate navigation solutions in real time. Incorporating AI-based image recognition, sensor fusion algorithms, and fault detection systems is essential to enable autonomous decision making when ground contact is unavailable [21].

IV. Low power consumption

Power is one of the most scarce resources in deep space. Solar power availability diminishes with the square of distance from the Sun, forcing spacecraft to operate on limited energy budgets. Navigation sensors must therefore be designed for ultra-low power operation, often using duty cycling, energy-efficient processing chips, and lightweight low-voltage components without compromising accuracy. Meeting these requirements allows deep space navigation systems to function reliably and independently over years-long missions, even when ground support is intermittent or unavailable [17].

9.3 Imaging Technologies for Navigation

9.3.1 Star Trackers

The operational principle of star trackers relies on utilizing stars as stable reference points to ascertain the exact orientation of a spacecraft. A star tracker comprises an optical device featuring a lens and a sensitive imaging detector, such as a CCD or CMOS sensor, which acquires digital images of the surrounding stellar field. The acquired image is further analyzed to identify stars, and their precise positions are determined by centroiding methods with sub-pixel precision. The observed star patterns are compared with an onboard star catalog utilizing pattern recognition techniques, including geometric voting or triangle matching, to identify the stars. Upon matching the seen stars with cataloged stars, the system calculates the spacecraft's attitude by juxtaposing the measured star vectors in the spacecraft's body frame with their established inertial frame coordinates. Mathematical estimating techniques, such as the QUaternion ESTimator (QUEST) algorithm, are employed to ascertain the spacecraft's orientation represented as quaternions or rotation matrices. This attitude data is always refreshed and provided to the spacecraft's navigation and control system, facilitating precise pointing, stabilization, and trajectory correction [22].

Star trackers use advanced algorithms to look at pictures of star fields to figure out which way the spacecraft is facing. These algorithms generally fall into two categories: star identification algorithms and attitude estimation algorithms.

Star identification algorithms look at the stars in an image and compare them to records in a catalog on board. Techniques based on angles, such as the pyramid algorithm and geometric voting, rely on the angular distances between stars. In contrast, vector-matching approaches, including the k-vector search, work by aligning the observed star vectors with pre-stored inertial vectors. Pattern recognition techniques, consisting of matching triangles or polygons, attempt to find various geometric arrangements of stars. On the contrary, magnitude-based filtering improves the brightness of stars.

Following star recognition, the orientation of the spacecrafts is estimated through attitude estimation algorithms which solve the Wahbas problem (minimize the difference between observed star vectors and known inertial vectors). Both the q method of Davenport and products eigenvalue analysis are known methodologies, and the very popular QUEST (QUaternion ESTimator) algorithm provides effective quaternion-based orientation estimations [13]. The singular value decomposition (SVD) technique offers higher accuracy through matrix decomposition, but at a higher computation cost. In the case of dynamical systems, filters such as the Extended Kalman Filter (EKF) and the Unscented Kalman Filter (UKF) are used to combine successive observations with a noise reduction. Taken together, these algorithms ensure that star trackers will deliver precise and reliable real-time attitude information for spacecraft navigation and attitude control.

9.3.2 Optical and Multispectral Cameras

I. Visible light cameras for terrain mapping

Visible light cameras are commonly used for navigation and terrain mapping as they operate in the visible spectrum (400–700 nm) and therefore provide clear high resolution images of the planetary surface. The process of daylighting using a CCD/CMOS camera includes an optical lens collecting sunlight that is reflected off an object. The sensor converts the light into digital images, which are used to analyze the various features like edges, landmarks, and surface textures [23]. Digital elevation models (DEMs) of changes in surface elevation can be obtained using visible light cameras using stereoscopic photography or multiple perspectives. DEM plays a vital role in navigation tasks that includes visual odometry, obstacle avoidance, path planning, georeferencing, and more, connecting images to maps or GPS data. Its advantages include high spatial resolution, cheap, and produce images that people can easily interpret [27]. However, it has its drawback that includes limited reliance on light conditions and effectiveness in low light or shaded areas. Despite these shortcomings, visible light cameras are still the most important in practical imaging technology for terrain mapping and navigation onboard spacecraft, flying drones, and planetary rovers.

II. Multispectral/hyperspectral imaging for planetary surface analysis
Multispectral and hyperspectral imagings are advanced techniques used in planetary surface investigations in which reflected radiation is measured across a number of wavelength bands, extending well beyond the visible spectrum into the near infrared and shortwave infrared [25]. While multispectral imaging extracts data of a few discrete spectral bands, hyperspectral imaging captures hundreds of narrow and contiguous spectral bands, making a spectral data cube for the spatial and spectral domains. This helps scientists and navigation systems to identify surface materials by their unique spectral signatures. This imaging is also required to mineral mapping, identifying water or ice due to their different absorption signatures and investigating geological processes, such as weathering, volcanism, and erosion [24]. In navigation, these sensors help rovers and landers to sense surface materials, recognizing landmarks, and to plan safe trajectories. Their main benefits include correct identification of materials, the detection of trace minerals and organic chemicals, and are used in scientific investigation and engineering projects. Challenges include large data volume, high processing and storage requirements, greater complexity than for visible cameras and sensitivity to lighting/atmospheric problems. Despite these limitations, multispectral and hyperspectral imaging tools are considered irreplaceable instruments for planetary exploration, enabling extensive compositional science and providing improved reliable navigation in extra-terrestrial environments.

III. Role in hazard detection and safe landing
High-resolution, real-time images of the landing site are crucial for sensing hazards and guaranteeing a safe landing during planetary exploration, and these can be captured using optical and multispectral cameras. Optical cameras records detailed surface features such as rocks, craters, slopes, and irregular terrain, whereas multispectral cameras improves hazard detection by differentiating materials like loose soil and solid bedrock. These cameras use stereoscopic photography to create digital elevation models (DEMs) that capture height changes on the surface and identify flat safe

landing areas [18]. On descent, on-board image processing systems scan the acquired terrain to automatically detect hazard zones and command the lander to make it to safe landing sites, reducing the need for Earth-based commands [27]. The use of visual information coupled with the inertial sensors improves the accuracy of navigation, and the hazard avoidance system adjusts the trajectory on the fly in response to unexpected obstacles. This ability is especially important in communication delay applications, such as Mars. As a result, optical and multispectral imaging substantially enhances mission safety and reliability through autonomous hazard detection, secure site selection and precise landing on uneven or uncharted planetary surfaces.

9.3.3 Stereo Vision and Depth Cameras

Structure from motion (SfM) and visual odometry (VO) are computer vision algorithms that use images of objects photographed with the camera to approximate three-dimensional structures and track motion, and are an essential component of navigation in the case of GPS or external reference unavailability.

I. Structure from Motion

Structure from motion (SfM) is a technique that can be used to obtain the 3-D shape of a scene through the analysis of a series of two-dimensional images taken of the same scene in different viewpoints. The relative camera position and orientation are calculated by identifying and matching the salient feature points among multiple photographs. When these camera postures are determined, the 3D coordinates of the feature points are determined using triangulation techniques and results in a sparse or rich 3D representation of the environment [26]. SfM is used in planetary exploration, remote mapping with airplanes, autonomous navigation, where the generation of accurate terrain maps is essential.

II. Visual Odometry

Visual odometry (VO) focuses on the movement of a camera (or vehicle) by focusing on the sequence of image frames. It works through the detection and tracking of features in successive photos, the measurement of their movement and the use of the information to estimate the camera motion in terms of translation and rotation as time progresses.

VO is a real-time posture estimator that is commonly found in rovers, drones, and autonomous vehicles to determine position during GPS unavailability or unreliability. SfM is used to construct global three-dimensional maps, and VO is used to conduct local motion tracking; together, they assist in reliable navigation *via* terrain mapping, localization, and obstacle avoidance. These techniques are advantageous because they are lightweight and passive and can offer detailed information on spatial measures, but face problems such as light sensitivity, featureless surfaces, and accruing drift error in visual odometry that are often reduced by combination with other sensors, such as inertial measurement units or LiDAR. As a result, SfM and VO represent a complementary pair to enable drone, rover, and autonomous systems navigation using vision.

9.3.4 Event-Based Cameras for Real-Time Perception

Neuromorphic vision sensors (or event-based cameras) represent a major breakthrough in the imaging technology designed to operate in real-time perception and navigation. Unlike the conventional frame-based cameras, which capture the entire images at pre-determined intervals, event-based cameras operate asynchronously with each pixel independently indicating change of brightness as individual events. This provides superb temporal accuracy and low latency as well as good data control since only changes in the scene are recorded [14].

I. Working principle

Every pixel of an event-based sensor is an independent detector which emits an event when it notices a change in the intensity of the light it receives. The output is a sequence of events which contain pixel position, time, and the polarity of brightness change, not continuous frames. This enables microsecond level awareness with much less redundancy of the data compared with the traditional cameras. The event-based cameras are very helpful to the navigation. Their high temporal resolution enables them to respond to changes in the scene in micro seconds making them fast to use in fast systems. Their efficiency in power consumption and bandwidth since they only record differences between brightness rather than full image frame makes them efficient in long-lasting missions. They also support a high dynamic range (HDR), which means that they can perform well under conditions with large differences in brightness or under low light, and that there is inherent resistance to motion blur, which is a common issue with the conventional cameras when one moves rapidly.

Event-based cameras can be well adapted to real-time obstacle avoidance due to its asynchronous data stream that enables an instant recognition of moving and immobile threats. They are applied in visual odometry and simultaneous localization and mapping (SLAM), which enables successful motion estimation and mapping and reduces computation requirements. This makes them essential to high-velocity robotics and drones, where agile navigation in variable environments is essential, and autonomous vehicles and space missions, where reliable perception in unfavorable lighting conditions, such as low light or glare, is essential.

Despite such advantages, event-based cameras have certain issues. It requires specialized event-driven algorithms since the result of their data output cannot be immediately applied using standard image processing methods [15]. The technology is also in the early stages of commercial viability compared to the traditional cameras, which prevents its further usage. Moreover, to be effective and reliable in their navigation, they often need to work together with other sensors, e.g., inertial measurement units (IMUs) or depth cameras to overcome their very specific weaknesses when isolated.

9.4 Sensor Technologies for Navigation

9.4.1 Inertial Measurement Units

MEMS-based accelerometers and gyroscopes are essential parts of modern IMUs. They are smaller in size and consumes less power, much useful for navigation in aerospace, automotive, and space applications. MEMS accelerometers measure linear acceleration along different axes by detecting movement of small mechanical structure when they are pushed by inertial forces. MEMS gyroscopes, on the other hand, measure angular velocity by using vibrating structures that respond to the Coriolis effect. These sensors work together to keep track of changes in speed and direction, which makes it possible to accurately estimate motion even when GPS and other outside signals are not available.

MEMS sensors with less cost and power are suitable in spacecraft, rovers, drones, and satellites where size and power efficiency are sensitive. In navigation, IMUs based on MEMS can be used to provide dead reckoning, control the attitude, and offer short-term tracking of motion, which can be combined with other sensor data, e.g., cameras or LiDar, to attain a higher precision. Nevertheless, MEMS devices drift and become noisy with time and these bias can build up to considerable errors when operating in long missions. To overcome this, they are usually used together with the external references or sophisticated filters like Kalman filters. Nonetheless, MEMS-based accelerators and gyroscopes continue to be a pillar of the modern navigation systems because of their versatility, durability, and capability to work effectively in extreme conditions.

I. Drift issues and correction methods

IMUs is a valuable control instrument that provides information about position, velocity and orientation using accelerators and gyroscopes, but with a major drawback referred to as drift. The error due to drift is that IMUs estimate motion in a constant combination of the acceleration and angular velocity measurements; a little error, noise, or bias in the sensor results, over time, causes the IMU to estimate the position and orientation substantially in error. This issue is of particular concern in missions which take long periods of time, or ones in which external reference signals (such as GPS) are inaccessible, because uncorrected drift may result in significant errors in navigation. Several correction methods are used in order to combat drift. Calibration of the sensor is used to minimize systematic errors like bias and scale factor errors. IMU data are combined with other sensors like GPS, cameras or LiDar to filter and estimate errors and correct them in real time using Kalman filters. Pedestrian or ground-based navigation Sometimes zero-velocity updates (ZUPTs) are utilized in pedestrian or ground-based navigation, the system sets the velocity to zero at times of rest to oppose the drift buildup. Furthermore, sensor fusion with vision or satellite data offers external references which anchor IMU measurements to stop the increase of errors. There is also exploration of advances in AI-driven error modeling to be used to predict and compensate dynamically on the drift patterns. All in all, although drift is an intrinsic problem in IMUs, through the combination of hardware calibration, algorithmic filtering and multi-sensor fusion, reliable and accurate navigation even in GPS denied or challenging conditions can be achieved.

II. Space-qualified IMUs

Special purpose IMUs are gadgets used in navigation that can be depended upon to perform well in a harsh space environment. An IMU is normally a combination of accelerators and gyroscopes, which detect the linear acceleration and angular velocity and continuously gives approximate values on the orientation, velocity, and position of a spacecraft. Space-qualified models are constructed to endure mechanical stresses, radiation exposure, vacuum, and large temperature changes both during launch as well as in orbit. In general, they are built to support radiation-hardened electronics, redundancy, and calibration methods to provide stability and uniformity. Such IMUs are important in autonomous navigation, especially in the event of inaccessibility of external references such as GPS, as in deep space missions or planetary exploration. Space-qualified IMUs with high performance allow the fine attitude control of spacecrafts during maneuvers, docking, and entry-descent-landing. Other sensors, such as star trackers and altimeters, are also sensor fused with them to reduce drift errors of inertial navigation. Space-qualified IMUs provide mission-critical navigation by providing a combination of robust design and resistance to space hazards and invaluable to the current spacecraft and planetary rovers.

9.4.2 Light Detection and Ranging

I. Principles of LiDar in planetary navigation

LiDar is another sensing technology that is very vital in planetary navigation as it gives a good three dimensional map of the immediate environment. The basic principle of LiDar is that short bursts of laser light are sent to a target and the time during which the reflected bursts reach the sensor is measured. The speed of light is always constant, so this time of flight measurement can accurately calculate distances to surface features using a high level of accuracy. LiDar generates 3D point cloud by scanning many points in various directions as fast as possible and capable of depicting the terrain, obstacles, and landmarks around a spacecraft. Applications in planetary navigation LiDar can be used to create real-time topographic maps, which are helpful in supporting localization, detecting hazards and safe landing missions. LiDar performs better than passive imaging sensors in low-light or visually impaired conditions, which are usually present on the surfaces of planets. It is also highly localized and accurate at different ranges, which, as a result, is useful in identifying rocks, slopes, craters, and other dangers that may pose a risk to the ability of the rover to move or to land the lander. LiDar can also be used in conjunction with other sensors like cameras and IMUs, when combined they will benefit the sensor fusion frameworks, gives both geometric accuracy and contextual information in autonomous navigation [19]. LiDar facilitated by high-resolution mapping, resistance to lighting conditions, and precise distance measurement has emerged as a fundamental technology in planetary missions, which is making exploration of complex extra-terrestrial landscapes safe and efficient.

II. Role in terrain mapping and hazard avoidance

LiDar is an important device in terrain mapping and hazard avoidance as it offers very precise and three-dimensional images of the surrounding environment. LiDar systems operate with laser pulses, which are used to determine the time of flight of light reflected by a given surface, which in turn allows the calculation of the distance to the object and ground features and features with a high level of accuracy. The possibility enables spacecrafts, rovers, and autonomous vehicles to build detailed DEMs of the landscape. In exploring the planets, this mapping is vital in the determination of slopes, craters, rocks and other irregularities that may cause danger on safe navigation. LiDar is used to avoid hazards by onboard systems, which can identify obstacles at different distances. LiDar is also mostly independent of the ambient light thus it can be applicable in difficult environments like darkness, shadows, or one with high glare like the space environment unlike cameras. With LiDar data combined with other navigation systems, such as IMU and cameras, unsafe landings and obstacle-clear mobility, and success in mission, are guaranteed with precision in spatial resolution and contexts of the mission. Examples Mars rovers, OSIRIS REx asteroid mapping. The LiDar technology has been essential to various historic space missions and has proved useful in navigation and mapping of harsh extraterrestrials. On Mars rovers, such as curiosity and perseverance by NASA, LiDar systems are employed to produce 3D high-resolution maps of the Martian surface, which allows to detect obstacles accurately and avoid hazardous paths and plan routes. With laser pulses, these rovers should be able to independently detect traversible routes, characterize geological features, and choose safe landing or exploration points through the scanning of the surface using laser pulses. Imaging and inertial sensors are combined with LiDar to provide a better localization and digital elevation of the Martian terrain.

The other prominent example is the OSIRIS Rex mission by NASA where the LiDar was extensively applied in mapping the asteroids. Examples of Mars rovers designed for planetary exploration and autonomous surface operations are depicted in Figure 9.5. A scanning LiDar system was used in the spacecraft to generate detailed 3D maps of the surface on asteroid. These maps were essential as they gave very important details on navigation within the microgravity space, safe maneuvering of the asteroid, and also in selection of a suitable sampling location. High accuracy and dependability of LIDAR allowed the mission group to counter the factors like rough topography of Bennu and the unstable surface characteristics. They demonstrate the importance of LiDar to precision mapping, safe navigation and scientific discovery in the exploration of planets and to discover asteroids.

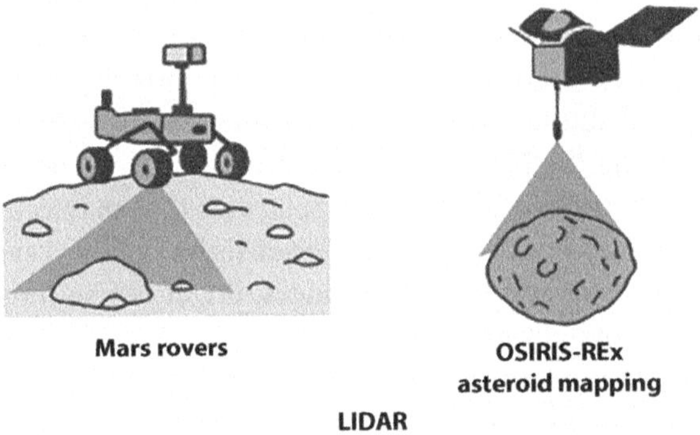

Figure 9.5 Mars rovers.

9.4.3 Short-Range Obstacle Detection

The proximity sensors, including radar, ultrasonic sensors, infrared detectors, etc., are essential due to the short range obstacle detection, which enables safe navigation in busy or moving environments. The method of operation of these sensors is the release of such signals sound waves in the case of ultrasonic, radio waves in radar, or pulses of infrared light and the time required by the reflected signal to come back to the sensor. This enables a clear way of determining the distance to the nearby obstacles, usually within a few centimeters to several meters, depending on the technology. Short-range detection is an important concept in navigation to avoid collisions, aid in fine maneuvering, and autonomous maneuvers such as docking a spacecraft, path correction of a rover or drone flying around an obstacle. Ultrasonic sensors have been shown to perform better at low speed and short distances whereas radar has more reliability in bad weather like dust, fog or dark nights. These sensors provide response to hazards by continuous checking the environment, which ensures safety and efficiency for mission critical operations. Proximity sensors, such as radar, ultrasonic, and other similar sensors, are very important in precision and safety in both space and planetary missions during docking and landing operations. Spacecraft docking in spacecraft docking, radar and ultrasonic detectors will give continuous measurements of the relative distance, speed, and orientation between two moving vehicles, and can fully align and approach them in conditions of low visibility, or where there is a high communication lag. These are critical to the autonomous docking systems where real-time feedback is important to achieve the smooth coupling without the threats of collision. In a similar vein, with planetary landing missions, the remotely operating proximity sensor help in the final descent, the altitude above the surface, and hazards including rocks or rough surfaces, and control the landing craft to a safe landing point. An example of these is radar altimeters, which provide more accurate altitude information, and ultrasonic sensors, which can give more fine-grained information about closeness to the ground. With such information incorporated into onboard navigation systems, spacecraft can autonomously change their descent path, throttle engines, and make safe touchdown landings. All in all, proximity sensors are important devices that would help to increase reliability and risk reduction and autonomous control during the most problematic parts of the docking and landing processes.

9.4.4 Radiation-Hardened Sensors

I. Radiation tolerance requirement

High-energy particles, cosmic rays, and solar radiation are highly harmful radiations that sensors in space environments are in constant exposure to. Such radiation may lead to severe problems like single-event upsets, corruption of data, the production of noise, and even the permanent damage of the hardware of the standard sensors. Given that the navigation systems require precise and consistent sensor data when performing functions, such as localization, obstacle detection, and path planning, the smallest errors brought about by the radiation can result in mission crucial failures. In the case of spacecraft, satellites, and rovers, which cannot be repaired or replaced, the ability to guarantee the reliability and longevity of sensors is of paramount importance. Radiation-hardened sensors are thus necessary to allow operation in sustained exposure ensuring performance integrity and eliminating malfunctions which may affect the navigation precision and the general mission safety. In addition, with missions that are going deeper into space beyond the protective magnetic field of the earth, the levels of the radiation become considerably larger and thus tolerance is even more crucial. So, this architecture is not only an option but also importance in navigation sensors in space exploration.

II. Methods: shielding, error correcting codes, redundant designs

Space missions require radiation-hardened sensors that are needed in navigation due to the presence of high-energy particles and cosmic radiation that can interfere or destroy electronic components. In order to be reliable, a number of protective measures are used. Shielding is the physical protection of radiation through the use of physical barriers, e.g., aluminum or hi-tech composite material to block or limit access of radiation to sensitive sensor electronics. Although shielding is efficient, it should be well tuned to give the optimal protection considering weight limitations in spacecraft designs. The application of ECCs at the data processing level enables the systems to detect and correct bit flips or corrupted information due to single-event upsets caused by radiation. This provides integrity of data and also stability in value even when operating in hostile environments. Also, redundant designs are also common in which several sensor units or parallel circuits are run at the same time such that in case one component fails, because of radiation, the other can instantly replace it. This redundancy increases fault tolerance and also greatly improves the mission reliability. Combining these techniques of shielding, ECCs and redundancy will constitute a complete plan to secure the navigation sensors so that their performance is accurate and has the full life span in the space environment which is full of radiations. Quantum sensing is a relatively new discipline which is taking advantage of quantum mechanical phenomena, including superposition, entanglement and tunneling, to sense physical quantities with unprecedented precision. The quantum sensors are also expected to be more accurate and reliable in navigation than it has ever been before, especially where other conventional methods of navigation, such as GPS is not available, jammy, or spoilt. It is the quantum sensing that is changing the way we undertake the process of inertial navigation, gravitational mapping, and the process of precision positioning by surpassing the constraints of the classical sensors.

Critical quantum concepts ensure superposition of atoms or particles being simultaneously in multiple states, making possible interference interactions which helps to detect small accelerations. Measurement sensitivity is enhanced in the entangled quantum particle beyond the classical threshold which reduces noise and increases precision. Atom interferometry matter waves of atoms are separated and brought together to form interference patterns. The phase change in these patterns is very much sensitive to acceleration, rotation, and change of gravity and is used in some high-precision magnetometers and field sensors in which very weak forces can be detected.

9.4.5 AI and Sensor Fusion in Navigation

Sensor fusion is an influential method in current navigation systems whereby data of various complementary sensors are fused together in order to initiate a more accurate, strong, and dependable perception of the surrounding. The imaging sensors, including cameras, can give detailed visual and informational information regarding the terrain, landmarks, and obstacles. IMUs provide accurate short-term motion information by tracking acceleration and angular velocity. Instead, LIDAR creates an in-depth

3D map of the environment with laser pulses at a high resolution, allowing detection of depth and distance. The shortcomings of each of these three sensing modalities are buffered when they are combined together, with cameras able to give semantic information that LIDAR is incapable of, and LIDAR able to give the correct depth where vision can wobble due to poor light or a featureless surface. This method allows strong localization, mapping, obstacle recognition, and trajectory planning. Conclusively, sensor fusion suggests a higher resilience of the system, uncertainty mitigation, and safer and more efficient navigation through the strengths of imaging, IMU, and LIDAR data in conjunction. The Kalman filtering, particle filters, and new fusion techniques, the sensor fusion process in navigation also heavily depends on the sophisticated filtering algorithms to integrate noisy and inaccurate data of several sensors to a coherent and robust approximation of position, orientation, and velocity [16]. One of the most popular approaches is known as Kalman filtering which is most effective with linear dynamics system and Gaussian noise. It offers the best recursive estimation algorithm, forecasting the state of the present, comparing it to sensor measurements and updating the estimate with the least error. The conventional Kalman filter is however not effective in complex and nonlinear environments and has given rise to Extended Kalman Filters (EKF) and Unscented Kalman Filters (UKF), which consider non-linearities in the motion and measurement models. Particle filters are used even to more complex and highly uncertain systems. They represent the probability distribution of state space by using a set of weighted samples (particles), and are therefore appropriate in nonlinear and non-Gaussian cases like visual inertial navigation or simultaneous localization and mapping (SLAM). Context-based weighting of sensors can be estimated using deep learning models, and their hybrids between neural networks and filtering approaches can be used to offer accessibility and flexibility. Advanced fusion methods use AI techniques to learn sensor correlations, adapt to changing conditions, and improve robustness. A combination of these tools of filtering and fusion guarantees the trustful navigation in complicated fields like autonomous car, drone and planetary rover where unpredictability, noise, and environmental fluctuation are critical issues. The use of artificial intelligence and especially deep learning image recognition has emerged as a major facilitator to step up navigation by classifying and detecting landmarks on the terrain. Through the analysis of high-resolution images of onboard cameras, AI models are capable of differentiating various types of terrain including rocky surfaces, sandy plains, slopes, or craters, which is vital in the safe planning of paths during planetary exploration missions or autonomous rover missions [28]. The use of convolutional neural networks (CNNs) and other sophisticated models is trained using massive data and is able to identify visual patterns to classify objects in real time even in high-visual or under-light conditions. In addition to the terrain, natural features like mountains, ridges or crater edges can be translated and tracked by AI systems, whether in the structured environment or not, and be used to position the object accurately when GPS is not available. By incorporating this image recognition feature into a navigation system, it becomes possible to make autonomous decisions, i.e., the rovers or drones will be able to choose safe paths, bypass dangers, and react to unfavorable environmental changes without the constant supervision of humans [29]. In addition, combined with the output of other sensors such as IMUs and LIDAR, AI-driven recognition can greatly enhance situational awareness and positional accuracy, and thus is a keystone technology in next generation autonomous navigation of space and on land.

9.5 Conclusion

With humanity exploring deeper into the solar system and plans of launching missions that could last a long time to planets, asteroids and deep space areas, the importance of having precise, autonomous and robust navigation systems cannot be underestimated. The main idea in such challenges is the focus on imaging and sensor technologies described in this chapter, which illustrates how star trackers, multi-spectral imagers, LIDAR, and MEMS-based IMUs are the key building blocks in the current deep-space navigation. The combination of radiation hardened hardware, AI-assisted image processing, and advanced sensor fusion is an indication that autonomy and precision can be advanced greatly in conditions where Earth-based communications are limited. Mars rovers and asteroid explorers case studies have given tangible examples of what these technologies are practically applicable and successful, and the discussion of new technologies like quantum sensing, neuromorphic vision, and onboard intelligence can be seen as promising a bright future. All these developments highlight the fact that the concept of Earth reliant navigation is gradually being replaced by systems in space that are progressively becoming self-reliant. The combination of the latest imaging, high-end sensors, and smart onboard processing will eventually become the hallmark of the new generation of deep space exploration. These technologies allow safer and efficient missions to be accomplished and have been a critical step toward sustainable interplanetary exploration by reducing dependence on networks of ground-based tracking and the increased autonomy of spacecraft.

AI disclosure statement: The author confirms that no artificial intelligence (AI)-generated content or AI-assisted writing tools were used in the preparation of this chapter (including summaries and the abstract). All text, analysis, and conclusions are the result of the author's original work.

References

1. Ma, Y., Gu, C., Jiang, J., Wei, X., Xie, D., Wang, G., Li, J., Review of autonomous optical navigation for deep space exploration. *IEEE Trans. Instrum. Meas.*, 2025.
2. Steffes, S.R. and Barton, G., Deep space autonomous navigation options for future missions, in: *Proc. AIAA SPACE and Astronautics Forum and Exposition*, p. 5369, 2017.
3. Gui, M., Liu, J., Zhang, C., Dai, M.Z., Advances in deep space probe navigation. *Appl. Sci.*, 15, 6, 3307, 2025.
4. Davarian, F., Asmar, S., Angert, M., Baker, J., Gao, J., Hodges, R., Israel, D., *et al.*, Improving small satellite communications and tracking in deep space—A review of the existing systems and technologies with recommendations for improvement. Part II: Small satellite navigation, proximity links, and communications link science. *IEEE Aerosp. Electron. Syst. Mag.*, 35, 7, 26–40, 2020.
5. Guinn, J.R., Riedel, J.E., Bhaskaran, S., Park, R.S., Vaughan, A.T., Owen, W.M., Ely, T., Abrahamsson, M., Martin-Mur, T., The deep-space positioning system concept: Automating complex navigation operations beyond the Earth, in: *Proc. AIAA SPACE 2016*, p. 5409, 2016.
6. Andreis, E., Panicucci, P., Ornati, F., Perico, D., Topputo, F., Autonomous vision-based navigation for deep-space CubeSats: Algorithm development and hardware validation, in: *Proc. 75th Int. Astronautical Congress (IAC 2024)*, pp. 1–11, 2024.
7. Khoroshylov, S.V. and Redka, M.O., Deep learning for space guidance, navigation, and control. *Space Sci. Technol.*, 27, 6, 38–52, 2021.
8. Nguyen, T.X.B., Rosser, K., Chahl, J., A review of modern thermal imaging sensor technology and applications for autonomous aerial navigation. *J. Imaging*, 7, 10, 217, 2021.
9. Parigi, F., Bassissi, E., Deva, L., Sanguineti, E., D'Alessandro, S., Fedele, A., Urban, D., Natalucci, S., Cicala, M., Morselli, A., Topputo, F., FUTURE (Fully aUtonomous feaTUre Recognition planetary Explorer): Revolutionizing nanosatellite autonomy through visual navigation, in: *Proc. Small Satellites Systems and Services Symp. (4S 2024)*, SPIE, 13546, pp. 454–468, 2025.
10. Stateczny, A., Kazimierski, W., Burdziakowski, P., Sensors and system for vehicle navigation. *Sensors*, 22, 5, 1723, 2022.

11. Wang, D., Li, M., Huang, X., Zhang, X., *Spacecraft autonomous navigation technologies based on multi-source information fusion*, Springer Nature, Switzerland AG, 2020.
12. Guthrie, B., Enhancing autonomous vision-based navigation in space with deep learning technologies, Ph.D., Thesis, University of Southampton, 2022.
13. Bi, Z., Yung, K.L., Ip, A.W., Tang, Y.M., Zhang, C.W., Xu, L.D., The state of the art of information integration in space applications. *IEEE Access*, 10, 110110–110135, 2022.
14. Jiancheng, F. and Xiaolin, N., Celestial navigation methods for space explorers. *Meas. Control*, 41, 3, 77–80, 2008.
15. Nesnas, I.A., Hockman, B.J., Bandopadhyay, S., Morrell, B.J., Lubey, D.P., Villa, J., Bayard, D.S., Osmundson, A., Jarvis, B., Bersani, M., Bhaskaran, S., Autonomous exploration of small bodies toward greater autonomy for deep space missions. *Front. Rob. AI*, 8, 650885, 2021.
16. AI-enabled autonomous navigation systems for space exploration. *IEEE Access*, 11, 105432–105445, 2023.
17. Castano, R., Rossi, F., Vaquero, T.S., Verma, V., Allard, D., Amini, R., Barrett, A., Choukroun, M., Davidoff, S., Dhamani, N., Hofstadter, M., Huffmann, B., Ingham, M., Jasour, A., Jorritsma, M., Van Wyk, E., Rabideau, G., Operating deep space autonomous spacecraft: Ground processes and tools for operability and trust, in: *Proc. 17th Int. Conf. Space Operations*, Dubai, UAE, pp. 1–15, 2023.
18. Ghanizadegan, K. and Hashim, H.A., Quaternion-based unscented Kalman filter for 6-DoF vision-based inertial navigation in GPS-denied regions. *IEEE Sens. J.*, 24, 12, 18245–18257, 2024.
19. Song, J., Rondão, D., Aouf, N., Deep learning-based spacecraft relative navigation methods: A survey. *IEEE Aerosp. Electron. Syst. Mag.*, 36, 8, 6–22, 2021.
20. Pauly, L., Rharbaoui, W., Shneider, C., Rathinam, A., Gaudilliere, V., Aouada, D., A survey on deep learning-based monocular spacecraft pose estimation: Current state, limitations and prospects. *IEEE Access*, 11, 64312–64330, 2023.
21. Turan, E., Speretta, S., Gill, E., Autonomous navigation for deep space small satellites: Scientific and technological advances. *Acta Astronaut.*, 196, 1–12, 2022.
22. Kumar, A., Patel, V., Sharma, R., Modeling considerations for developing deep space autonomous systems, in: *Proc. IEEE Aerosp. Conf., Big Sky*, MT, USA, pp. 1–10, 2024.
23. Ma, Y., Gu, C., Wei, X., Autonomous optical navigation: Fundamentals, challenges and missions. *IEEE Trans. Aerosp. Electron. Syst.*, 60, 3, 2890–2902, 2024.
24. Mahendrakar, T., Ekblad, A., Fischer, N., White, R.T., Wilde, M., Kish, B., Silver, I., Performance study of YOLOv5 and Faster R-CNN for autonomous navigation around non-cooperative targets, in: *Proc. IEEE Int. Conf. Robot. Autom. (ICRA)*, London, U.K., pp. 4432–4438, 2023.

25. Andreis, E., Panicucci, P., Topputo, F., An image processing pipeline for autonomous deep-space optical navigation. *IEEE Trans. Aerosp. Electron. Syst.*, 59, 4, 7211–7222, 2023.
26. Andreis, E., Panicucci, P., Topputo, F., An autonomous vision-based algorithm for interplanetary navigation. *J. Guid. Control Dyn.*, 47, 1, 115–128, 2024.
27. Pugliatti, M., Rizza, A., Piccolo, F., Topputo, F., The Milani mission: Overview and architecture of the optical-based GNC system, in: *Proc. IEEE Aerosp. Conf., Big Sky*, MT, USA, pp. 1–12, 2022.
28. Bhaskaran, S., Autonomous navigation for deep space missions, in: *Proc. AIAA Guidance, Navigation, and Control Conf.*, Minneapolis, MN, USA, pp. 1–9, 2012.
29. National Aeronautics and Space Administration (NASA), Gateway autonomy for enabling deep space exploration, NASA Technical Report, Document ID 20230014588, 2023.

10

Cryogenic Electronics for Lunar and Deep Space Missions

G. Loganathan[1]*, S. Sivasakthiselvan[2], M. Palanivelan[1] and Ahmed A. Elngar[3]

[1]Department of Electronics and Communication Engineering, Rajalakshmi Engineering College, Chennai, Tamil Nadu, India
[2]Department of Computing Technologies, SRM Institute of Science and Technology, Chennai, Tamil Nadu, India
[3]Computer Science Department, Faculty of Computers and Artificial Intelligence, Beni-Suef University, Beni Suef, Beni Suef Governorate, Egypt

Abstract

The future of the human race in the prospect of long-term lunar habitation and the extension of human spacecraft into deep space, the need to have electronic systems that can be utilized in these extreme cryogenic conditions has emerged as a priority. This chapter reviews the concepts, design techniques, and new innovations that support the development of cryogenic electronics to mission to the outer space. The temperature in these environments is often less than −150°C, which makes it extremely challenging to the traditional semiconductor devices and methods of packaging. This chapter also discusses the methods through which power efficiency, radiation tolerance, and thermal stability can be attained, and it also covers how cryogenic circuits can be integrated with spacecraft infrastructure. It emphasizes the way in which engineering innovation and multi-disciplinary teamwork can turn theoretical knowledge into practice using the chosen case studies of lunar programs and interplanetary programs. The conclusion parts describe the research trend in the future and potential of developing new technologies that have the prospect of increasing the performance limit of the space-borne electronics in extreme temperature conditions.

***Keywords*:** Cryogenic electronics, deep space, spacecraft, cryocooling techniques, radiation hardening

**Corresponding author*: logeshece57@gmail.com

N. Sathish, Prasanth Aruchamy, Rajesh Kumar Dhanaraj, Nithya Rekha and Karthik Palani (eds.) Space Electronics for Deep Space Exploration, (229–246)

10.1 Introduction

The deep-space and the exploration of the Moon subject the electronic systems to more challenging environmental conditions than one would experience on Earth. The ambient temperatures of these regions may vary between the extent of –250°C and +120°C [1], and the high vacuum and high radiation field exposures are prolonged. In this case, the typical electronic components perform poorly, their switching nature cannot be relied upon, and the material tends to fatigue. This problem is address by cryogenic electronics systems. Cryogenic operation has got its own advantage. First, low thermal noise, which increase sensitivity of devices. CMOS and SiGe Leakage currents are significantly reduced, resulting in a high signal-to-noise ratio and analog front-end stability improvements [2, 3].

Second, it can directly thermalize cryogenic electronics with low-temperature sensors. In cases where the readout electronics are in the same cryogenic environment, parasitic capacitances are reduced and the signal quality is also enhanced [4]. This type of co-located design has been found to be important in high precision missions including the Lunar Gateway and Europa Clipper as well as the James Webb Space Telescope (JWST) where sensitivity and low power dissipation are mission vital criteria. However, these benefits are accompanied with serious issues. At temperatures lower than about 100°K, the ionization efficiency of the semiconductor dopants reduces leading to carrier freeze-out and threshold voltage changes in MOSFETs [5]. Furthermore, recurrent thermal cycling gives rise to mechanical stress which causes fatigue to interconnects and packaging materials. As a result, the achievement in this area is based on the creation of radiation-hardened, thermally stable, and mechanically resistant technologies of materials and packaging. Cryogenic electronics is becoming the core technology of the future of lunar and interplanetary exploration systems with the combination of sophisticated device physics, materials science, and thermal management practices.

10.1.1 Motivation

The cryogenic electronics design and implementation are not based solely on theoretical interest but also on strong functional requirements of contemporary space exploration. Space is a place that continually puts a test on engineering capacity. The lack of atmosphere and long-term radiation and excessive thermal gradient cause a situation where the normal electronics used on the earth do not behave in a predictable manner, or at all. At the lunar night, when the period is about fourteen days on earth, the surface temperatures may drop to less than −180°C. In this case, the carrier mobility of standard silicon transistors is massively reduced, and the transistors become almost non-conductive. Additionally, the accuracy standards of modern space missions have been multiplied several times. Contemporary forms of observatories, like the James Webb Space Telescope (JWST) and the high-resolution devices of interplanetary probes, are dependent on ultra-sensitive detectors that are able to detect small signals, such as infrared photons of the faraway galaxies, or the electromagnetic reflections of the ice layers beneath the surface. At these levels of sensitivity, even low-level thermal noise will mask useful information. Sensitivity is not the only way of looking at the significance of cryogenic systems. They allow long-term stability in deep-space missions where the maintenance or replacement of the components is not possible.

Cryogenic conditions assist in reducing leakage currents, reducing power requirements, and stabilizing bias requirements all of which enhance increased operating life. In missions requiring continuous operation, e.g., planetary landers, orbital observatories, and relay satellites, this reliability is a key to mission success. Simply put, the need to create cryogenic electronics can be driven by both scientific curiosity and a need to do so. The ability to collect accurate measurements, extend the mission life, and venture into areas of space that were once deemed to be hostile are all possible by ensuring that electronic systems are stable, energy-efficient as well as responsive in cryogenic environments, thanks to the efforts of the engineers. Along with the success of lunar and interplanetary missions, these technologies also support the overall frontier of human capability in the mission to the extreme environments.

10.1.2 Environmental Challenges in Lunar and Deep Space

Figure 10.1 presents the schematic representation of the lunar thermal environment and its impact on on-board electronics. The thermal and radiative issue of lunar environment to degrade the performance of electronics systems. In comparison to the conditions on the Earth, temperature and pressure levels here are manageable, on the contrary, the conditions in the extraterrestrial environments are very dynamic with extremes that may occur in the environment and deteriorate the traditional electronic devices and subsystems within a short period.

I. Extreme temperature variations

Temperature on the surface of the moon fluctuates to an average of +120°C at the time of the day and less than −180°C at night. These giant thermal

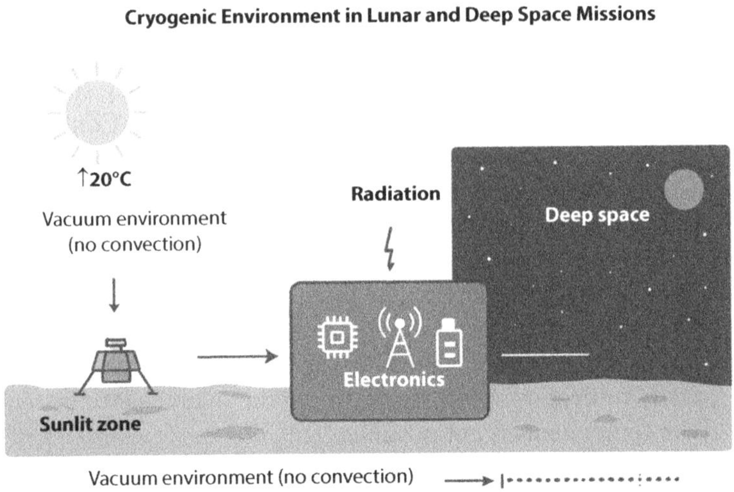

Figure 10.1 Schematic representation of the lunar thermal environment and its impact on on-board electronics.

swings happen in a cycle of 2 weeks in which the atmosphere is not substantial to mediate heat exchange. Components that are subjected to such environments undergo cycles of expansion and contraction numerous times, and this may lead to mechanical fatigue, micro-cracking, and eventual breakdown of solder joints and interconnects. Deep-space missions travelling many radii beyond the orbit of the earth (e.g., missions to the outer planets or the interstellar medium) can experience ambient temperatures down to below 40°K. Such cryogenic environments cause problems in the carrier dynamics of semiconductor, threshold voltages, and mobility properties, requiring re-architecture systems and new material systems that operate best at sub-ambient temperatures.

II. Radiation and high-energy particles

The other significant issue is the constant exposure to cosmic rays, solar wind, and gamma radiations. Without the magnetic field of the earth in deep space, the electronics of spacecrafts is not shielded against these high-energy fluxes. Effects caused by radiation including single-event upsets (SEUs), total ionizing dose (TID) damage, and displacement damage have the potential to cause changes in logic states, to degrade the properties of transistors and ultimately leave devices useless [6].

In a few cases, cryogenic temperature may alleviate radiation effects by slowing down defect movement, as well as inhibiting annealing mechanisms, but in other cases cryogenic temperature causes trap related anomalies in some semiconductor materials.

III. Vacuum and outgassing effects
Space has a hard vacuum that removes convective heat transfer and radiative and conductive processes become the only thermal regulating processes. Moreover, as a result of the vacuum environment, polymeric materials, adhesives, and insulators tend to outgas, resulting in molecular contamination of optical and electronic surfaces.

IV. Micrometeoroid impacts and mechanical stress
In longer missions, the electronics of spacecraft are susceptible to impacts of micrometeoroids and vibration during launch and orbit insertion. Most materials become less ductile and brittle at cryogenic temperatures, and this increases the potential of fracture under mechanical stress. Mechanically resistant packaging, redundant shielding, and stress-resistant interconnects are hence important to the reliability of a system.

10.2 Semiconductor Behavior at Cryogenic Temperatures

Using electronic systems under cryogenic temperatures has an extensive performance, reliability, and efficiency factor that most cannot do without in space and planetary exploration. Whereas standard electronics focus on the ability to keep the apparatus at room temperature, the cryogenic conditions offer the chance to harness the physical processes that appear only under low temperatures and result in better device property and improved system performance.

10.2.1 Enhanced Carrier Mobility and Performance

In low temperatures, carrier scattering through phonons is highly suppressed, and this leads to a significant enhancement in mobilities of electrons and holes. In semiconductor devices, this would translate to high switching speeds, low on-resistance, and less power dissipation. As an example, cryogenic CMOS and GaAs devices exhibit a higher transconductance and lower leakage currents, and can operate at high frequencies with a low energy consumption, which is a significant benefit of space missions in which data is to be acquired and processed, and power is a finite resource [7].

10.2.2 Superconductivity and Quantum-Level Advantages

On reaching sufficiently low temperatures, some materials become superconducting, i.e., zero electrical resistance and magnetic fields are expelled (Meissner effect). Superconducting electronics such as Josephson junctions and Rapid Single Flux Quantum (RSFQ) circuits are ultra-fast and low-power dissipation switching [8, 9]. The properties of superconducting circuits also render them useful in quantum sensors or magnetic field measurements, as well as very high-precision timing systems all of which are critical in navigation, scientific research, and interplanetary communication.

10.2.3 Lower Power Consumption

Cryogenic operation by nature minimizes both the statical and dynamic power losses. This power efficiency is directly converted to a long mission life and less mass needed by power systems and cooling systems in space where all the watts of power are produced using scarce sun or nuclear energy.

10.3 Design Considerations and Techniques

Figure 10.2 presents the active cryocooling setup for spacecraft payload electronics. The design of electronic circuits operating in low temperatures requires extreme care in energy efficiency, reliability, and stability at very high or very low temperatures [10]. Cryogenic electronics are in contrast to more traditional ambient-temperature systems in that device physics, material properties and thermal dynamics no longer follow conventional body behavior. Circuit designers are therefore required to embrace special techniques in order to guarantee predictable performance and durability of such an environment.

10.3.1 Biasing Strategies

Compensation of variation in threshold voltages and transconductance changes caused by low temperatures is one of the most significant problems

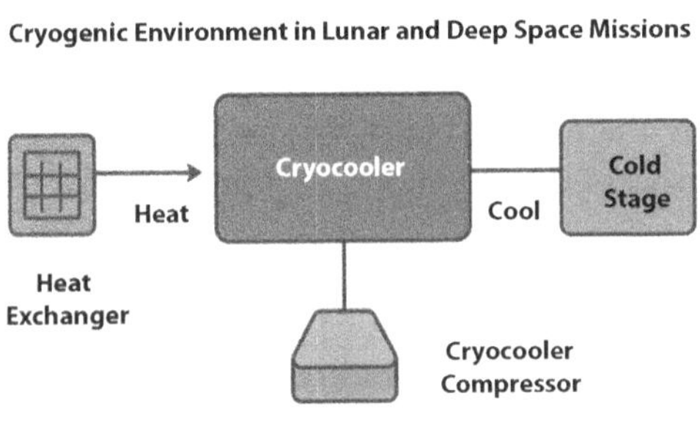

Figure 10.2 Typical active cryocooling setup for spacecraft payload electronics.

in the design of cryogenic circuits. In order to stabilize biasing, designers use temperature-adaptive reference generators [11]. Combinations of circuits like proportional-to-absolute-temperature (PTAT) and complementary-to-absolute-temperature (CTAT) references are usually used in order to have constant bias currents over an extended temperature range.

10.3.2 Device Modeling

The traditional SPICE models that have been designed to operate at room temperature are not reliable at cryogenic temperatures because they do not consider carrier freeze-out, bandgap-widening, and changes in mobility [12]. As such, correct modeling of circuit behavior must involve physics-model device models, which incorporate temperature-dependent models and quantum effects. Such sophisticated models enable the engineer to model a response more accurately to the circuit resulting in optimal performance to be realized before manufacture [13].

10.3.3 System Integration

Lastly, the cryogenic system needs to be coupled to the room temperature electronics, which normally deals with the data collection, control, or telemetry capabilities. In order to achieve compatibility, the designers use level-shifting circuits, and the use of differential signaling, and this enables the signal to be translated between tempos in the signal without the need to lose its fidelity. Efficient integration will make cryogenic subsystems play a positive part in the overall system performance in the spacecraft architecture.

10.4 Integration with Spacecraft Systems

The subsystems that comprise cryogenic electronics have to be integrated (seamlessly) with spacecraft power, telemetry, and data buses so that they can operate reliably and synchronously over large temperature ranges. Some strategic integration strategies are: Low-voltage differentiating signaling (LVDS), onboard calibration routines, miniaturized cryogenic system-on-chip (SoC) platforms, digital signal processors (DSPs), and power distribution networks use low-dropout regulators (LDOs).

10.5 Cryogenic Materials, Interconnects, and Packaging

One of the most significant aspects of performance and reliability of the cryogenic electronics that will pass through lunar and deep-space flights is the choice of the materials and the packaging technologies. Cryogenic temperature material behavior, including thermal-expansion coefficient, elastic modulus, electrical conductivity, and fracture toughness, at low temperature are quite distinctly different compared to their behavior in the high temperature. When the temperature falls, the materials become tighter and lower in ductility, and their behavior in terms of crystal-structure changes. As such, what would be the case with parts that are not critical and therefore can behave well under earth conditions may fall apart calamitously under cryogenic space conditions unless they are carefully chosen and highly qualified.

The electronic systems are repeatedly exposed to the high temperature differential differences in the space missions on the launch process, lunar evenings, eclipses, and the deep-sea cruise process. Temperature variations result in mechanical forces in the electronic assemblies, due to varying contraction of the incompatible materials. Stresses may accumulate during extended periods of missions and lead to micro-cracks, delamination, and electrical discontinuities. Thus, the propensity toward the engineering of the cryogenic materials is not restricted to the electrical performance, since it implies the long-term mechanical stability, radiation resistance, and capacity to endure the vacuum.

10.5.1 Substrate and Structural Materials

The types of semiconductor substrates employed are: silicon (Si), silicon-on-insulator (SOI), silicon carbide (SiC), and sapphire due to their mechanical properties and predictable low temperature behavior. Conventional bulk silicon is in widespread use as it is mature and well understood but at cryogenic temperature SOI substrates have superior electrical isolation capability, and reduced parasitic properties, which can be exploited to enhance circuit stability and noise characteristics. The oxide buried layer of the SOI devices also suppresses the leakages and latch-up phenomena when the gadgets are exposed to radiation.

SiC has been found to be a highly promising material in cryogenic and space applications. It has a large bandgap and can be used at a wide range of temperatures and using small leakage currents. SiC is also very thermal and thus a good conductor of heat even at low temperatures. In addition, its inherent resistance to radiations renders it applicable in missions where radiations are that of high energy particles, such as missions around Jupiter or in the interplanetary space. These attributes make SiC particularly attractive to high-power front-end electronics and cryogenic mixed-signals.

Sapphire substrates have also found application in superconducting circuitry as well as microwave cryogenic electronic circuitry due to low dielectric loss, high thermal stability, and even high lattice integrity at extremely low temperatures. They are also essential in such applications as superconducting detector, quantum sensor, and high-frequency communication components, in which signal integrity and reduction of loss are of paramount concern.

Housing structural materials: Frame structural materials, mechanical supports and housing structural materials must be dimensionally stable in thermal cyclic operating conditions. The reasons why titanium alloys are very popular are the high to-weight ratio, corrosion resistance and relatively low coefficient of thermal expansion (CTE). Invar is an iron-nickel alloy which is extremely low CTE, and so is often used where stability is critical. Silicon-based electronics: The stability between light weight and CTE capability: aluminum-silicon composites offer a trade-off between the two. Proper CTE alignment of substrates, interconnects as well as structural materials significantly reduces the stresses that are precipitated by the thermo effects, and these can cause delamination, warping or even fracture during cooldown and warm-up cycles.

10.5.2 Interconnect Technologies

Another problem that is also a major concern during cryogenic temperatures is the interconnect reliability because as an electronic system, electrical connections are the most vulnerable components. Classical solder joints typically based on tin alloys turn brittle at low temperature and fracture under thermal loads. Moreover, the difference in contraction between metals and substrates can induce shear stress which can enhance mechanical failure along solder interfaces.

Ductile behavior, permanence of electrical properties, and oxidation resistance still make the gold and aluminum wire bonds, the workhorses of cryogenic electronics. They are the materials that are also flexible in low temperatures such that they can contract in case of heat without rupturing. The wire bonds, however, introduce longer electrical paths, and it increases the parasitic inductance and might be a limitation in high-frequency/high-speed operations.

To counter the weaknesses, complex interconnect technologies are gaining popularity, such as flip-chip bonding and copper pillar interconnects. Flip-chip bonding enables the die and substrate to be in electrical contact with either solder bumps or conductive pillars, which enables far shorter lengths of interconnects, as well as parasitic effects, are reduced by an enormous extent. The copper pillars interconnects possess superior current carrying capability and mechanical strength as compared to the traditional solder bumps.

Through-silicon vias (TSVs) allow miniature three-dimensional integration, through which electrical contacts are allowed to be made through the silicon substrate. Cryogenic systems have reduced interconnect density, increased signal integrity, and increased functional density than TSV integration. Nonetheless, it is important to note that reliability of TSV at cryogenic temperature must be critically sharpened on via-fill materials, barrier layers and stress relief structures to prevent contraction and delamination of the cracking, an outcome of thermal contraction.

10.5.3 Packaging and Encapsulation

The cryogenic-compatible packaging must possess the capability to provide mechanical protection, radiation shielding, and good thermal conductivity and inhibit outgassing in vacuum conditions. Even the most minute amount of the outgassed material in space may condense onto fragile optical or electronic surfaces and affect performance. Therefore, packaging materials should be chosen and processed in a manner to achieve a high vacuum and contamination level.

The ceramic package like alumina and aluminum nitride (AlN) is extremely popular due to its great thermal stability, mechanical strength and the hermetic sealing ability. Good thermal conductivity is of particular interest to aluminum nitride because this material has guaranteed that the heat is easily transported within the cryogenic electronics and cooling systems. They are also ceramic packages, which are not degraded by radiations.

The new packaging techniques are progressing to phases that need thermal straps on their packaging, isolations of vibrations, and electromagnetic interference in a single multipurpose package. The integrated designs provide system-level reliability, reduced mass, and spacecraft easy integration.

10.6 Passive and Active Cryocooling Techniques

Passive and active cryogenic cooling methods form one of the focal points of research in the modern study of low-temperature electronics, in part because such systems have found use in spacecraft where environmental conditions differ radically with those found in terrestrial laboratories. The necessity to control thermal energy in such platforms puts much emphasis on the development of solutions that are not only light in weight and energy efficient but also able to stabilize temperature, reduce thermal noise, and in the process protect the integrity of delicate electronic components.

10.6.1 Passive Cryogenic Cooling

This passive cooling is a zero-power and non-moving process which is based upon radiative heat transfer. It uses multilayer insulation (MLI), which is an array of reflective foils with low-conductivity materials in between to thermally isolate the warmer parts of the payload to the cooler parts. MLI is not new to solar and other thermal penetration prevention since it has been used in prior missions to space successfully. Other passive mechanisms, which include deployable radiators and high-emissivity sunshields, operate using the concept of extending large surfaces, which radiate heat into the cool space of the void. Experimental experience shows that with suitable positioning of these radiators in respect of the Sun, it is possible to achieve a low temperature of less than 50°C of the instruments. The James Webb Space Telescope is an example of such a method: its five-layered sunshield creates a strong temperature gradient that keeps its scientific instruments alive and in cryogenic temperature with no active cooling system involved. Even though active systems, unlike passive methods, do not have the mechanical complexity, they still require careful control of the spacecraft attitude, path, and sun direction.

10.6.2 Active Cryocooling Systems

Active cooling systems are required when temperature drops below 20°K or when the exact temperature control is necessary. Standard exemplary versions are stirling-cycle coolers, pulse-tube coolers, and Joule-Thomson (JT) systems. These machines work by compressing, expanding and distilling heat on the payload sequentially *via* mechanical means. Pulse-tube coolers are of interest to space operation due to the fact that all the cold end has no moving components and thus, minimal vibrational noise and a long operational life. Stirling coolers are more likely to be efficient although they require a high level of vibration isolation. Pre-cooling stages are usually integrated in JT coolers to obtain the lowest temperature levels. An effective active cooler design should also be vibration dampened, electromagnetic suppressed, and power reduced. Magnetic bearings, dynamic balancing and active vibration cancellation technologies are used to guarantee long term reliability and power consumption over multi years deep space missions.

10.7 Radiation Hardening at Cryogenic Temperatures

Radiation effects cannot be ruled out even in cryogenic temperatures. Single-event upsets, ionizing dose absorption, and displacement damage are phenomena which are observed with time. Low temperature can alter electrical behavior of devices and hence tough design and test procedures are required to predict and counter such effects.

10.7.1 Radiation Effects in Cryogenic Environments

When the temperature is low defect annealing is inhibited, and radiation-induced defects accumulate and do not disappear. Oxides trap charges which increases voltage passes and noise levels and deteriorates the performance of the devices. Reduced carrier mobility on the contrary can reduce the severity of some transient radiation events, such as transient events with single events.

10.7.2 Shielding and System-Level Mitigation

Protective layers made of aluminium, tantalum, or composite materials can absorb radiation though they are bound to add weight. As a result, the issue of weight optimization is a crucial factor. System-level protection uses error-injection and error-correcting codes, watchdog timers, and recovery code-routines, techniques (which, together, provide a greater level of confidence in the systems) that the system will be able to withstand radiation glitches and continue running.

10.8 Reliability, Fault Tolerance, and Redundancy Strategies

Space is a hostile place and when a spacecraft leaves the earth, it could almost be impossible to make changes or repair it. Therefore, the design paradigm is based on reliability. Cryogenic electronics should be robust to missions that are subjected to thermal, mechanical and radiations.

10.8.1 Thermal Cycling and Mechanical Fatigue

Periodic changes of extreme sub-cutting conditions with ambient temperature cause mechanical loads of materials and wiring, which causes wear in the long term. Examples of prediction lifespan tests are accelerated-life tests, finite-element thermal-mechanical tests, and failure-mode tests, which can be used to estimate mission viability.

10.8.2 Fault-Tolerant Architectures

Cutting is spread throughout the architecture. Triple-modular redundancy (TMR), dual-strings and graceful-degradation techniques are among the techniques that guarantee the survival of the system even when some of the subsystems have failed.

10.9 Case Studies and Mission Applications

Several pioneering space missions have been done using cryogenic electronics and subjected to harsh thermal conditions, such systems require creative design, material, and system integration techniques. The case studies below underline the real-life aspects of the application and difficulties of cryogenic technologies to aerospace missions.

10.9.1 James Webb Space Telescope

The James Webb Space Telescope (JWST) is one of the most recent examples of cryogenic electronic systems in the contemporary space exploration. Its Mid-Infrared Instrument (MIRI) has a very low temperature of just 7°K with the help of a closed-cycle cryocooler coupled with cryogenic CMOS readout circuits to allow extra-sensitive infrared detection. The cryocooler was carefully designed to supply the telescope with the necessary constant cooling potential and overcome mechanical vibrations, which may alter the optical positioning of the telescope. Extensive thermal gradient analysis throughout the optical bench was done by the engineers to make sure material expansion and contraction during cooldown did not lead to a misalignment of critical components.

The giant ground-based environmental testing was used to test the structural stability, electrical continuity, and signal fidelity in cryogenic conditions by simulating thousands of thermal cycles. The MIRI electronics needed radiation-hardened, low-leakage CMOS devices which could operate reliably at 7°K where the properties of transistors differ radically to those at room temperature. Each subsystem, including the compressor units of the cryocooler up to the digital read out electronics, needed to work flawlessly over a distance of 1.5 million kilometers in space The lessons of MIRI cryogenic performance are now used as reference systems in space observatory systems of the next generation and quantum communication payloads and define the future of deep-space sensors.

10.9.2 Lunar Reconnaissance Orbiter

Lunar reconnaissance orbiter (LRO) satellites have their thermal mapping systems based on SiGe-based cryogenic amplifier. The last SiGe amplifier modules were subjected to rigorous thermal-vacuum and vibration testing to ascertain the electrical and mechanical integrity of the modules following hundreds of thermal cycles. Parallel radiation tests: Recurring radiation tests ensured that transistor gain and noise figures did not change when exposed to ionizing radiation and cryogenic temperatures. Finally, these amplifiers showed strong signal fidelity, making radiometers and thermal imagers by LRO to provide high-resolution lunar temperature maps with a high degree of accuracy. The LRO experience highlights the importance of considering microscale mechanical mismatches at cryogenic temperatures with mission critical consequences, and the need to adopt an integrated fabric, microelectronic dependability, and systems level validation strategy of space missions on long duration space missions.

10.10 Future Directions: Quantum Electronics and AI-Assisted Cryogenic Systems

The future of cryogenic electronics is now on the border of quantum technologies and artificial intelligence on one side and the newest materials science on the other. Since the orbital industry is becoming more independent, the exploration of space and space exploration, and the heightened scientific demands, the old ways of electronic design are no longer effective. In the past, the crystal ice state conditions were considered as rather an operational constraint, today, however, operational studies actively exploit such conditions to find new physical phenomena, and to improve system behavior. These new space-grade cryogenic platforms do not only accommodate instrumentation, they are also proving to be smarter, adaptive, and mission-critical subsystems and will radically alter spacecraft architectures.

One potential application of quantum electronics is quantum electronics with AI-assisted cryogenic temperature control where quantum electronics may deliver transformational performance in sensing performance, energy efficiency, system data processing, and system resilience. It is of interest especially to deep space missions that are long range and to which the unavailability of power, communications with long latency, and other unfavorable environmental conditions demand highly autonomous and efficient electronic systems.

10.10.1 Quantum and Superconducting Electronics

Analysis of the development of the design of cryogenic systems involves a critical curve of quantum and superconducting electronics. Some materials pass through a superconducting transition at cryogenic temperatures which is electrically resistant to zero and the MeissnerOchsenfeld expulsion of magnetic fields. The effect allows the construction of electronic circuits having very low power dissipation and very low noise levels—not possible in the conventional semiconductor technology.

Josephson-based superconducting circuit quantum electronic architectures, including the quantum computer, are based on quantum computers. The junctions can be used to precisely control quantum states and have been applied in superconducting qubits, ultrasensitive magnetometers, and single-photon detectors. These devices are inherently useful in instruments of science that demand high sensitivity in the space domain, such as the far-infrared astronomy, cosmic microwave background, gravitational-wave, and the observations of the dark-matter.

The cryogenic quantum sensors can detect small distortions of electromagnetic fields, temperature or radiation with high precision never observed before. The superconducting transition-edge sensors (TES) and microwave kinetic inductance detectors (MKIDs) have already been applied in space-based telescopes to enhance spectral performance and the detection limit. The immediate benefit of a cryogenic working of such sensors is that the signal fidelity and accuracy of measurements is enhanced by the decrease in thermal noise. Unlike conventional CMOS logic, the SFQ logic encodes data using quantized magnetic flux pulses, can support functions with extremely high frequency with extremely low energy consumption. The technologies are yet to be mature to be deployed to space, though they can potentially be utilized to offer real-time onboard data analytics to space missions that generate significant scientific payloads.

Nonetheless, there is a lack of means to imagine how to employ quantum and superconducting electronics on orbital platforms. The severe technical issues are the ultra-stable cryogenic conditions, the painstaking electromagnetic shielding, and the shipping, which endures the vibration of the launch and radiation.

10.10.2 AI-Assisted Adaptive Cryogenic Systems

Artificial intelligence and machine learning are being embraced by cryogenic electronics to enhance flexibility, efficiency and reliability. Conventional methods of thermal control and system-management need pre-programmed control laws and intervention on the ground, neither of which can be applied in deep-space missions where communications delays may take minutes or hours and the environment cannot be predicted.

AI-based cryogenic systems contribute to the optimization of thermal control, biasing conditions, and power distribution in real-time with sensor feedback and predictive models. Based on machine-learning techniques, temperature gradients, vibration signal, and electrical performance indicators are used to dynamically control the actions of cryocooler, thermal routing, and electronic biasing. This responsiveness is flexible and consumes less power without impairment of optimum operating limits of sensitive cryogenic devices.

In addition to thermal optimization, AI is the determining component of the fault detection and recovery. Monitoring system parameters on a continuous basis, AI-based algorithms will have an opportunity to detect subtle anomalies that are the antecedents of component failures—such as a progressive increase in leakage current or shifting vibration signature. Early warning will enable the system to initiate the corrective procedure, i.e., reconfigurating the circuit, turning on of the backup elements or reconfiguring the operating points and therefore, prolonging the life cycle of the total mission assets.

10.11 Conclusion

Cryogenic electronics are an important technology to the future of lunar and deep space exploration. They represent the integration of the human imagination, scientific creation, and multidisciplinary cooperation, which allows a stable operation of electronic performance in those conditions, when the traditional system would stop its activity. Continued advances in materials science, semiconductor device modeling, and system-level integration have continued to improve the performance and stability of electronic systems in very low cryogenic conditions. Such developments increase the reliability of operation of such missions into space, as well as widen the horizons of what can be done scientifically and technically in alien environments. The history of cryogenic electronics is a part of a larger story of human endurance, of making physical limits into discovery space. With the current pace of advancement in this field, not only will new scientific knowledge be unlocked as never before but also the future generation of engineers and scientists will be inspired and the future that a continued human and robot presence will explore into our solar system and further, even beyond our solar system.

AI disclosure statement: The author confirms that no artificial intelligence (AI)-generated content or AI-assisted writing tools were used in the preparation of this chapter (including summaries and the abstract). All text, analysis, and conclusions are the result of the author's original work.

References

1. Chen, W., DiPirro, M., McKinley, I., Cho, C., Tseng, H., Active cryocooling needs for NASA space instruments and future technology development. *Cryogenics*, 141, 103877, 2024.
2. Eastoe, J., Noah, G.M., Dutta, D., Rossi, A., Fletcher, J.D., Gomez-Saiz, A., Method for efficient large-scale cryogenic characterization of CMOS technologies. *IEEE Trans. Instrum. Meas.*, 2024.
3. P Pepe, C., Development of superconducting single-particle detector transition-edge sensor, 2024.
4. Buzio, M., Di Capua, V., Fiscarelli, L., Hernandez, U.M., Cryogenic tests of electronic components and sensors for superconducting magnet instrumentation. *Meas. Sens.*, 38, 101436, 2025.
5. Anjankar, S.C. and Dhavse, R., Design and optimization of MOS capacitor based radiation sensor for space applications. *Arabian J. Sci. Eng.*, 49, 5, 7013–7028, 2024.
6. Vozenin, M.C., Alaghband, Y., Drayson, O.G., Piaget, F., Leavitt, R., Allen, B.D., Doan, N.L., Rostomyan, T., Stabilini, A., Reggiani, D., Hajdas, W., More may not be better: enhanced spacecraft shielding may exacerbate cognitive decrements by increasing pion exposures during deep space exploration. *Radiat. Res.*, 201, 2, 93–103, 2024.
7. Sebastiano, F., Fundamentals of Cryo-CMOS Circuits and Systems for Quantum Computing: Challenges and prospects in interfacing future quantum computers. *IEEE Solid-State Circuits Mag.*, 17, 3, 61–71, 2025.
8. Wei, J., Zhang, P., Yi, X., Hong, M., Fu, X., Tang, X., Qian, K., Zhang, X., Liao, W., Zang, J., Zhang, L., Cryogenic Total Ionizing Dose Effects and Annealing Behaviors of SiGe HBTs. *IEEE Trans. Device Mater. Reliab.*, 2025.

9. Duran, E.A., Pulgar, A., Izquierdo, R., Koblischka, D.M., Koblischka-Veneva, A., Koblischka, M.R., Zadorosny, R., Bridging superconductors with UN development goals: perspectives and applications, arXiv preprint arXiv:2409.17307, 2024.
10. Xiao, K., Wang, Y., Wang, H., Weng, F., Wu, Q., Li, G., Liu, S., Zhang, H., Miao, J., Progress and Prospects of Cryogenic Thermal Control Technology for Space Power Transmission Systems Based on High-Temperature Superconducting Cables. *Microgravity Sci. Technol.*, 37, 3, 32, 2025.
11. Penanen, K. and Weilert, M., Overview of First Two Years of JWST MIRI Cryocooler On-Orbit Performance, 2024.
12. Pillet, X., Interplanetary Rail Launcher: Open-Source White Paper (Hardware & Software)–20 km Electromagnetic Catapult, LiF Cryogenic Shoes and Retractable MHD Shield (TRL, Bottlenecks, Roadmap 2025–2040), 2025.
13. Srinivasan, J.M., Barltrop, C., Berman, S., Bushman, S., Dickson, J., Drain, T., Engelbrecht, C., Fugett, D., Gruel, D., Hill, S., Horner, M., Europa Clipper flight system overview. *Space Sci. Rev.*, 221, 1, 1–30, 2025.

11

Revolutionizing Fault Management with AI: From Detection to Self-Repair

P. Shree Nandhini[1*], P. Rithika[1], M. Vinodhini[1] and K. K. Devi Sowndarya[2]

[1]*Department of Computer Science and Engineering, Sree Sakthi Engineering College, Coimbatore, Tamilnadu, India*
[2]*Department of Electronics and Communication Engineering, DMI College of Engineering, Chennai, Tamilnadu, India*

Abstract
Tracking and anticipating errors continuously is an absolute necessity for modern technical systems in order to manage effectively technical problems and keep systems running smoothly. As the amount of data and data environments continues to grow, including real-time, high-volume, and multiple types of network systems, relying on traditional human-supported troubleshooting methods has become too ineffective and slow. automated fault management is emerging as the vehicle through which companies, such as healthcare and transportation, can utilize artificial intelligence (AI) to not only highlight potential problems but also create and implement solutions immediately, as they arise. In these cases, self-repaired AI takes the process a step further by allowing software to automatically identify, diagnose, and address any issues from beginning to end without people needing to intervene at any point. Combining AI-based analytics and artificial neural networks specifically created for hybrid network architectures, these methods enable companies to create an extremely flexible, adaptable, and efficient means to maintain and protect critical infrastructure systems.

Keywords: Artificial intelligence, fault detection, machine learning, deep learning, self-repair

**Corresponding author*: shreenandhinip@gmail.com

N. Sathish, Prasanth Aruchamy, Rajesh Kumar Dhanaraj, Nithya Rekha and Karthik Palani (eds.) Space Electronics for Deep Space Exploration, (247–260) © 2026 Scrivener Publishing LLC

11.1 Introduction

Despite the complexity and challenges that occur with the introduction of failure and performance-related errors with respect to nodes, multiple fault management strategies have been proposed within reviewed literature for improving the overall dependability and resiliency of edge computing systems.

11.1.1 Adoption Challenges in Fault Management

I. Constrained devices and computation offloading

Edge computing is that more and more computation has been removed from high-performance computing platforms (HPCPs) located in cloud data centers. In addition to having fewer computing resources available (cores, RAM, HDD space) than HPCPs, edge devices must also create an app that fits within a wide range of application constraints (e.g., size, power consumption). For instance, a wearable device may utilize a microcontroller (e.g., STM32F411VE, GAP8) instead of an industrial-style computer. In this regard, it should be noted that edge devices will have a variety of physical form factors, from small chips and boards to more traditional configurations (e.g., RPi, Jetson Nano). The user through the way the edge devices are connected and their physical attributes have created a scenario that designers can exploit in both ways—design opportunities and challenges. A lot of emphasis is placed on the user's side in terms of needs, constraints, and objectives so that the most effective product can be provided.

In edge computing, it is vital that the workloads are optimized and made to fit the processing capacity of each edge device. Each edge device has unique processing power, and this could lead to a situation where the accuracy or performance is diminished when the device is tried on processing of large amount of data. One way to help the edge device's limited processing power is to use the method of computation offloading. The method allows division of tasks or portions of workloads among various edge devices or cloud platforms, thus providing faster and potentially greater processing capabilities [1, 2].

II. Device fleet management

One of the significant issues for edge computing is the need to manage the heterogeneous collection of dispersed edge devices. In cloud computing, all edge devices operate from a common data centre which has the same specifications, are located at one geographical site, and are likewise associated. Supporting various types of devices, operating systems, and software that are occasionally utilized globally is often required when it comes to edge computing.

To mitigate the problems arising from this issue, implementing systems like MQTT, CoAP, AMQP, DDS, or OPCUA that provide standardized communication methods that allow for cross-device and cross-system interoperability can be beneficial. Beyond that, there are existing studies [3] that focus on machine learning (ML) to improve the available protocols and address the existing issues. Additionally, because edge devices may be positioned in difficult-to-reach places, handling hardware issues is more difficult than in a clouded environment. This is especially true when it comes to software issues, especially when it comes to the brittle relationship between edge devices and the cloud. Research is being done to provide more dependable implementation methodologies for edge-focused environments because it is more difficult to distribute new software to the edge fleet than to the cloud [4, 5].

III. Security and privacy

The problem of reducing stress through intelligent sleep monitoring using Internet of Medical Things (IoMT) devices is covered in Rachakonda *et al.* [8]. The most advanced cryptography is used to protect the system's integrity and confidentiality of data, records, and transfers of foi IoMT devices through blockchain technology.

Authorization mechanisms help protect personal health records from unauthorized parties' access. Another important risk is that of data privacy regarding the system's operational state. This is particularly relevant in instances where the device in question could be lost, stolen, or where someone might attempt to gain unauthorized access by direct physical means (i.e., to users' personal information linked to an edge device).

User privacy and data management are key components of edge computing to prevent abuse and/or misuse of sensitive and confidential customer data. The use of consumer data is especially prevalent within the Internet of Things (IoT) space, so providing users with trust due to the ability to assure data integrity and protection of their personal data while stored and during transferring is critical.

Role-based access control are paramount in regulating the access authorization for data to prevent unauthorized access to data by anyone other than those who are supposed to have it (i.e., authorized personnel). Additionally, because edge devices may be positioned in difficult to reach places, handling hardware issues is more difficult than in a clouded environment cross device and cross system. The concept of data minimization is also an important principle associated with privacy, whereby organizations collect and process only the minimum amount of data necessary to meet the goals of the documentation or use for which it was originally collected. User privacy should always be a top priority when designing edge computing solutions, as well as complying with local laws. Different jurisdictions have different laws concerning users' privacy. The General Data Protection Regulation (GDPR), the California Consumer Privacy Act (CCPA), and the Health Insurance Portability and Accountability Act (HIPAA) are typical examples.

In some instances, host countries may impose restrictions on when, where and how certain types of data are locally hosted. This includes protecting Internet of Medical Things (IoMT) devices from cybersecurity-related risks, enabling secure communication between IoMT device and the cloud, maintaining data integrity throughout a decentralized network, and effective management of electronic health records (EHRs), all of which fall under the expansive umbrella of healthcare security technologies. Because of the growing use of IoMT and the increased volume of data exchanged with traditional healthcare systems IoMT devices are susceptible to many kinds of cyberattacks (DPR). including impersonation attacks (i.e., identity/threat deception), denial of service (DoS) attacks, and remote device take-over (hijack).

IV. Data privacy

Dealing with large distributed datasets without loss of utility of the data is one of the identified problems by Chamikara *et al.* [6] in preserving data privacy in distributed ML. The suggested distributed privacy-preserving algorithm is to violate privacy by disrupting the privacy *via* data disruption and avoid privacy leakage in the data exchange in distributed systems. The technique enhances both data utility and privacy by utilizing the parameters made at a centralized location to perform perturbation on the edge nodes.

As Attaullah *et al.* [7] mention, "F-Classify" is a new privacy preservation method that classifies Multiple Sensitive Attributes (MSA) as well as the quasi-identifiers using Fuzzy logic. The proposed method, which makes use of fuzzy logic and multi-dimensional partitioning for classification and anonymization, aims to overcome the major limitation that conventional privacy preservation methods have because of the correlation of the attributes.

Rachakonda *et al.* [8] worked on tackling the stress problem using the smart sleeping monitoring system. The paper leverages IoMT devices with state-of-the-art encryption for ensuring data integrity, as well as confidentiality and securely logs and monitors the data transfers between IoMT devices using blockchain technology. In this method, private health data are ensured not to be seen without approval.

11.1.2 Adoption Challenges in Artificial Intelligence

I. Bias and fairness

Artificial intelligence (AI) urges to develop from biased datasets may lead to the magnification of social biases present in our current system, thus resulting in discriminatory actions.

II. Privacy and security

AI technology will require the collection and use of extremely large amounts of personal data to perform its function, leaving room for questions regarding the security of an individual's private data and risks associated with an individual's privacy violations if that private data is not adequately secured.

III. Data quality

While AI employs highly sophisticated algorithms, if the input training data used by the AI model is inadequate or poor quality, the resulting AI model may provide false or unreliable results.

IV. Explainability and transparency

AI model development and training typically results in complex models (especially "deep learning" [DL]) which results in challenges for human

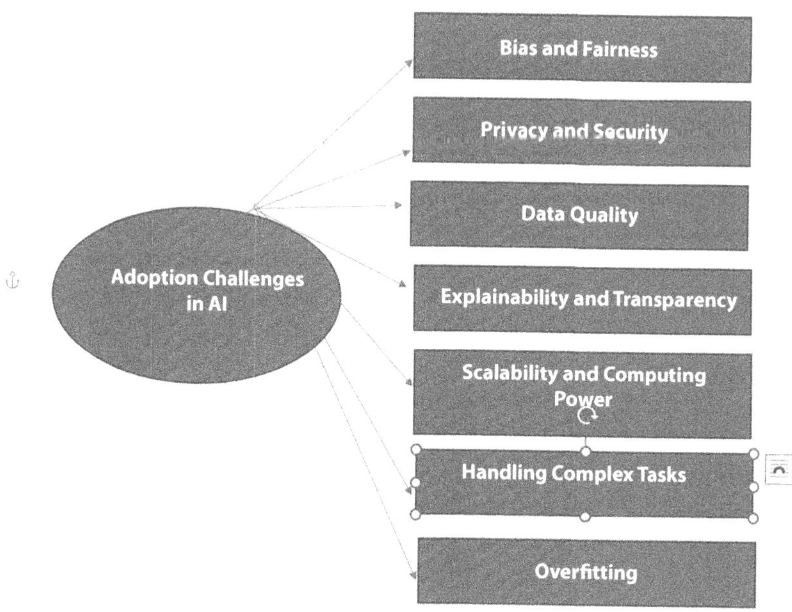

Figure 11.1 Adoption challenges in AI.

beings to understand the fundamental reasoning behind the AI model's decision-making process, making it difficult to defend the model's.

V. Scalability and computing power
Advanced AI model training and execution requires a vast amount of computing power and energy resources, resulting in increased costs and an increase in carbon emissions associated with the use of such large amounts of energy.

VI. Complexity in Tasking
Some AI models are unable to complete extremely difficult or uncertain tasks due to their lack of versatility, which can produce errors or non-consistent performance.

VII. Overfitting
AI models can become overly specialized to their initial training set (leading them to overfit) and thus may not be able to generalize to a new, previously unseen input dataset. Figure 11.1 shows that the organizational, cultural, and technical obstacles that businesses encounter when implementing AI at scale are known as artificial intelligence adoption issues.

11.2 Fault Management with AI

11.2.1 Fault Management Using ML Techniques

Dealing with large distributed datasets without loss of utility of the data is one of the identified problems by Chamikara *et al.* [6] in preserving data privacy in distributed ML. The suggested distributed privacy-preserving algorithm is to violate privacy by disrupting the privacy *via* data disruption and avoid privacy leakage in the data exchange in distributed systems. The technique enhances both data utility and privacy by utilizing the parameters made at a centralized location to perform perturbation on the edge nodes.

Using a neural network (NN) with a support vector machine (SVM), genetic algorithm back propagation neural network (GA-BPNN), RF, Gaussian process regression (GPR), logistic regression (LR), and long short-term memory recurrent neural network (LSTM-RNN), another study [10] demonstrated how to use ML models in the battery management system of a lithium-ion battery for the prediction of faults in the remaining useful life (RUL), charge state, and health state. In a different paper [11], the authors utilized a medium Gaussian support vector machine (MG-SVM) on a motor bearing dataset to forecast bearing defects for electric motors.

11.2.2 SVM-Based Fault Type Identification

The support vector machine (SVM) is one supervised ML technique that can be applied to regression or classification issues. Each feature is linked to a particular position, and each data point is represented as a point in an n-dimensional space, where n is the number of features. The next step in the classification process is to identify a hyperplane that separates the data into distinct categories. Both linear and nonlinear issues can be resolved with SVM [12]. In order to identify the fault categories, this study contrasts the recommended method with the SVM methodology.

11.2.3 KNN-Based Fault Type Identification

The KNN is a type of supervised learning algorithm that determines the distance between each training point and the test data in an effort to predict the proper class of test data. The KNN method selects K points that are reasonably near to the test data. The test data are assigned to the category with the highest probability once the probabilities of the different training data categories have been established [13]. The KNN is used in this work to identify and compare faults.

11.2.4 Fault Management Using DL Techniques

Many researchers have utilized DL methodologies to improve error forecasting [14]. Jiang and Norlund [14] developed a new methodology to predict fault through the use of Seismic Data. Convolutional neural networks (CNNs) were used to extract images from seismic data and segmented *via* the U-net architecture. The extraction of the most prominent attributes used a random forest. The location of the Fault was predicted *via* a fault reconstruction approach using GANs.

Fault probability maps were more accurate once the "discontinuity along dip" was developed as a primary seismic characteristic. Additionally, the authors of the paper [15] suggested a structure-base-based data augmentation technique to broaden the diversity of the semi-real and semi-synthetic seismic dataset, which originated from numerous working locations in China's Tarim Basin. Enhancing U-Net's performance in fault detection and deep neural networks' performance in fault prediction was the aim of this data augmentation strategy.

Using DL techniques to predict failures in complex manufacturing processes. A second recent paper used DL-based approaches to predict and determine the causes of manufacturing process failures in very complex processes. These approaches used attention mechanisms to identify faults in sensor data along production lines and determined fault predictions based on different fault types and predicted the fault events using DL. The authors used long short-term memory (LSTM) models to train the neural networks to learn the process structure of the plant. By applying LSTM models to the data, the authors calculated anomaly scores and were able to determine the accumulation of past fault scenarios into a single incident.

11.3 Revolutionizing AI with Health Care

AI, a powerful tool for digitally evaluating and diagnosing X-ray pictures of the thorax, is being used by cardiologists to detect various ailments earlier, lower diagnostic errors, and do away with needless medication. AI has also been developed as a novel and effective way to analyze electrocardiogram (ECG) and echocardiogram (ECHO) data, providing cardiologists with information that assists them in making well-informed decisions regarding patient management. A recent study from an Oxford-based hospital states that using AI, the ultromics platform (a three-dimensional echocardiographic technology) analyzes echocardiographic images to uncover characteristics of beats that correlate with ischemic heart disease.

AI has shown it can really help with finding diseases on, like skin cancer and breast cancer and problems with your eyes and pneumonia. It does this by using machines that take pictures of the inside and outside of your body. AI has also made programs that can find people who have problems with their brains like Parkinson's disease. These programs listen to the way people talk and look for patterns in their voices. They can even tell when someone might have a problem with their brain like a psychotic episode by looking at the patterns in the way they talk. AI is really good, at looking at these patterns and helping doctors figure out what is going on with a person's body. Finally, one recent study used ML models to estimate the likelihood that individuals would develop type 2 diabetes.

11.3.1 Medical Research and Drug Discovery

The research community generates an overwhelming amount of data through their research efforts, and AI will facilitate researchers on how to interpret the data generated from their research projects; therefore, facilitating the organization of the data that the researchers have collected. As such, there are many benefits that researchers can achieve from the use of AI; thereby changing how they conduct research and also enabling research efforts around the world.

Pharmaceutical companies leverage AI to support the rapid identification of potential drug candidates and the more efficient delivery of the complete cycle R&D process. With AI's assistance, pharmaceutical companies will also have access to all of the data from a variety of research sources; thus allowing them to bring together that information with other types of research to create and identify the best and most likely successful drug candidate based on the available data from previous research evaluations. With the implementation of AI, pharmaceutical companies will create and develop new drugs based on information derived from many different research methodologies.

Compared to their pharmaceutical industry competitors, pharma companies that utilize AI have realized substantial benefits in their capability to identify and formulate and develop new drug candidates. When utilizing ML in clinical trials, it can improve the quality, quality of care, quality of participant experience, and the chance of success. ML will have a more complicated role than other technology tools in a clinical trial, and ML will need to be evaluated more in-depth with regard to philosophical and logistical implications than other technology tools. Like ML, natural language processing (NLP) has demonstrated promise in addressing many of the same issues associated with managing clinical trial participants, but the application of both technologies to improve the clinical trial experience and quality of clinical trials, is still being studied. A comparison of the various methods for managing clinical trial participants will continue to be an important area of research [26].

11.3.2 Revolutionizing AI with Quantum Computing

Numerous technology firms are engaged in research to develop computing devices based on quantum mechanics and are beginning to investigate a branch of new technology called quantum computing. Quantum computing employs enough mathematics and computer science to implement computer algorithms which rely upon quantum mechanical principles for the calculation of the database; Quantum computing possesses an inherent ability to manipulate data in a manner unlike any existing computing technology in use today, thus providing unprecedented computational capability.

Quantum computing represents an immense opportunity for the transformation of multiple industries; building on the advancement of quantum computing will not only enhance existing AI technology but create additional advancements in the scientific arena.

Transforming the traditional way of addressing some of the most complex problems that have been too difficult for traditional computers to resolve (such as optimizing supply chains) is one of many opportunities for quantum computing technology. Quantum computers are capable of addressing problems that were once considered impossible, including optimization of logistics for supply chains, and simulating the interaction of many complex chemical compounds when researching drugs sensitive data, new kinds of secure encryption methods are required due to quantum computing. Quantum computing and AI have a mutually beneficial relationship; quantum-enhanced ML can take advantage of quantum speed and huge data sets that were previously thought unimaginable for traditional methods.

Since AI technology is still in a nascent stage, the fusion of these technologies has profound implications within the health care, finance and many others, as solutions to complex issues within each industry can be provided at a speed and accuracy not achievable in previous generations of devices. At this point in time, however, it is important to be cognizant that quantum computing itself remains in the early stages of development and has significant technical issues, including error correction and qubit stability.

To get over the coming problem and fast track quantum technology all over the world, big money is being put into quantum hardware and software. For the full realization of quantum computing, firms and researchers across the globe are competing.

11.3.3 Revolutionizing AI with Education

By offering individualized learning experiences *via* adaptable platforms, AI is revolutionizing education. The automation of administrative tasks to free up teachers, the improvement of student support through chatbots and AI tutors, and the improvement of assessments through automated grading and data analysis to identify learning gaps.

I. Custom learning pathways

Intelligent Tutoring Systems: AI-powered tutoring programs offer individualized, one-on-one personalized clinician/patient care that identifies an individual's learning gaps and offers support to address those needs.

II. The streamlined administration of schools

Scheduled Tasks: Even though teachers are busy, they are still responsible for managing their classrooms, and all of the administrative tasks (i.e., grading homework, taking attendance, writing reports, etc.) that go along with the classroom. By automating these administrative tasks, teachers can dedicate their time to instruction and improving student learning and participation.

III. Greater student accessibility and support
Assistive Technology: AI-based technologies, such as speech recognition and customized learning games for children with disabilities, provide additional access to assistive technology and create an inclusive learning environment.

IV. Improved student engagement and immersion
Augmented and virtual reality combined provide students with greater ease and comfort in learning and gaining understanding of areas where they may struggle/experience challenges. With such technology, new avenues for teaching/learning have opened (i.e., subjects that might be tough for students in traditional education).

V. Improved assessment and feedback
Measuring student performance in real-time: AI enables teachers to assess and monitor student academic performance as it occurs, find learning behaviours that create unique patterns for each student, and tailor their teaching technique based upon the needs of the individual.

11.4 Fault Management and Revolutionizing AI

AI is moving the management of issues from a reactive to a proactive approach by automating the diagnosis of complex infrastructure and production equipment as well as enabling predictions on when equipment will fail based upon data obtained *via* big data. The application of big data in a similar manner as well as using ML or DL allows AI to continuously improve system reliability, performance, and uptime by identifying unique data patterns, predicting equipment failures prior to their occurrence, and making recommendations based upon previous failures.

11.5 Conclusion

The automated fault management of IT systems, networks, and infrastructures are changing due to ML and DL in AI. Predictive maintenance is one of the most important advancements in fault management because it saves time and makes it easier to detect, diagnose, and predict faults. AI will also enable organizations to provide service-oriented strategies, thus using their resources most effectively (e.g., 5G). AI has also automated many repetitive tasks that require little human involvement and allowed workers to concentrate on more complicated tasks. There is also ongoing research to create explainable AI, and using these models can be combined with sources of sustainable energy, to improve transparency.

AI disclosure statement: The author confirms that no artificial intelligence (AI)-generated content or AI-assisted writing tools were used in the preparation of this chapter (including summaries and the abstract). All text, analysis, and conclusions are the result of the author's original work.

Bibliography

1. Wang, J., Pan, J., Esposito, F., Calyam, P., Yang, Z., Mohapatra, P., Edge cloud offloading algorithms: Issues, methods, and perspectives. *ACM Comput. Surv.*, 52, 1–23, 2019.
2. Mach, P. and Becvar, Z., Mobile edge computing: A survey on architecture and computation offloading. *IEEE Commun. Surv. Tutorials*, 19, 1628–1656, 2017.
3. Donta, P.K. and Dustdar, S., Towards intelligent data protocols for the edge, in: *Proc. IEEE Int. Conf. on Edge Computing and Communications (EDGE)*, Chicago, IL, USA, 2–8 July, 2023.
4. Li, B., He, Q., Cui, G., Xia, X., Chen, F., Jin, H., Yang, Y., READ: Robustness-oriented edge application deployment in edge computing environment. *IEEE Trans. Serv. Comput.*, 15, 1746–1759, 2022.
5. Wang, E., Li, D., Dong, B., Zhou, H., Zhu, M., Flat and hierarchical system deployment for edge computing systems. *Future Gener. Comput. Syst.*, 105, 308–317, 2020.
6. Chamikara, M.A.P., Bertok, P., Khalil, I., Liu, D., Camtepe, S., Privacy-preserving distributed machine learning with federated learning. *Comput. Commun.*, 171, 112–125, 2021.
7. Attaullah, H., Anjum, A., Kanwal, T., Malik, S.U.R., Asheralieva, A., Malik, H., Zoha, A., Arshad, K., Imran, M.A., F-Classify: Fuzzy rule-based classification method for privacy preservation of multiple sensitive attributes. *Sensors*, 21, 4933, 2021.
8. Rachakonda, L., Bapatla, A.K., Mohanty, S.P., Kougianos, E., SaYoPillow: Blockchain-integrated privacy-assured IoMT framework for stress management considering sleeping habits. *IEEE Trans. Consum. Electron.*, 67, 20–29, 2021.
9. Shahbazi, Z. and Byun, Y.-C., Smart manufacturing real-time analysis based on blockchain and machine learning approaches. *Appl. Sci.*, 11, 3535, 2021.

10. Samanta, A., Chowdhuri, S., Williamson, S.S., Machine learning-based data-driven fault detection/diagnosis of lithium-ion battery: A critical review. *Electronics*, 10, 1309, 2021.
11. Lin, S.-L., Application of machine learning to a medium Gaussian support vector machine in the diagnosis of motor bearing faults. *Electronics*, 10, 2266, 2021.
12. Jeong, K., Choi, S.B., Choi, H., Sensor fault detection and isolation using a support vector machine for vehicle suspension systems. *IEEE Trans. Veh. Technol.*, 69, 3852–3863, 2020.
13. Zhou, Z., Wen, C., Yang, C., Fault isolation based on k-nearest neighbor rule for industrial processes. *IEEE Trans. Ind. Electron.*, 63, 2578–2586, 2016.
14. Jiang, F. and Norlund, P., Seismic attribute-guided automatic fault prediction by deep learning, in: *Proc. EAGE Annual Conf. Exhibition, Online*, 8–11 December, European Association of Geoscientists & Engineers, Utrecht, pp. 1–5, 2020.
15. Wang, S., Si, X., Cai, Z., Cui, Y., Structural augmentation in seismic data for fault prediction. *Appl. Sci.*, 12, 9796, 2022.
16. Li, Y., A fault prediction and cause identification approach in complex industrial processes based on deep learning. *Comput. Intell. Neurosci.*, 2021, 6612342, 2021.
17. Yang, H.-S. and Kim, Y.-S., Design and implementation of machine learning-based fault prediction system in cloud infrastructure. *Electronics*, 11, 3765, 2022.
18. Ghosh, P., AI early diagnosis could save heart and cancer patients, BBC News, 2018. Available at: http://www.bbc.com/news/health-42357257.
19. Wang, D., Khosla, A., Gargeya, R., Irshad, H., Beck, A.H., Deep learning for identifying metastatic breast cancer, arXiv preprint., arXiv:1606.05718, 2016.
20. Esteva, A., Kuprel, B., Novoa, R.A., Ko, J., Swetter, S.M., Blau, H.M., Thrun, S., Dermatologist-level classification of skin cancer with deep neural networks. *Nature*, 542, 115–118, 2017.
21. Rajpurkar, P., Irvin, J., Zhu, K., Yang, B., Mehta, H., Duan, T., Ding, D., Bagul, A., Langlotz, C., Shpanskaya, K., *et al.*, CheXNet: Radiologist-level pneumonia detection on chest X-rays with deep learning, arXiv preprint, arXiv:1711.05225, 2017.
22. Bedi, G., Carrillo, F., Cecchi, G.A., Slezak, D.F., Sigman, M., Mota, N.B., Ribeiro, S., Javitt, D.C., Copelli, M., Corcoran, C.M., Automated analysis of free speech predicts psychosis onset in high-risk youths. *npj Schizophr.*, 1, 15030, 2015.
23. IBM Research, IBM 5 in 5: With AI, our words will be a window into our mental health, 2017. Available at: http://www.ibm.com/blogs/research/2017/1/ibm-5-in-5-our-words-will-be-the-windows-to-our-mental-health/.
24. Academy of Royal Medical Colleges, Artificial intelligence in healthcare, 2019. Available at: http://www.aomrc.org.uk/wp-content/uploads/2019/01/Artificial_intelligence_in_healthcare_0119.pdf.

25. O'Mara-Eves, A., Thomas, J., McNaught, J., Miwa, M., Ananiadou, S., Using text mining for study identification in systematic reviews: A systematic review of current approaches. *Syst. Rev.*, 4, 5, 2015.
26. Weissler, E.H., Naumann, T., Andersson, T., Ranganath, R., Elemento, O., Luo, Y., Freitag, D.F., Benoit, J., Hughes, M.C., Khan, F., *et al.*, The role of machine learning in clinical research: Transforming the future of evidence generation. *Trials*, 22, 1–15, 2021.
27. Martonosi, M. and Roetteler, M., Next steps in quantum computing: Computer science's role, arXiv preprint, arXiv:1903.10541, 2019.
28. Losert, M., How quantum computing could change the world. *DUJS*, 6, 2015.

12

Robotics and Autonomous Systems for Planetary Exploration

K. Naresh Kumar Thapa[1]*, R. Krishnaprasanna[1], V. Yokesh[1] and Weiwei Jiang[2]

[1]*Department of ECE, Sathyabama Institute of Science and Technology, Chennai, Tamilnadu, India*
[2]*School of Information and Communication Engineering, Beijing University of Posts and Telecommunications, Beijing, China*

Abstract
This book looks at the significant changes in planetary exploration robotics that happened between 2020 and 2025. It shows how systems that can work on their own and make scientific discoveries on their own came about. It looks at cutting-edge technology like artificial intelligence, improved navigation systems, edge computing, and machine learning that have made it possible for robots like NASA's perseverance rover and ingenuity helicopter to work on Mars with more freedom than ever before. This chapter looks at multinational lunar expeditions, new ways to collect samples, and how robots can work together with people. It also talks about new technologies like 3D printing in space and bio-inspired robotics. This chapter describes how autonomous systems have fundamentally transformed space exploration, allowing robots to make intricate decisions millions of miles from Earth without human intervention, thereby expediting scientific discovery and laying the groundwork for sustainable planetary exploration and eventual human settlement beyond Earth.

***Keywords*:** Planetary exploration robotics, autonomous system, improved navigation system, lunar expeditions, artificial intelligence

**Corresponding author*: k.nareshkumarthapa@gmail.com

N. Sathish, Prasanth Aruchamy, Rajesh Kumar Dhanaraj, Nithya Rekha and Karthik Palani (eds.) Space Electronics for Deep Space Exploration, (261–292) © 2026 Scrivener Publishing LLC

12.1 Introduction

Robotics in space exploration has transformed human exploration of celestial bodies, providing new opportunities to explore new environments that are hostile or otherwise inaccessible to humans. Robotics and autonomy are inherently linked to the development of knowledge regarding the formation, the state of habitation and geologic evolution of the solar system. In the last 5 years, the leading space agencies and industry partners have increased the pace of introducing robots to explore the planets, making bold moves not only on known terrain, such as Mars, but also exploring the possibilities of having greater missions to the Moon, Venus, Titan, and remote asteroids. This introduction sums up the reasons behind this field of study, its history, and the changes that have been transforming the field ever since the past 5 years.

12.1.1 Motivation for Robotic Planetary Exploration

The evolution of the simple rovers of the 1990s to the advanced autonomous systems of the modern days is one of the greatest technological advancements in the engineering field. The initial planetary robots, such as the NASA Sojourner rover that landed on Mars in 1997, were in essences remote controlled cars, which had to be continuously supervised by mission controllers on the Earth. All movements, all scientific observations, and all operational decisions needed to be planned, sent, and made in a considerate manner concerning the time lag involved in communication over interplanetary distances.

This was initiated by the Mars Exploration Rovers Spirit [1] and Opportunity in 2004 which made some of the simplest autonomous navigation possible. Nevertheless, the curiosity rover [13] in 2012 was the first in the modern era, with its advanced autonomous driving system and basic machine learning terrain analysis features. Recent generations of planetary robots, such as the perseverance rover and the ingenuity helicopter by NASA, can be seen as a giant leap in the right direction. These systems combine various types of artificial intelligence, high-order sensor fusion, and highly-level decision-making algorithms into systems that can work more autonomously than ever before.

12.1.2 The Autonomous Revolution

The tendency toward actual autonomy defines the modern planetary robotics. There are several ways in which modern systems can operate autonomously, analyze their surroundings, discover scientifically interesting targets, plan intricate multi-day operations, and respond to unanticipated events autonomously. The navigation autonomy enables rovers to navigate difficult terrain with the use of computer vision and machine learning to decide what places are in danger or safe and what routes to take. Scientific autonomy allows instruments to autonomously identify interesting phenomena and control the parameter of observation to optimize the quality of the data. Operation autonomy gives systems freedom to manage their resources, diagnose, and carry out recovery processes themselves.

The revolution has been made possible by the addition of edge computing. The current generation of the planetary robots bear the heavy computer capability and can execute advanced AI procedural systems on board, thus not requiring to transmit the raw data to Earth to process them. This has made real-time decision making possible and has also made robotic systems able to react instantly to changing circumstances or opportunities enabling the system to have a better performance as time progresses and to respond well to the previously unforeseen conditions that were not initially anticipated during the design process. Image and visual exploration by using deep learning algorithms can enable robots to detect the presence of geological features, evaluate possible biosignature, and move successfully through complex terrain with human or even super human accuracy.

12.1.3 Key Drivers in Recent Years: Autonomy, AI, Multi-Agent Systems

Probably the most significant reorientation in planetary robotics between 2020 and 2025 is toward increased autonomy, which is driven by artificial intelligence, machine learning, and multi-agent interaction. Onboard artificial intelligence systems have learned to make sense of sensor data, map terrain, plan the best path, and identify unexpected occurrences of scientific interest and do so with minimal oversight by ground crews. This is particularly useful considering the extended time lag of communication between the earth and the other planets that are very far. Simultaneously, the multi-robot systems not only coordinated but also swarming have started displaying the promise of coverage of an area, distributed sensing and collective problems solving. Such teams are capable of self-organizing to investigate hazardous or complex habitats, and members may even specialize at various tasks (ex: scouting, sampling, repairing).

The missions made possible through the creation and combination of these technologies have been able to realize more scientific benefits, navigate a more hazardous environment, and respond to the changing conditions on the planet by itself. AI-controlled independence has also led to unexpected findings since it has been able to discover previously ignored features and phenomena, at a significantly faster rate than the human analyst.

12.1.4 Current Challenges in Planetary Exploration

There is a distinctive cluster of challenges of modern planetary exploration and these issues facilitate a technological innovation. The inherent problem of the time delay in communication will always remain the primary problem, messages taking between Earth and Mars can be delivered one way between 4 and 24 minutes based on the position of the planets. Such delay precludes control in real-time and requires great autonomy. The environmental challenges have also been on the rise with missions being directed to more difficult areas. The Martian climate is characterized by severe changes in temperature, months-long dust storms, and crater walls covered with boulders and smooth plains.

Recent expeditions have ventured into polar areas and underground caves among other places that were inaccessible have required sophisticated autonomous functions. The limitations on resources still contribute to the innovation of efficiency and reliability. Robotic systems in the present day are required to extend their scientific productivity by several years after being designed to operate. Limited power budgets have prompted the creation of intelligent power management systems and adaptive operation modes because of the necessity to gain the maximum scientific payback.

12.2 NASA's Perseverance Rover: A Technological Marvel

The Mars 2020 perseverance rover [2] is the culmination of modern planetary robotics technology which includes the innovations of autonomous navigation, scientific instrumentation, and collection of samples which seemed unbelievable only 10 years ago. Offered in July 2020 and landed in Jezero Crater in February 2021, Perseverance has turned around our thought process on what will be accomplished by robotic explorers.

12.2.1 Advanced Autonomous Navigation

The navigation system in perseverance is a step toward autonomy in mobility that is several generations ahead. Enhanced AutoNav system enables it to scan the terrain of up to 50 m in the future and route safe paths automatically. The system is capable of detecting and evading rocks, sand traps, and steep slopes and still maintain progress on scientific goals using not only a stereo camera but also by having hazard cameras plus even greater computer vision algorithms.

Machines used in the rover were programmed through machine learning algorithms, which were trained with millions of images of past missions to Mars. Such algorithms are able to classify terrain type, predict the interaction between the wheel and the terrain and also estimate the traversability with a high degree of accuracy. Experience may also be learnt by the system and its abilities to assess the terrain are enhanced with time.

Among the most important innovations, there is the fact that the rover can think and drive at the same time. The earlier rovers were required to stop, take pictures, scan the landscape, strategize a route, and engage in a lengthy process that curtailed the amount of the daily traveling. As long as it is on a trajectory, perseverance can examine terrain, and it has a great boost to its operation.

12.2.2 Ingenuity Helicopter: Martinian Powered Flight

The successful operation and deployment of the Ingenuity Mars Helicopter [7] is, perhaps, one achievement that indicates the technological revolution in the realm of planetary exploration better than others. Initially planned as an experimental technology test; with a projected 30-day lifespan and five flights, Ingenuity has significantly surpassed any imaginative projection, flying dozens of times and still operating long into 2024.

The difficulty of flight to Mars is not something that can be exaggerated. The atmosphere of the planet Mars contains lower than 1 time less oxygen than the atmosphere of the earth and hence the creation of lift is extremely challenging. The design of the helicopter demanded the innovation of several fields:

I. Rotor technology

The contra-rotating rotors on Ingenuity rotate at a rate of about 2,400 RPM which is almost five times the speed of normal Earth helicopters. The carbon fiber blades are carefully designed to ensure that they do not undergo structural failure if a high level of rotational force caused by rotational forces is experienced and to be as aerodynamically efficient as possible in the thin Martian atmosphere.

II. Autonomous flight control

Every flight is autonomous because the time lag between it and Earth makes real-time control infeasible. Travelling is performed with the help of the combination of inertial measurement units and cameras and altimeter sensors to control the stable flight state and arrive at predetermined waypoints.

III. Thermal management

Martian evenings may go up to −80°F (−112°C) which would ruin a standard electronics. There is a complex thermal management system, including heaters and insulation, in ingenuity to ensure that the important components remain within operation ranges of temperature.

IV. Edge computing

The helicopter is equipped with a special flight computer that is able to work on sensor information and execute control actions in real-time. This system has to ensure a tradeoff between computation power and rigid requirements of weight and power.

12.2.3 Sample Collection and Caching Technologies

One of the most advanced robotic mechanisms that had ever been used in another planet is the sample collection system of perseverance. The system is able to excavate deep into the rock formations, sample the rock formations, process the samples to control contamination and attach them into ultra clean titanium tubes which will be returned to Earth in future. The drilling system has adaptive control algorithms that have the capability to change the drilling parameters in response to rock properties that were metered in actual time. The sensors are used to track the drilling progress, wear of the bits and the quality of samples and the system can be adjusted to maximize the techniques used by the system to collect samples in different geological formation. The sample processing system comprises of high-end contamination control provisions. The samples are sterilized under a controlled setting by each sample tube to avoid contamination which can disrupt subsequent scientific examination of the sample. The system is also able to prepare sample of witnesses tubes that will record the amount of pollution at the stage of collection.

12.2.4 Terrain-Relative Navigation Systems

Perseverance used terrain-relative navigation (TRN) [6], a ground-breaking technology in the entry, descent, and landing (EDL) sequence. In the descent phase, the spacecraft had cameras to take pictures of the Martian surface and compared them in real-time with pre-loaded terrain maps to permit the vessel to navigate toward a safe landing site in an unprecedented precision. This technology allowed the landing of the site of the scientifically interesting, and hard to reach Jezero Crater, which would have been too risky to have been landed on before. The system had precision at a range of 7 m of the planned landing point an extraordinary feat considering the complexity of an autonomous guidance throughout a speedy descent through the Martian atmosphere.

12.2.5 Scientific Instrumentation Revolution

The revolution of scientific instrumentation was an era during which scientific instrumentation achieved significant breakthroughs in a scientific domain.

Perseverance is a conceptual framework of seven scientific devices that signify great breakthroughs on miniaturization, automation, and analytical ability:

I. MASTCAM Z

The Mast Camera-Zoom (MASTCAM Z) [4] is a modified way of perceiving and researching Mars. The advanced dual camera system has the capability to zoom the two cameras three to one in that the zoom allows close images of things that are distant without necessarily having to move the rover an important factor considering the amount of energy and time available to carry out rover operations. The cameras can record the true-color and multispectral images in nine different wavelengths giving mineralogical data that is extremely important to scientists in identifying the types of rocks and other possible biosignatures at a greater distance. The high-resolution sensors used in the system with 1600 × 1200 pixels are capable of resolving detail as fine as 0.15 mm at the distance of 1 m, bringing out the structures and textures which could have not been captured by earlier Mars cameras. Higher autofocus functions mean shots at any range and under any lighting are produced with a sharp, clear image, and panoramic photography features allow the scientist to make the 360-degree image that would take his/her billions of miles, to stand virtually next to the rover. The video progressive recording feature was, perhaps, most dramatically used when it was used to record the historic flights of the ingenuity helicopters, the first fuel powered flight conducted to another planet. The system has machine learning algorithms that are used to identify targets autonomously and optimize imaging parameters so that MASTCAM-Z can change exposure times, focus distances and spectral filters depending on what is in the scene without Earth commands. Such freedom maximizes the number of scientific returns and minimizes the data transmission and complexity in operation in prolonged autonomous operations, which would be a feature of the new generation of intelligent scientific instruments which are revolutionizing planetary exploration.

II. MEDAz

The Mars Environmental Dynamics Analyzer (MEDA) is a unique weather station, an environment analyzer that measures the atmosphere conditions continuously and with a high level of precision never seen before. It is a complex set of instruments that are used to measure the velocity and direction of wind, temperature, humidity, pressure level and characteristics of dust particles at different altitudes. Working around the clock on a Martian day, night, MEDA offers the essential information regarding weather patterns and changes in climate by the season. Its measurements were really necessary in planning the flights of the Ingenuity helicopters associated with the characterization of wind conditions and density of the atmosphere. The radiation sensors of the system also detect the surface environment, which is important in future human missions to know the radiation levels on the surface and the most appropriate time to venture out.

III. PIXL

The Planetary Instrument for X-ray Lithochemistry the PIXL technology is a new chemical analysis technique that is based on microscopic imaging and X-ray fluorescence spectroscopy. This new tool [5] can identify the inner structure of rocks and soils at the finer scales than a grain of sand, limiting chemical differences at 120 μm. The system used by PIXL, robotic positioning system, allows sensors to be placed against rock surfaces with less than a millimeter accuracy and allows chemical mapping of the details of textures, veins, and crystal boundaries in detail. The high scanning rate of the instrument produces detailed chemical maps in hours, not days and embedded algorithms that run onboard processes them automatically, discovering scientifically interesting compositional abnormalities, which may reflect previous water action or have the ability to form biosignatures.

IV. SuperCam

SuperCam [3] is the most multifaceted remote sensing instrument that has ever been sent to Mars which has been able to offer five complementary devices in one package. Its high laser energy has a vaporizing capability of seven meters of rock sample, which forms a plasma with whose spectral signature can reflect the composition in terms of elements through the spectroscopy of laser-induced breakdown (LIBS). The device utilizes Raman spectroscopy to identify minerals, infrared spectroscopy to detect organic substances, and high-resolution color imaging to provide the context and to study acoustic data with its onboard microphone the first ever to send tracks on Mars. AI software automatically chooses the best targets and manipulates the analysis parameters, allowing SuperCam to describe dozens of samples each day and the most promising of them interested in taking a closer look with contact instruments. These tools are capable of working in coordinated actions where information on some of the tools will be used to deactivate others. Machine learning algorithms are used to filter through initial data and determine which targets are the most potentially productive to spend more time and power going into scientific space.

12.3 Lunar Robotics and Commercial Partnerships

A renewed interest in lunar exploration has taken place in the last 5 years due to an immortal blend of government initiatives and a developing lunar economy. The novel paradigm has radically transformed the way in which robotic lunar missions are done and focused on cost efficiency, prototyping, and novel technological solutions.

12.3.1 Artemis Program Robotics Integration

The Artemis program by NASA is the most ambitious lunar exploration mission since the Apollo program and the role played by robotic systems is very vital to the upcoming proposals of human missions. This is in a contrast to the overall approach of the program in robotics, with a push and focus being on modularity, business relationships, and speedy technology demonstration.

The Lunar Gateway, the first lunar orbit space station of humankind will also be used as a functional test of intelligent robots. The station will also have robotic arms that will handle cargo, external maintenance robots, and autonomous docking systems, which will prove to be vital in aiding the activities on the lunar surface.

Artemis has resource prospecting robots as one of its priorities. These systems have to work in the permanently shadowed areas of lunar polar craters, where the temperature goes down to less than –230°C (–382°F) and there is no solar power. High-technology power systems, such as radioisotope heater units and long-duration battery technologies, allow it to work in such harsh environments.

12.3.2 Commercial Lunar Payload Services

The collaboration with commercial firms leads to deliver scientific payloads to the lunar surface, the Commercial Lunar Payload Services (CLPS) program has transformed the economics of lunar exploration. This technique has lowered cost by a factor of 100 to factor of 5 over conventional missions as formulated by the government, as well as escalating the number of missions and diversifying technologies.

Such corporations as astrobotic, intuitive machines, etc. have already worked out innovative lunar landers with the use of the latest autonomous technologies. These systems are also required to go through the difficult lunar approach and landing process with minimum ground support and thus demands an advance guidance, navigation and control system.

The CLPS model has increased development of technology through establishing market incentives in the area of innovation. Inter-provider competition has led to improvement on propulsion efficiency, landing accuracy, and flexibility in integrate payload.

12.3.3 VIPER Rover: Water Ice Prospecting Technology

The Volatiles Investigating Polar Exploration Rover (VIPER) [12] is an upcoming category of lunar rover exploiter with a focus on resource investigations. VIPER, planned to be sent to lunar south pole, integrates various advances in the field of technology that will be essential in the future to use lunar resources.

I. Advanced mobility systems

The mobility system of VIPER will be made to work on steep slopes and loose regolith and be stable in the low gravity lunar conditions. The rover will have a sophisticated wheel structure and optimization of the system to survive on the moon.

II. Special scientific instruments
The rover contains a complete set of instruments to detect and study the presence of water ice. These are neutron spectrometers to detect the hydrogen labels, near infrared spectrometers to detect the direct water and drilling systems to sample under the ground.

III. Power management under extreme environments
Power under permanently shadowed areas need to be innovated. VIPER installs radioisotope heater modules to preserve important systems and high-tech battery technologies that can work in extremely low temperatures.

12.3.4 Autonomous Landing and Hazard Detection

The recent missions to the moon have involved more elaborate autonomous landing systems [14] that are able to detect and avert the risks encountered on the latter part of the descent. Such systems portray vital technology toward future missions to difficult parts of the lunar surface. The current lunar landers implement hazard detecting systems on the basis of LIDAR, which recognize boulders, craters, and steep slopes, during descent and automatically turn toward safe landing sites. These systems should be able to work in real-time with very tight computational and power limits of loading precision at landing. Computer vision algorithms determine the features of the surface and categorize the terrain suitability at real time. The machine learning systems that have been trained on images of the lunar orbits are capable of determining safe landing sites and directing the spacecraft to the most appropriate touchdown.

12.3.5 International Lunar Exploration

The stakes of the lunar exploration have been on a new level all over the past 5 years. Chang'e missions by China have shown capabilities of return sample in advanced ways whereas India Chandrayaan program has challenged the limits of cost effective exploration of the moon.

I. Chang'e sample return missions
In December 2020, the Chang'e 5 mission [11] was the first to bring lunar samples to Earth in a successful mission in more than 40 years, a feat that had not been realized since January 1979. The mission displayed outstanding autonomous skills in its multistage activity. Having landed in the Mons Runker area, the robotic sampling system on the lander retrieved about 1.73 kg of substance using surface scooping and the subsurface drilling

technique, being fully automated because of communication limitations. The ascent vehicle then conducted an autonomous docking experiment in lunar orbit and the unprecedented autonomous rendezvous maneuver of the return spacecraft was a critical technology in which precise navigation and control of the ascent vehicle was necessary, and real-time ground communication was unavailable. The use of sophisticated computer vision systems allowed the space vehicle to find and meet the orbiter and dock with one when it was moving at a speed of thousands of kilometers/hours around the moon. The mission sampler transport system was autonomously operated in which the sealed sample transport system was transferred between vehicles demonstrating highly intelligent robotic manipulation in the hostile space environment. The success of Chang'e 5 confirmed that China can execute challenging autonomous operations which will be needed in the sample return missions of Mars and asteroids later on. These capabilities were further expanded in the mission of Chang'e 6, which was the first mission to make a successful return to the far side of the Moon in 2024, necessitating relay satellite communications and even more advanced autonomous operations. These missions have made China a leader in the field of robotic sample returns technology and have proved that very complex multi vehicle activities can be effectively executed without human control on a regular basis.

II. Chandrayaan innovations

Chandrayaan, India lunar exploration project has impacted the world of cost effective space robotics with innovative methods of engineering designs capable of maximizing scientific pay off at minimal cost of mission. The mission Chandrayaan 2 and the fact that the lander has successfully landed chose to land in 2019 showed an unprecedented technological success with the orbiter still offering valuable scientific data. Through the mission, India demonstrated its ability to come up with state of the squadron spacecraft systems, such as high resolution cameras, radar gadgets to identify underground water ice, and more advanced communications units at a small fraction of the conventional mission expenditures. In August 2023, Chandrayaan 3 [15] became the first spacecraft to land successfully in the lunar south pole and test improved autonomous landing algorithms, allowing it to make a precise touchdown on uneven terrain. The Vikram lander and Pragyan rover of the mission had developed new thermal management technologies that enabled them to operate during the day on solar energy and survive the harsh −200°C lunar nocturnal environment with custom-designed insulation and phase-change materials. The six wheeled rocker-bogie suspension system of the rover was designed, assembled locally in India and proved to be highly effective in finding its way through lunar regolith and rocks. The system created by Pragyan of laser induced breakdown spectroscopy and alpha particle X ray spectrometer offered *in situ* chemical analyses on lunar soil, which confirmed that the surface of the moon contained a variety of different elements, such as sulfur, in regions near the south pole. This focus of the mission on ruggedized commercial off the shelf products, reduced mission architecture, and effective project management has resulted in a new paradigm of affordable planetary exploration of interest to other countries and agencies. The Indian strategy demonstrates that advanced space research does not have to be an expensive undertaking, which can be accomplished through clever thinking, narrow goals, and creative solutions.

III. Collaborative missions

Over the last 5 years, there has been greater international collaboration in the development and exploration of the moon with countries increasingly sharing resources, technological know-how and combining missions. The Artemis Accords that were signed by many countries are set to provide the principles of peaceful lunar exploration and the use of resources, as well as access to share technologies and jointly plan missions. One of the examples of how the collaboration with the global community can bring things that nations could not do on their own is the European Service Module, which supplies power to the Orion spacecraft, which is part of the Artemis program developed by NASA with the contributions of the European Space Agency. Russia and China proposed the building of the International Lunar Research Station where international countries were welcomed to join in the permanent robotic and, finally, human presence in the lunar south pole. This partnership uses the Chinese landing and sample return experience and the Russian space station experience. In early 2024, the Smart Lander for Investigating Moon (SLIM) mission of Japan showed accurate landing technology as it landed within 10 m of its destination, although it had a technology called the last-minute engine failure that will be valuable to the development of international landing systems. South Korea has their Korea Pathfinder Lunar Orbiter (KPLO) satellite with the unique NASA ShadowCam instrument on board, and this is an example of bilateral collaboration with one country developing the spacecraft and the other one a notable influence of scientific equipment. The Rashid rover is being developed by the United Arab Emirates, under its Mohammed bin Rashid Space Centre, in collaboration with partners in Japan, and demonstrates how a fast rate of capability development can be attained by emerging space countries through international relationships. This level of cooperation does not confine itself to the physical aspect but also to data sharing deals, shared scientific analysis groups, and co-ordinated observation campaigns where two or more spacecraft observe the same process at the same time. Communication relay satellites are being perceived as common international infrastructure and countries are aligning to ensure that there is constant coverage to support all lunar missions irrespective of origin. It is this collaborative method that spurs innovations in technology in less time, offers individuals programs backup operation since there is a backup route for essential faculties, and creates a cordial relationship between the international community due to collective objectives and interdependence in space-building endeavors.

12.4 Advanced Autonomous Navigation Systems

12.4.1 Visual Inertial Odometry in Space Environments

The invention of high-performance visual-inertial odometry (VIO) has revolutionized planetary robot autonomous navigation. Visual inertial odometry is an in-camera information that is used in conjunction with motion information that is given by inertial measurement devices to give the accurate and real-time location of the object or object group when the GPS becomes unavailable. This sensor fusion design approach allows planetary rovers to monitor their position and orientation at all times, both on flat featureless terrain and in sub-optimal lighting. Up-to-date VIO systems have positioning errors of less than 1% of range to be covered and rovers can autonomously travel hundreds of meters without needing specific knowledge of the location. The technology has turned out to be essential to the planetary exploration mission where remotely operated robots can continue working even when communication goes dead and make immediate navigation decisions without requiring input response over the ground controller. These systems integrate visual data taken by cameras and inertial data of accelerators and gyroscopes to assist in giving precise real-time positioning and navigation functions even in GPS denied conditions.

12.4.2 Technical Implementation

Present VIO systems to space uses are complex configurations that include a number of advanced components that do not only work together but also create reliable navigation functionality. Stereoscopic depth perception, where multiple synchronized cameras are used in an array at positions to give add redundancy in terms of camera failure or to block the lens with dust, is possible using multi camera arrays when multiple cameras are used. These camera systems are normally running at 10 to 30 frames/sec rate and capture synchronized stereo images pairs, which are then processed by an algorithm to analyze the images, draw out the three-dimensional environmental form. Inertial measurement units have a large rate with a frequency range of 200 to 1000 Hz, which is used to give continuous motion measurements to make it possible to fill in the gaps between camera frames and offers precise velocity measurements. SQIMUs shall be required to be accurate throughout a very high and low temperature range and be able to survive launch transverse forces as well as decades of constant service. Extremely sophisticated sensor fusion methods are, in particular, Extended Kalman Filters (and more recently factor graph optimization) methods, which smartly combine both visual and inertial data streams to generate robust state estimates. These algorithms take into account the sensor noise properties, temporary visual errors when the cameras are covered or when they observe terrain with no features and correct the accuracy even when the manoeuvers are dynamic. The continuously updating fusion process compares predicted motion provided by the inertial sensors to observed motion provided by the visual tracking, and therefore gives more accurate estimate compared to the predictions made by the inertial sensors themselves.

12.4.3 Environmental Adaptations

The space environment comes with peculiarities that demand special adoptions in order to make VIO systems to operate effectively. The variations in lighting are extreme since the planetary surfaces undergo extreme variations when vehicles travel through shadowed and bright illuminated areas, and ranges of differences are over six orders of magnitude. Contemporary VIO systems also add adaptive exposure control algorithms [8, 9], which automatically change camera settings to ensure that at these transitions the imagery remains usable, and high dynamic range techniques of image create richer detail in dark and bright areas at the same time. The aura and obscuration caused by the Martian dust storms and lunar regoliths is a constant problem to camera lenses and can seriously impair the performance of visual tracking. The unnecessary arrangement of the sensors, having cameras in various positions and direction will be used to ensure that at least some sensors have clear views that do not suffer any interference as the others are not completely visible. Automated clean systems such as protective covers, gas jets and vibration devices serve to keep the optical cleanliness of the systems during the mission which may last through time. Hundreds of degrees Celsius of temperature change have an influence on sensor operation and mechanical stability of mounting structures. The thermal compensation algorithms are continuously aimed at keeping temperature dependent sensor biases and drift and advanced environmental control systems operate heaters, radiators, and insulation to ensure that sensors operate within operational temperature limits. Parameters needed to represent camera intrinsics, position of sensors relative to space should be preserved during thermal cycling, and thus mechanical design and regular recalibration steps have to be performed.

12.4.4 Machine Learning for Terrain Classification

Artificial intelligence has caused a revolution in the way robotic vehicles perceive and communicate with planetary terrain, allowing superhuman accuracy in the analysis of millions of pictures. With the current systems, it is possible to classify surface materials, predict traversability, and detect scientific targets with superhuman uniformity and speed. Algorithms used in machine learning when trained on massive data sets are more likely to spot patterns that cannot be spotted by human eye, and they are able to spot minor deviations in textures that identify that this could be a different type of rock or geological landmarks of geological interest. These AI systems [10] can be resolved an order of magnitude more quickly than humans and allow assessing the terrain in real time when operating different rovers. The technology has transformed into more basic rule-based classifiers to more advanced deep learning networks that can have a fine sense of comprehending complex geological scenes and has radically altered the way planetary robots see and engage with their worlds.

I. Deep learning architecture

Modern ways of classifying the terrain [17–20] are based on complex neural network structures that are tailored to process planetary images. Convolutional neural networks are the core of their systems, as the majority of them operate based on the multi-layer CNNs, analyzing high-resolution images to determine the type of rocks, soil type, and surface textures with impressive precision. These networks scan various spectral bands concurrent with visible light, near-infrared, and thermal data retrieving finer geological data that would have taken up much human work to accomplish. The contemporary CNNs are capable of recognizing dozens of individual terrain classes (rock types, sand dunes, bedrock exposures, and even science-related items such as the hydrated minerals, etc.). Recurrent neural networks allow sequential analysis of data with RNN models that use time-dependent input data to enhance classification, by considering dynamic content in either the changing nature of terrain features or after changing viewpoint. Such networks are able to follow the variations of lighting, viewing geometry, and changes in season on the appearance of the terrain, training to perceive identical things with different conditions. Transformer models are the state of the art, and updated transformer-based architectures with attention demonstrating their recent progress and improved results in more complicated scene understanding exercises. The models are able to study relations among far flung sections of pictures and they comprehend the context of images and spatial correspondence, which are absent in simpler structures. Transformers are also very efficient at the detection of very subtle patterns and anomalies that could potentially represent a scientifically valuable target making the analysis of the terrain and detection of hazards more subtle than before.

II. Training and adaptation

The application of machine learning systems to planetary missions needs to perform due diligence on the methods of training so that the products can depend on the application in the unheard-of conditions. High fidelity simulations of the environment on the surface of a planet generate synthetic data that provides training datasets with a wide diversity of conditions the robots could face without it being on a planet. The photo realistic images are formed as the result of advanced rendering methods and models of physics-based lighting and elaborate terrain databases when the conditions of light and atmospheric effects and the geometries of the view are varied. Such artificial datasets may contain edge cases, which would otherwise be hard to test to with real missions making sure algorithms are able to deal with edge cases. Transfer learning can utilize the models that are already trained on the terrestrial datasets with backgrounds of millions of labeled images on the Earth by means of geological survey and the analysis of the terrain. Such models acquire more generic concepts of features and textures on the terrain and space relationships, which are generalized across planetary bodies and are then refined with limited planetary information to Mars-specific or Moon-specific properties, such as various mineralogy, atmospheric variations, and gravitational influences. This method leads to a very large reduction in the quantum of data needed on the planetary training. There are also online learning capabilities in some advanced systems, which allow the models to update themselves on new data they have occurred during missions, and further increase their classification accuracy as additional planetary data is made available. Such systems are able to recognize when they meet terrain types outside the well represented training information and put the uncertain classifications underground review as the system learns on any corrections made by the operator to perform better in future.

III. Performance validation

To test machine learning systems in a space environment, the testing procedures must be strict and reliable so as to validate the reliability of the effective implementation of such mission in a poor less environment. The high-fidelity virtual worlds offered by simulation environments replicate the entirety of operating conditions anticipated, and thus allows a great deal of testing of AI systems prior to deployment, without jeopardizing a costly physical system. Such simulations simulate lighting changes, dust storms, equipment failures and other conditions so that programmers can check system behavior under thousands of conditions. The use of terrestrial analog, such as desert habitats, polar regions, and volcanic landscape on Earth as test grounds to be able to predict conditions on other planetary bodies is validated with real hardware and sensors. Field tests in places, such as the Atacama Desert in Chile, dry valleys in the Antarctic, and the Iceland lava fields, are exposing algorithms to the geographical features, photo weather, and power challenges analogous to Mars or the moon. The methods of cross-validation use many independent AI systems that analyze the same data to estimate confidences and to identify possible errors of the automated classification. Where various algorithms concur in their classification, one has a lot of confidence; where there is disagreement, this serves as a good sign that there is more to look at. Statistical validation measures do have accuracy, precision and recall, and other performance measures that are also quantified to ensure that the systems are weighed against the high standards of reliability.

12.4.5 Simultaneous Localization and Mapping

The simultaneous localization and mapping (SLAM) technology allows robotic vehicles to create maps [22] of the unknown spaces and simultaneously provide information of their position inside of the map representing a major chicken and egg dilemma in robot navigation. This is necessary in order to be able to have autonomous exploration of planetary surfaces in cases where existing maps might be incomplete, too old to be used in the light of changes that occur on the surface or may not be fine enough to navigate safely. SLAM systems create increasingly robust and accurate maps until the robot has covered the area, with measured landmarks used to correct the cumulative errors in positioning, as well as resolving the ability to go back to previously passed locations. The technology has developed to the level of a dramatic point in the last decade in that it has ceased to be a computationally intensive research algorithm but one that can actually be run in real time on space-qualified processors.

I. SLAM implementations in the modern world

Recent SLAM systems to investigate the planets have multiple advanced functions that allow firm operation in adverse conditions. The camera-based systems that are used to develop visual maps of the environment in detail and monitor the position of the vehicle are used in visual SLAM, which may be conducted in real-time and gain highly detailed information about the environment to be used in path planning and identify scientific targets. These systems recognize and trace visual features that are unique, such as rocks, crater rims, and geological boundaries forming maps, that describe correlation of three dimensional locations and visual descriptors. Visual SLAM works well in feature-rich environments and extreme light, but fails in the featureless ones or at the very bright ones. LIDAR SLAM uses the laser-based ranging systems which offer the most accurate distance readings that can be used to map the terrain features with a lot of accuracy with a range of more than 100 m with an accuracy of a few centimeters. LIDAR generates geometric maps indicating the shape and the position of obstacles of the terrain as exceptionally precise and not subject to light conditions and visual texture. The systems do not malfunction in the dark, dust or fog which would fog cameras. Multimodal SLAM is the most advanced method incorporating the visual, LIDAR, and the inertial data to develop strong maps that would be valid even when each of the sensors experiences poor functioning. Fusion algorithms are smart enough to assign weights to various sensor data based on the present-day reliability of the sensor and smoothly switching between sensor modalities as conditions alleviate and combining the strengths that each mode of sensing has in lieu of one another.

II. Loop closure detection

The detection of when a vehicle is back to a location previously explored is one of the most important problems in SLAM because it will be able to rectify cumulative drift errors. Visual vocabulary techniques hold the databases of independent visual features that make it possible to recognize the previously visited places even when the conditions of viewing have altered greatly based on the angle of the sun, change of season, or the amount of dust that has settled. These are systems that utilize bag of words representations of images wherein images are defined over the visual features present in them so that they can be easily searched in big databases of maps to get matches. Geometric verification is further confirmation that spaces of any feature are as originally expected at the location that is being theorized. When a robot believes that it has reached a former location, geometry consistency tests are used to ensure that the noticed features are present in their proper relative locations, which filters out all false positive detection, which might corrupt maps. Global optimization algorithms engage in global closures and ensure global consistency of the entire map by adjusting all the loops and optimizing them to correct cumulative drift errors. These optimization algorithms spread out corrections to the map systematically adjusting the estimated locations of all previous features that were already mapped to make the best possible fit. The concept of successful loop closure allows enhancing map accuracy by several aspects, fixing up the errors that compounded during a long trip.

III. Multi-sensor fusion technologies

The combination of data obtained through the use of more than one sensor gives it strength and precision that is not possible through any one sensor system and produces navigation possibilities that are more than the aggregate limitations of the individual sensors. State-of-the-art planetary space robots [16] employ advanced fusion designs, which integrate dozens of diverse sensors, ranging between wheel odometry and star trackers, hazard cameras, and spectrometers. Sensor fusion provides solutions to some of the fundamental shortcomings of any of its sensors: cameras cannot work in the dark, LIDAR is subject to dust, and inertial sensors tend to drift and provide inaccurate state estimations. State-of-the-art fusion algorithms consider various sensor properties, such as update rates, accuracy, reliability and failure modes, and combines the contribution of each sensor optimally through the present operating conditions.

IV. Sensor architectures

Modern multi-sensor devices make use of a variety of sensing modalities that give their full awareness of their environment and the state of the vehicle. Proprioceptive sensors measure the internal state of the vehicle, such as wheel odometry, to determine distance covered by, for example, rotating the wheel and joint position to determine manipulator arm positions, motor current to detect heavy loads, and health monitors to measure temperatures, voltages, operating state. These sensors give a continuous feedback on vehicle condition and activities. Exteroceptive sensors monitor [23] the surroundings, such as more than one camera system to give visual control in various views, LIDAR and radar, which detects the distance and the topography of the terrain, spectrometers to study the surface structure, and detectors of radars to check the space climate. These sensors inquire data on the environment that can be used to enable navigation, obstacle evasion, and scientific exploration. Environment sensors detect the condition of the environment, temperature, pressure, humidity, wind speed, dust density, radiation, and other factors related to vehicle performance and scientific processes. The data can be used in making operational decisions, such as when to drive, when to seek shelter due to dust storms, and the best time to carry out scientific observations. All this is as a result of proprioceptors, exteroceptors, and environmental sensors that allow full situational awareness so that they can operate intelligently and autonomously.

V. Fusion algorithms

The advanced algorithms combine data of multiple sensors with advanced mathematical models. Bayesian Networks utilize the probabilistic model of uncertainty in sensor values and offer principled generalization of combining information across a variety of sources, and consideration of the sensor noise properties and reliability. These networks spread the indicators of uncertainty, as they do not only give the state approximations but also disclose confidence that would refer to reliability. Particle filters is a Monte Carlo algorithm to pursue a set of hypothesis regarding the condition of the vehicle and the environment to give strong estimation in Gabling scenarios that simple models cannot perform. Representing probabilities, these filters are sets of samples or particles making it possible to represent complex and multimodal distributions. Particle filters are best used when one can have several meaningful interpretations of sensor data which are kept alive until more information is available to disambiguate the situation. One of the measures is the selection of sensors, and data fusion strategies quantifiable as information content is done through information theory measures to optimize the limited computational resources. The methods are focused on sensors that present the most information in the presence of existing uncertainty, and the intelligent provision of processing capabilities in order to maximize state estimation accuracy without violating maximum computation limits. Entropy-based mechanisms determine which of the available sensors to query, and how often, based on the tradeoff between the information accumulated and power usage or processing resources.

VI. Fault tolerance

The space missions demand high levels of reliability which propels the evolution of fault tolerant sensor systems that assure that they remain operational even after the failure of parts. Sensor redundancy offers essential capabilities to have multiple independent sensors that supply important functions in case a number of sensors fail due to attaining old age, radiation effects, or other mechanical issues. The most common examples of redundant configurations [21] are several identically different sensors in different locations and varied types of sensor with different physical principles measuring similar quantities. To ensure that sensor delivery is in good shape, the aberration detection algorithms ensure that sensor performance is constantly checked through machine learning to detect deterioration or fault before it influences the operations of a mission. These systems adapt patterns of normal sensor behavior in normal operation, and use these patterns to identify deviations that indicates problems such as drift, increase of noise or systematic biases. Early warnings can be used to switch to backup sensors or in other instances, adjust the fusion algorithms to offset such impaired performance. The graceful degradation design approach will ensure that the systems have the fundamental functionality, even in the case in which they are under reduced sensor capacities. Should cameras fail, LIDAR and inertial will provide entry-level navigation: should LIDAR fail, visual and inertial system systems will still operate. The implementation methods of mission-critical functions possess numerous independent channels, which require that failure of a sensor will not affect the success of the mission.

12.5 Secure Data Transmission and Storage

12.5.1 Edge Computing for Real-Time Decision Making

The application of the massive computing capabilities on planetary robots has allowed autonomous decision making, which has changed missions into being remotely controlled missions to intelligent missions. In modern space missions, edge computing system is integrated whereby intricate algorithms are computed locally without necessarily relying on Earth analysis, but instead responding to opportunities and risks instantly billions of miles away.

12.5.2 Hardware Architectures

In the modern planetary robots [25], advanced computing architectures that maintain a balance between processing power, coupled with rigid mass, volume and power limitations, and radiation limitations have been adopted. Radiation-hardened processors are based on error correcting memory and radiation resistant designs to attain billions of operations per second at low power consumption. GPU acceleration is able to do parallel processing of machine learning inference and computer vision, making trillions of computations per second. The implementation with FPGA offers power-efficient resource reconfigurability resources that are fine tuned to do specific tasks and can be reprogrammed mid-mission. Neuromorphic computing systems are designed to replicate the biology of neural systems meaning that future applications of this technology will provide AI applications with ultra-low power requirements.

12.5.3 Real-Time Processing

Edge computing facilitates milliseconds decisions on diverse essential functions. Computer vision algorithms detect objects at camera frame rates (over 30 fps) to process hazards, scientific targets, and navigation objects such that they can be used on a continuous autonomous basis. The predictive analytics are used to track hundreds of parameters of the system and identify cases of the nuance of anomaly before failures do happen. Independent science systems analyze data in the form they are being gathered and automatically modify instruments and observation plans to optimize scientific reward when the instructions are not relayed to the Earth.

12.5.4 Thermal Control and Power Management

Running of the powerful computers in space needs advanced management systems. Dynamic power scaling is a scaling system that scales the performance of the computational unit to the available energy and thermal limits, reducing speeds or going into low-power states when required. Thermal-aware computing allocates the computational load among the processors to avoid over-heat as it can only be lost to space *via* radiation as well as maximizing the throughput. Power limited missions' thermoelectric generators are used to recover waste heat in energy harvesting technologies additional energy enhancement in the systems.

12.5.5 Autonomous Science Operations

The combination of AI and scientific devices yields systems that theorize, devise experiments, and interpret findings without human participation, acting instantly on the discoveries that do not need days on Earth to get a command. These systems integrate scientific knowledge in the form of rules in addition to machine learning model, thus allowing robotic explorers to reason scientifically and make hundreds of investigations each day.

12.5.6 Intelligent Target Selection

Technologically advanced robots are able to recognize and prioritize scientific goals with advanced AI. A spectral scan at dozens of wavelengths not only identifies minerals and atmospheric effects but also successively identifies potentially biosignificant phenomena at the level of an expert. Anomaly detection algorithms ramify atypical attributes that should be looked into by comparing observations and what is expected to happen. Multi-objective optimization trades the science, accessibility, resource utilization, and even mission requirements to come up with the best observation plans based on the limited budgets with the greatest scientific output.

12.5.7 Adaptative Experimental Design

The modification of protocols by AI systems, in turn, adapts the scientific method to the initial outcomes independently. Bayesian optimization determines parameters that give informative results most likely to increase the returns of the limited measurements. Sequential analysis modulates later measurements using initial data to emphasize eventualities, which may be productive, such as the time to add more integration when identifying organic compounds. Hypothesis testing facilitates the independent formulation and testing of theories, the development of measurements to discriminate among rival explanations.

12.5.8 Data Quality Optimization

With automated systems, maximum collection in different conditions is made possible. Adaptive sampling varies rates, integration times and parameter depending on signal properties increasing integration of weak signals and sampling rates of dynamics. Advanced signal processing is used to extract weak signals of noisy data using noise reduction algorithms.

Calibration management keeps track of the performance integrating data on drift and automatically applies the necessary corrective actions to ensure the accuracy over a long mission.

12.5.9 Predictive Maintenance Systems

The AI-based maintenance systems identify issues in advance of failures and plan checks to achieve maximum missions since it is impossible to do any physical maintenance in space. Such systems are constantly observing the health of the spacecrafts, anticipating failure of certain parts and therefore responding before the component fails instead of responding to failure.

12.5.10 Health Monitoring

Rigorous monitoring is done on key subsystems. Vibration analysis identifies bearings wear, imbalance, and mechanical issues using machine learning algorithms that are abnormally extracted signatures. Thermal monitoring involves the use of thermal models together with an endless reading of the condition of the thermal to reveal any anomaly that signifies insulation, electronics, and mechanical issues. Analysis of the electrical system monitors power generated, power stored, and distributed detecting degradation of solar panels, battery capacity loss and power electronics problems so that compensating adjustments can be made.

12.5.11 Failure Prediction

AI makes predictions of the probability of failures. Machines based on prognostic models that have been trained on historical failure data to predict any remaining useful life, with associated confidence intervals, using a history of usage, exposure to the environment, and performance measures. The progression tracking of anomalies tracks the development of problems over time, and forecast when minor problems can turn into major failures. Integration of environmental factors considers the temperature cycles, radiations and mechanical loads, and understands that the components in severe environments wear out sooner than those in mild environments.

12.5.12 Maintenance Optimization

The maintenance activities are optimized by automated systems. Resource-constrained scheduling trades between maintenance and scientific goals and resource availability to produce schedules that produce the most total mission value. Preventive and corrective trade-offs analyze the possible options of conducting preventive maintenance or waiting to fail, depending on the probability of failure, complexity of repair, and potential effects on the mission. Spare parts management perform the optimization of allocation to maximize the mission lifetime, as well as minimizing mass and volume requirement.

12.5.13 Deep Learning of Image Analysis

Deep learning has transformed planetary science in a way that allows it to be analyzed faster and at scale efficiency that was never achievable by humans. Contemporary AI recognizes images better than humans, detecting smaller details and methodically using learned patterns in millions of images. This human and machine sameness is an incredible enhancement of scientific productivity.

12.5.14 Advanced Neural Architecture

Modern systems use the latest model that will have an optimized design and help in the imaging of the planet. Transformer architecture attention mechanisms emphasize computational resources on parts of an image they consider relevant and enhance accuracy and efficiency whilst offering interpretability. Multi-scale processing using hierarchical networks compares images at many resolutions of the image simultaneously and identifies the small and large-scale patterns. Domain adaptation applies transfer learning to specialize planetary networks trained on Earth images by using millions of labeled pictures on Earth, but only thousands of pictures on the planet.

12.5.15 Scientific Applications

The analysis which is AI-based makes new possibilities. Rock types, mineral compositions, and rock structures are identified by the geological classification, which is expert in identifying one out of thousands of images they have to work on every day to map them in detail. The detection of atmospheric phenomena makes it possible to track real-time weather patterns, dust storms, and dynamics and use the information as important science and mission planning. The image difference technique is applied to detect changes over time to spot seasonal change, erosion, and dynamic changes such as changing gullies, moving dunes and new impact craters.

12.5.16 Quality Assurance and Validation

To be reliable, there has to be advanced validation. Quantification of uncertainty gives confidence measures that allow the measurement of automated results—systems give the percentage of confidence that allows scientists to concentrate on uncertain cases. The human–AI interaction, a combination of efficiency and expertise in human–AI interaction, can identify the ambiguous processes, which can be verified by the expertise, and routine processes, which are executed with the help of analytical machinery. Cross-validation involves use of many independent systems that analyze same images and the agreement of the systems strengthens confidence and where there is disagreement it specifies the problem. Single algorithm, ensemble methods, co-occurring methods: A combination of two or more methods, although it is not always the case, tend to be more effective than individual algorithms.

With the incorporation of these AI technologies, planetary robotics have become literally free to make their own scientific findings. With the further development of technologies, future missions will be even more scientifically independent, and the speed and breadth of the space exploration will boost greatly [24].

12.6 Conclusion

This chapter encapsulates an essential revolution in the exploration of the planets concerning the remotely controlled machines up to an ordinary autonomous intelligent system that possesses the ability to conduct a scientific discovery on its own. This is a revolution, which has taken place in the past 5 years and is one of the greatest developments in space exploration history.

12.6.1 The Transformation

Current planetary robots are not only advanced remote-control vehicles anymore but also advanced systems with edge computing, machine learning, and advanced sensors. Minute- or hour-long communication delays turn autonomy into one of the operational requirements and not only a desire anymore. The use of systems such as Perseverance Enhanced AutoNav, which allows autonomous terrain analysis, path planning, execution, as well as thinking while driving make a huge difference in the operational efficiency.

The flight of ingenuity on Mars is an indication of this new era. All flights are fully autonomous, the helicopter senses the surrounding environment, it keeps airborne and finds the way without human control in some of the critical situations. This was made possible by the dramatic developments in lightweight structures, self-control, and heat dissipation.

The AI is intertwined with making discoveries by scientific tools on Perseverance MASTCAM-Z, MEDA, PIXL, and SuperCam. These systems automatically detect targets, optimize observation and follow-up studies without commands on Earth and make significant contributions to scientific productivity with a number of machine learning-enabled systems.

12.6.2 Global Expansion

The world space exploration of the moon has encountered a new space age renaissance. The Artemis program incorporates the Gateway station to VIPER polar prospecting advanced robotics. As proven in Commercial Lunar Payload Services, commercial relationships drastically lower the prices of missions and improve the frequency thereof. The Chang Yu-5 Indian Chandrayaan-3 south pole landing and the Chang Yu-5 autonomous rendezvous and docking accomplishments demonstrate that sophisticated exploration is now universally available with the help of innovativeness in the engineering application.

12.6.3 Technical Foundations

Under VIO systems have a positioning error of less than 1% of the distance covered. Terrain classification in the style of deep learning with CNNs, RNNs, and transformers can achieve superhuman consistency, which is facilitated by machine learning. SLAM technology builds rich maps as positions are tracked and the multi-modal approach involves the use of visual, LIDAR, and inertial measurements. Multi-sensor fusion provides a combination of dozens of sensors using Bayesian networks and particle filters, so that it can be successfully operated in even cases of component failures.

12.6.4 AI as Enabler

Edge computing takes serious processors, GPUs, and FPGAs into space, which allows real-time decision making. Independent science functions enable the autonomous robots to come up with hypotheses, experiment design, and results analysis. Predictive maintenance is a system that keeps track of the health of spacecraft and is used to identify issues prior to the failure process by analyzing vibration, thermal sensors, and prognostic models. Deep learning is also an analysis of planetary images with expert, eye-level accuracy, and it can process thousands of images a day to classify geologically and identify changes.

12.6.5 Looking Forward

These are the working technologies that have demonstrated their value in the current missions. Perseverance is an autonomous mission studying Jezero Crater. Ingenuity is flying across the skies of Mars, and rovers studying the moon around the world improving our insights. The autonomous revolution has demonstrated the capability of robots going out on their own and discovering things without the continuous supervision of human beings. It is no longer a question of whether or not it is possible to autonomously explore but how far we will get by discovering more about our solar system and our position in the universe through these capabilities.

AI disclosure statement: The author confirms that no artificial intelligence (AI)-generated content or AI-assisted writing tools were used in the preparation of this chapter (including summaries and the abstract). All text, analysis, and conclusions are the result of the author's original work.

References

1. Farley, K.A., *et al.*, Mars 2020 mission overview. *Space Sci. Rev.*, 216, 8, 1–41, Dec. 2020, doi: 10.1007/s11214-020-00762-y.
2. Maki, J., *et al.*, The Mars 2020 engineering cameras and Microphone on the Perseverance Rover: A next-generation imaging system for Mars exploration. *Space Sci. Rev.*, 216, 8, 1–48, Dec. 2020, doi: 10.1007/s11214-020-00765-9.
3. Maurice, W., *et al.*, The SuperCam instrument suite on the Mars 2020 rover: Science objectives and mast-unit description. *Space Sci. Rev.*, 217, 3, 1–89, Apr. 2021, doi: 10.1007/s11214–021-00807-w.
4. Bell, J., *et al.*, The Mars 2020 Perseverance Rover Mast Camera Zoom (Mastcam-Z) multispectral, stereoscopic imaging investigation. *Space Sci. Rev.*, 217, 1, 1–61, Feb. 2021, doi: 10.1007/s11214-020-00755-x.
5. Hecht, M., *et al.*, Mars Oxygen ISRU Experiment (MOXIE). *Space Sci. Rev.*, 217, 1, 1–73, Feb. 2021, doi: 10.1007/s11214–020-00782–8.
6. Tao, Y., *et al.*, Terrain-relative navigation for Mars landing using landmark tracking. *J. Guid. Control Dyn.*, 44, 6, 1119–1132, Jun. 2021, doi: 10.2514/1.G005674.

7. Balaram, T., *et al.*, Ingenuity Mars Helicopter: From technology demonstration to extraterrestrial scout, in: *2021 IEEE Aerospace Conference*, Big Sky, MT, USA, pp. 1–19, 2021, doi: 10.1109/AERO50100.2021.9438165.
8. Gao, G., *et al.*, Deep learning for autonomous planetary exploration. *IEEE Trans. Aerosp. Electron. Syst.*, 57, 5, 3123–3138, Oct. 2021, doi: 10.1109/TAES.2021.3074139.
9. Parra, L., *et al.*, Artificial intelligence for autonomous space systems: Challenges and opportunities. *IEEE Aerosp. Electron. Syst. Mag.*, 36, 8, 24–39, Aug. 2021, doi: 10.1109/MAES.2021.3072407.
10. Colagrossi, A., *et al.*, CubeSat formation flying: Enabling technologies and mission concepts. *Adv. Space Res.*, 66, 6, 1291–1315, Sep. 2020, doi: 10.1016/j.asr.2020.05.035.
11. Li, C., *et al.*, The Chang'e 5 mission: Sample return from a young mare on the Moon. *Natl. Sci. Rev.*, 9, 2, 1–14, Feb. 2022, doi: 10.1093/nsr/nwab188.
12. Takir, D., *et al.*, VIPER rover and the volatiles investigating polar exploration rover mission, in: *2020 IEEE Aerospace Conference*, Big Sky, MT, USA, pp. 1–10, 2020, doi: 10.1109/AERO47225.2020.9172323.
13. Rankin, A., *et al.*, Driving curiosity: Mars rover mobility trends during the first seven years, in: *2020 IEEE Aerospace Conference*, Big Sky, MT, USA, pp. 1–14, 2020, doi: 10.1109/AERO47225.2020.9172469.
14. Golombek, M., *et al.*, Selection of the Mars 2020 landing site. *Space Sci. Rev.*, 216, 1, 1–60, Jan. 2020, doi: 10.1007/s11214-019-0624-7.
15. Soman, S., *et al.*, Chandrayaan-3: India's lunar landing mission. *Acta Astronaut.*, 212, 45–58, Nov. 2023, doi: 10.1016/j.actaastro.2023.07.028.
16. Zhang, J., *et al.*, Autonomous visual-inertial odometry and mapping for planetary rovers. *IEEE Trans. Rob.*, 38, 4, 2456–2473, Aug. 2022, doi: 10.1109/TRO.2021.3139619.
17. Otsu, K., *et al.*, Supervised autonomous driving for planetary rovers using machine learning, in: *2020 IEEE International Conference on Robotics and Automation (ICRA)*, Paris, France, pp. 9386–9392, 2020, doi: 10.1109/ICRA40945.2020.9197083.
18. Hewitt, R., *et al.*, Deep learning for terrain classification and mapping from planetary imagery. *IEEE Rob. Autom. Lett.*, 6, 2, 2958–2965, Apr. 2021, doi: 10.1109/LRA.2021.3062360.
19. Ono, M., *et al.*, MAARS: Machine learning-based analytics for automated rover systems, in: *2022 IEEE Aerospace Conference*, Big Sky, MT, USA, pp. 1–12, 2022, doi: 10.1109/AERO53065.2022.9843379.
20. Viswanathan, P., *et al.*, Multi-robot coordination for planetary exploration: State-of-the-art and future directions. *Rob. Auton. Syst.*, 145, 103865, Nov. 2021, doi: 10.1016/j.robot.2021.103865.
21. Burleigh, K., *et al.*, Delay-tolerant networking for deep space communications, in: *2021 IEEE Aerospace Conference*, Big Sky, MT, USA, pp. 1–11, 2021, doi: 10.1109/AERO50100.2021.9438523.

22. Nesnas, T., *et al.*, Autonomous exploration and mapping for planetary surface missions. *J. Field. Rob.*, 39, 5, 651–679, Aug. 2022, doi: 10.1002/rob.22073.
23. Wettergreen, D., *et al.*, Robotic sample acquisition and caching for Mars sample return. *Acta Astronaut.*, 186, 405–420, Sep. 2021, doi: 10.1016/j.actaastro.2021.06.008.
24. Chien, S., *et al.*, Onboard autonomy for Mars 2020 Perseverance rover operations. *J. Spacecr. Rockets*, 59, 3, 1046–1062, May 2022, doi: 10.2514/1.A35168.
25. Abdi, H., *et al.*, Edge computing and AI for real-time decision making in space robotics. *IEEE Access*, 11, 28456–28474, Mar. 2023, doi: 10.1109/ACCESS.2023.3259187.

13

Flexible, Printed, and Wearable Electronics for Space Applications

V. Yokesh[1]*, B.P. Bhuvana[1], C.P. Koushik[2] and Nouhaila El Koufi[3]

[1]Department of ECE, Sathyabama Institute of Science and Technology, Chennai, India
[2]School of Computing Science Engineering and Artificial Intelligence, VIT Bhopal University, Bhopal, India
[3]Department of Computer Science, Hassan First University, Settat, Morocco

Abstract

Recent advances in flexible, printed, and wearable electronics are poised to transform space technology by offering lighter, more adaptable, and more energy-efficient alternatives to traditional rigid hardware. This chapter explores how these emerging systems can strengthen spacecraft functions, support continuous astronaut health monitoring, detect structural problems, and make better use of available resources. It highlights the major benefits of additive and printed manufacturing—such as reduced material waste, compatibility with curved or irregular surfaces, and on-demand production in microgravity. Wearable technology equipped with multifunctional sensors can continuously track astronauts' health and safety, enhancing their well-being throughout space missions. Similarly, adaptable sensor networks can be integrated into the surfaces of spacecraft and within living areas to detect environmental fluctuations and recognize potential structural problems at an early stage. This chapter emphasizes critical material concerns such as radiation protection, heat resistance, and the long-term durability of materials in demanding space environments. Utilizing recent studies and real-world examples, it examines the latest advancements, challenges, and emerging opportunities in flexible and wearable electronic systems designed for space missions. Ultimately, these technologies are positioned to support more sustainable, human-centered space exploration.

**Corresponding author*: yokesh.inba@gmail.com

N. Sathish, Prasanth Aruchamy, Rajesh Kumar Dhanaraj, Nithya Rekha and Karthik Palani (eds.) Space Electronics for Deep Space Exploration, (293–326) © 2026 Scrivener Publishing LLC

Keywords: Flexible electronics, printed electronics, wearable technology, space mission monitoring, human spaceflight systems

13.1 Flexible & Wearable Space Electronics

Space applications could be revolutionized by recent developments in flexible, printed, and wearable electronics, which provide lightweight, versatile, and energy-efficient substitutes for conventional rigid systems. The use of these new technologies in space is discussed in this chapter, with particular attention to how they will enhance spacecraft systems, monitor astronaut health, identify structural issues, and make use of existing resources. Some of the primary advantages of additive and printed manufacturing are discussed in this article, including the fact that minimal material is wasted, that it can be utilized on non-flat surfaces, and that it can be completed on demand in microgravity. By enabling small, flexible, and multipurpose electronic platforms, these technologies have the potential to completely transform the architecture and operation of next-generation satellites, spacecraft, and astronaut systems.

13.1.1 Evolution of Space Electronics

Early satellite missions in the 1950s and 1960s, which depended on large, heavy, and power-hungry components, are where the history of space electronics begins. Developments in material science, microelectronics, and miniaturization over the years have produced radiation-hardened and lightweight systems.

However, new possibilities for integrating intelligence into unusual surfaces, like spacecraft exteriors, astronaut clothing, and deployable structures, have been made possible by recent advancements in flexible substrates, nanomaterials, and additive manufacturing. A paradigm shift away from traditional circuit-board-based architectures is represented by soft, stretchable, and reconfigurable electronic systems intended for space use.

13.1.2 Limitations of Conventional Rigid Systems

Traditional rigid electronic systems are dependable, but they have a number of disadvantages in space environments:

- ***Mass and Volume Constraints:*** Rigid boards impact launch costs and efficiency because they limit design flexibility and increase payload weight.

- ***Mechanical and Thermal Stress:*** Rigid interconnects may deteriorate or break as a result of abrupt temperature changes and mechanical vibrations.
- ***Integration Challenges:*** Conventional systems do not work well for embedding sensors and electronics on curved or irregular surfaces (such as helmets, suits, or inflatable modules).
- ***Restricted Wearability:*** Wearing big electronic devices inside astronaut suits causes problems. It makes it hard for astronauts to move freely, stay comfortable, and track their performance effectively over a long mission.

These challenges clearly show we need electronic devices that are lightweight, can do multiple things, and conform (or blend) into the suit's shape. This is crucial if we want them to work well in the constantly changing environment of a space mission.

13.1.3 Motivation for Flexible Wearable Tech

These difficulties really highlight that we must have electronics that are light, can do multiple tasks, and mold to the shape of the suit. This way, they can properly integrate and keep up with the continuously shifting demands of a space mission:

- ***Lightweight and Compact Design:*** It's easier and quicker to set up the system once it's in space (increases system deployability).
- ***Mechanical Flexibility and Conformability:*** They can be built right into curved or uneven areas, woven into the fabric of the suit (textiles), or even become a part of the structure or frame itself (structural components).
- ***Enhanced Reliability:*** When we use these new electronics, it helps spread out the physical stress that normally concentrates in one spot. This is important because it makes the equipment much tougher and longer-lasting, especially when it goes through big temperature changes (thermal cycling) and operates in the zero-gravity environment of space (microgravity).
- ***Multifunctional Integration:*** This capability is great because it allows us to put things like sensors (for monitoring),

communication tools, and power components (batteries or energy collectors) all together in one thin, printed layer.
- ***Real-Time Health Monitoring:*** Wearable gadgets can keep tabs on an astronaut's body while they are outside the spacecraft doing spacewalks (EVAs). Specifically, they can monitor their vital signs (like heart rate), their stress level, and how hydrated they are.

These kinds of technologies are really important because they effectively bridge the gap between how well a device can functionally adapt to new situations and how much intelligence is built into the material itself. This connection is absolutely vital for the success of future missions, both those involving humans and those using robots.

13.1.4 Objectives and Scope of the Chapter

This chapter is essentially a look at how we're developing, making, and using electronics that are printed, flexible, and wearable, specifically for space missions.

The main goals it covers are:

- Highlighting material breakthroughs: Looking at new materials that are both reliable and flexible enough to handle the harsh conditions of space.
- Discussing manufacturing methods: Exploring printing and fabrication techniques that can actually work well in a microgravity environment.
- Sharing real-world examples: Showing current ways these electronics are used, like collecting energy, monitoring spacecraft, and tracking astronaut health.
- Examining the practicalities: Studying the challenges, testing steps, and necessary requirements for actually sending these systems into space.
- Describing cutting-edge tech: Detailing the newest concepts for space exploration, such as self-managing sensor networks, "smart skins" for spacecraft, and integrating AI.
- Building the future foundation: By blending material science, electronics, and aerospace engineering, the chapter lays the groundwork for creating the next generation of electronic systems designed to thrive in space.

13.2 Materials for Space Flexible Electronics

Flexible, printed, and wearable electronic systems for space missions are highly dependent on material choice and tunings for truly effective performance in the harsh space environment. Due to the particular characteristics of space as indicated by high radiation intensity, temperature variations, vacuum, and also continuous mechanical vibrations. Thus, the selected materials must have the most crucial mechanical adaptability, high electrical abilities, and demonstrated environmental endurance at the same time [1, 2].

In this section, we shall study specifically four critical material constituents: substrates, conductors, dielectrics, and encapsulants. The longevity and integrity of the space-grade flexible electronics directly rely on these materials. The substrate is essentially the backbone of any multifunctional electronics platform. In terms of space applications, these substrates have rigorous specifications based on being resistant against atomic oxygen (AO) erosion, having little outgassing (the release of gases in a vacuum), and exceptional dimensional stability.

13.2.1 Substrate Materials

The substrate is the cornerstone of any multifunctional electronic device. Space substrates should have low outgassing, be dimensionally stable, and be impervious to atomic oxygen erosion.

- **Polyimides (e.g., Kapton®):** It is extensively used in spacecraft because of its exceptional mechanical flexibility, compatibility with vacuum, and thermal stability (up to 400°C).
- **Parylene Coatings:** They are suitable for biocompatible wearable applications and encapsulation because they have superior dielectric and moisture barrier properties.
- **Graphene and Carbon-Based Films:** Offer conductive, ultra-thin, and lightweight substitutes with the extra advantages of mechanical durability and radiation resistance.
- **Emerging Substrates:** Hybrid polymer composites and flexible ceramics are being investigated for EMI-immune and radiation-shielded designs.

These materials make it possible to integrate sensors and circuits on inflatable habitats, astronaut suits, and curved spacecraft surfaces.

13.2.2 Conductive Inks for Printed Circuits

Functional inks made of conductive, semiconductive, or dielectric materials are essential to printed electronics [3]. These inks must remain conductive and adherent under vacuum and heat stress in order to be used in space electronics.

- **Metallic Nanoparticle Inks (Ag, Cu, Au):** High electrical conductivity is provided, but surface stabilization is necessary to stop oxidation, particularly in systems that use copper.
- **Carbon Nanotubes (CNTs) and Graphene Inks:** Provide radiation-tolerant, flexible, and lightweight substitutes that work well with strain sensors and antennas.
- **Conductive Polymers (PEDOT: PSS, polyaniline):** Because they are stretchable and compatible with textiles, they are useful in wearable biosensors.
- **Hybrid Nanocomposites:** Combine metal and carbon fillers for greater flexibility, conductivity, and environmental endurance.

These inks allow interconnects, antennas, and sensing layers to be additively manufactured, eliminating the need for rigid wiring or heavy soldering.

13.2.3 Thermal Stability in Harsh Environments

The encapsulation must have long-term resistance to atomic oxygen, ultraviolet light, thermal cycling, and mechanical abrasion in flexible circuits.

- **Silicone-Based Encapsulants:** Provide wearable components the elasticity and UV protection they require.
- **Fluoropolymers (e.g., Teflon®):** The low friction and chemical inertness of fluoropolymers, such as Teflon®, make them perfect for deployable structures.

- **Ceramic Nanocomposite Coatings:** Enhance mechanical strength and radiation protection without compromising flexibility.
- **Thermal Management Materials:** To improve heat dissipation during high-power operations, phase-change materials (PCMs) and graphene-enhanced composites are combined.

Proper encapsulation guarantees device dependability and safety during prolonged orbital or terrestrial missions.

13.2.4 Radiation-Tolerant and Low-Outgassing Materials

When it comes to electronics in space, radiation poses a significant challenge. That is why selecting materials that can withstand these harsh conditions is so important. Here are some key options:

- **Radiation-Hardened Polymers**: Materials like polyimides and fluoropolymers are great choices because they maintain their electrical properties even after being exposed to cosmic radiation for long periods.
- **Carbon-Based Nanomaterials**: Carbon nanotubes and graphene are impressive because they offer excellent charge mobility and can resist damage from radiation, making them highly durable.
- **Low-Outgassing Adhesives and Coatings**: These materials are vital for protecting sensitive spacecraft sensors and optics from contamination in the vacuum of space.
- **Shielding Strategies**: To enhance radiation protection while keeping flexibility, layers made from metal oxides or boron nitride can be added to the design.

By incorporating these innovative materials, both manned and unmanned space missions can enjoy improved durability, safety, and adaptability. The result is a new generation of flexible electronics designed to thrive in the demanding environment of space, combining materials that are specifically engineered for functionality, stability, and resilience against the elements. From polyimide substrates to graphene conductors and radiation-resistant encapsulants, these innovations enable the realization of lightweight, resilient, and high-performance electronic systems suitable for the extreme demands of outer space.

13.3 Printing and Fabrication Techniques

Novel, accurate, and scalable procedures are required to develop lightweight but high-performance devices by fabricating flexible and printed and wearable electronics to be used in space [4]. In contrast to traditional microfabrication methods, which require photolithography and etching, printing methods permit additive, maskless, and material efficient microfabrication onto flexible or curved substrates.

When applied to space systems, such methods need to be able to provide sound adhesion, fine resolution, radiation survivability, as well as reliability under vacuum and thermal conditions. This section presents some of the most important printing and fabrication schemes, schemes of integration, and the problem of modifying such schemes to microgravity and extra-planetary space.

13.3.1 Inkjet & Aerosol Printing for Space Tech

The next generation of printed electronics is inkjet printing and aerosol jet printing (AJP) that provides a high level of design freedom and material efficiency.

- **Inkjet Printing:** Employs a controlled droplet ejection of functional inks on substrate. It is perfect in the creation of conductive traces, antenna, and thin-film sensors that have a resolution of approximately 20-50 um. It is a computer-controlled process that facilitates quick prototyping and *in situ* design adaptations which will be important in space missions where fabrication on demand is needed.
- **Aerosol Jet Printing (AJP):** Has much better resolution (down to 10 μm) and can work with a large range of inks, such as metallic nanoparticles, CNTs, and polymers. The capability of AJP to print on non-planar and 3D surfaces means that it is very sensitive to curved spacecraft modules or interfaces on spacesuits.
- **Space Adaptation:** The in-orbit manufacturing platforms can be miniaturized with inkjet and aerosol-based systems allowing electronic components to be repaired or customized in-orbit.

Both strategies reduce material waste, and no vacuum-based processing is used, which is compatible with sustainable and portable fabrication systems of the future space habitation.

13.3.2 Roll-to-Roll Additive Manufacturing

Roll-to-roll (R2R) fabrication enables continuous production of flexible electronics over large areas—an essential feature for scalable deployment in satellites, deployable solar arrays, or inflatable space modules.

- **R2R Printing:** Enables high-throughput multilayer deposition of conductive, dielectric, and encapsulation layers.
- **Integration with Additive Manufacturing:** Sensors and interconnects can be seamlessly integrated into lightweight polymeric or composite structures in hybrid systems created by combining R2R with additive manufacturing, or 3D printing.
- ***In Situ* Fabrication:** Reliance on Earth-based supply chains can be decreased by using additive methods for on-demand production on board spacecraft or in lunar habitats.

Automation, adaptability, and self-fabrication of spacecraft and intelligent surfaces that can alter their configuration in response to mission demands are enabled in such programs.

13.3.3 Integration with MEMS, Sensors, and Antennas

A versatile approach to create integrated (printed and flexible) electronics in conjunction with micro-electro-mechanical systems (MEMS) and supplementary sensor or communication modules has been an important element of a multifunctional and intelligent space platform. Such hybrid integration improves the system's overall performance, sensing accuracy, data communication efficiency while reducing weight and structural complexity and is therefore important for space borne applications.

- **Hybrid Integration:** Combines flexible interconnects with rigid ICs or MEMS dies using low-temperature bonding or anisotropic conductive films.
- **Embedded Sensors:** On a single flexible substrate, printed strain gauges, temperature sensors, and biosensors can be co-fabricated with power and communication modules.
- **Printed Antennas:** Lightweight, conformal antennas composed of graphene layers or conductive inks can be incorporated into spacecraft skins or astronaut suits to decrease bulk and improve communication efficiency.

This system-level integration method creates small, multifunctional platforms that can monitor, communicate, and make decisions in real time while in space.

13.3.4 Challenges in Microgravity-Compatible Fabrication

Engineering the incorporation of existing terrestrial fabrication procedures into microgravity environments has significant engineering challenges.

- **Material Flow and Deposition:** Fluid behavior drastically alters in zero gravity, influencing substrate wetting and ink droplet formation.
- **Curing and Sintering:** Laser or photonic sintering provide controlled, energy-efficient alternatives that are appropriate for in-space applications, as traditional thermal curing may not be as effective.
- **Contamination Control:** Closed-loop systems are necessary for the fabrication of printed electronics in spacecraft in order to stop nanoparticle dispersion in cabin environments.
- **Automation and Robotics:** Robotic arms and AI-powered vision systems will be essential components of future in-orbit manufacturing systems for tasks like alignment, inspection, and precision printing.

In order for autonomous on-demand fabrication systems to support and repair electronic infrastructure during extended missions, these issues must be resolved. Technologies of additive fabrication and printing [5] are both destroying the conventional definition of how electronics could be developed and implemented in space. The use of inkjet and aerosol jet, roll-to-roll printing, and hybrid integration with MEMS and sensors are new techniques that enable lightweight, customizable, and repairable systems, and increase the scope of their use. As they develop to work in microgravity, they promise to provide *in situ* manufacturing ecosystems—one of the enabling factors of sustainable space exploration and extraterrestrial colonization.

13.4 Wearable Electronics for Astronauts

In the space missions, continuous health surveillance of astronauts in long space missions is very essential to optimize the mission performance and to ensure the safety of the mission [6]. There are physiological and psychological stresses associated with space travel, which include muscle atrophy due to micro-gravity, radiation, cardiovascular, and thermal stress. Wearable electronics arise due to flexible and printed technologies, providing non-invasive real-time monitoring of vital parameters as biosensors on suits, in textiles or patches on the skin. One of these systems comprises sensing, wireless communication, and power modules to form autonomous biomedical monitoring networks which are capable of working in the harsh environment of space.

13.4.1 Physiological Parameter Tracking

While traditional rigid biomedical instruments are impractical in micro-gravity and constrained cabin environments, flexible, printed, and conformal sensors offer lightweight, low-power, and unobtrusive alternatives that seamlessly integrate into astronaut garments or skin-mounted patches, ensuring real-time bio signal acquisition, wireless transmission, and AI-assisted health assessment. These systems are the cornerstone of next-generation space medicine and are essential for maintaining astronaut health during long-duration space missions.

Numerous physiological parameters that can be tracked with flexible and conformal sensors are essential to modern astronaut health systems:

- **Electrocardiogram (ECG) Sensors:** Heart signals are recorded by printed graphene or Ag/AgCl electrodes incorporated into clothing or skin patches, facilitating the early identification of stress levels and arrhythmias.
- **Pulse Oximetry (SpO_2):** Blood oxygenation and perfusion are measured using flexible optoelectronic sensors that use printed LEDs and organic photodiodes.
- **Temperature Sensors:** To avoid overheating during extra-vehicular activities (EVAs), localized skin temperature mapping is provided by printed resistive thermistors or carbon nanotube-based sensors.
- **Electromyography (EMG) Sensors:** Integrated into suits to monitor muscle activity and fatigue, helping to control microgravity exercise.

These sensors' wireless interface with on-board health systems enables telemetric health analysis and AI-assisted diagnostics, as illustrated in Figure 13.1.

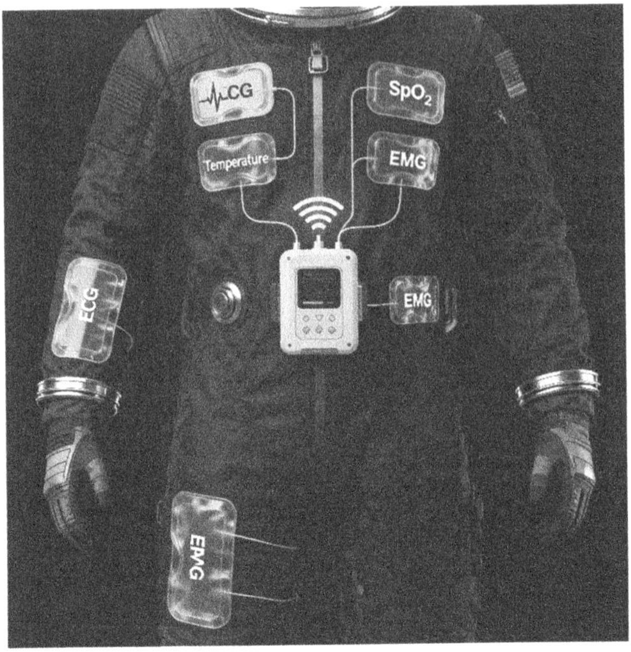

Figure 13.1 Conceptual illustration of wearable biosensors for astronauts.

13.4.2 Flexible Biosensors and E-Textiles for Space Suits

E-Textile technology forms the foundation of wearable sensing systems. Sensors and interconnects can bring connected conductive yarns, fibers, and fabrics into their smart textiles of multifunctional sensing, allowing conventional suits to become smart textiles. To get a comprehensive assessment of health, multimodal data streams must be integrated.

- **Conductive Textiles:** Silver-coated nylon, carbon-fiber yarns, and PEDOT: PSS coatings provide electrical pathways without compromising comfort or breathability.
- **Textile-Integrated Sensors:** Stretchable fiber sensors are crucial for ergonomics and fatigue management because they can identify changes in motion, respiration, and pressure.
- **Smart Suit Integration:** In multilayered astronaut suits, printed sensor arrays and e-yarns are seamlessly laminated to provide environmental and biomechanical monitoring.
- **Wireless Modules:** Real-time communication with spacecraft medical systems is made possible by textile antennas and NFC/BLE nodes.

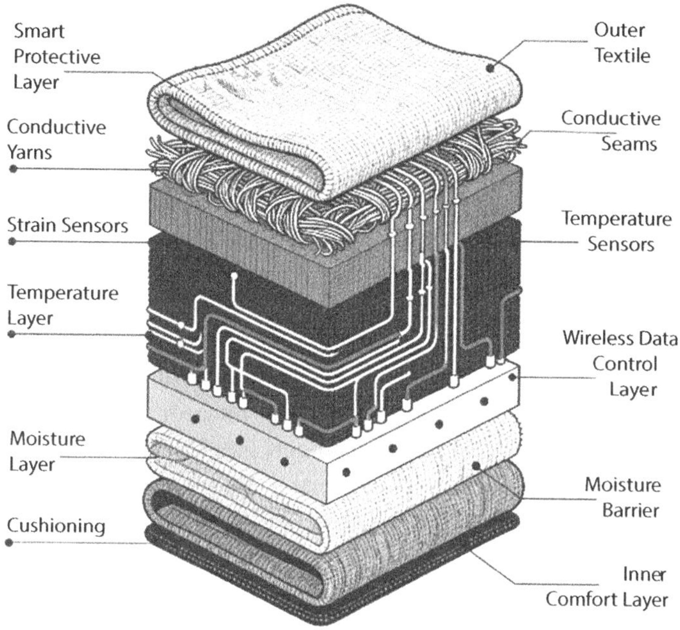

Figure 13.2 Cross-sectional view of a smart e-textile space suit.

Because of this integration, heavy biomedical harnesses, as seen in Figure 13.2, can be replaced with lightweight, discrete, and non-invasive physiological monitoring.

13.4.3 Real-Time Telemetry and Data Analytics

Advanced intra-body communication (IBC) protocols, Bluetooth Low Energy (BLE), or Near Field Communication (NFC) are used to wirelessly transfer data obtained from wearable physiological systems [7]. In the electromagnetically complex environment of a spacecraft, these wireless frameworks guarantee transmission that is low-power, short-range, and resistant to interference. After being gathered, the biosignals are processed by AI-driven analytics platforms, neuromorphic processors, or embedded microcontrollers, allowing for on-board health assessment without exclusively depending on ground communication.

- **Edge computing:** Using edge computing techniques, the wearable module itself can extract features and process signals locally. Even with limited connectivity to mission control, this distributed computation guarantees continuous operation, minimizes latency, and lowers the amount of communication bandwidth needed.
- **Data fusion:** Data fusion frameworks correlate environmental and motion data (pressure, acceleration, radiation levels, for example) with physiological signals (ECG, EMG, SpO_2, temperature, etc.) to formulate a comprehensive physiological profile of the astronaut.
- **Integration of machine learning:** To identify complex physiological patterns and predict potential health anomalies, machine learning algorithms including deep neural nets and temporal analysis models like LSTM networks are used. Predictive analytics in real time considers aspects like cardiovascular stress, fatigue, dehydration, and circadian rhythm anomalies.
- **Data security:** In the area of space health informatics, data confidentiality and integrity are necessary. To maintain the integrity of data during transport and storage, wearables employ secure communication protocols (e.g., AES, ECC, post-quantum variants) and lightweight cryptographic modules.

As visualized in the in the Figure 13.3, the processed data is forwarded to mission control and to astronauts. In this way it allows ongoing, adaptive

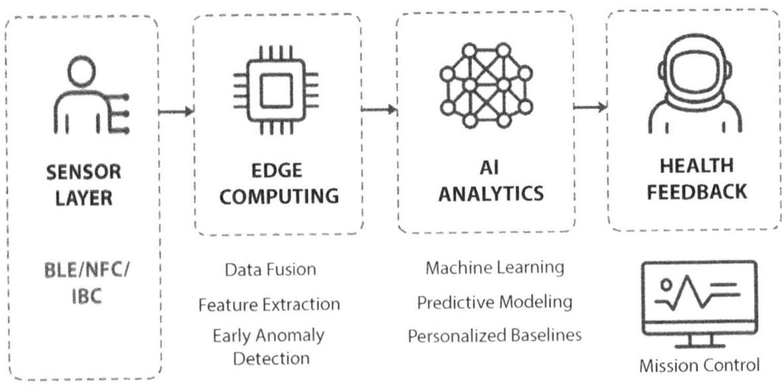

Figure 13.3 Block diagram for intelligent data processing and health analytics.

health feedback. These closed-loop monitoring systems enable automated notifications, personalized exercise recommendations and on the spot intelligent regulation of oxygen or heat based on physiological information in real time. Such systems—that combine edge AI, flexible sensors and secure wireless telemetry—are key to operational resilience and next-generation astronaut healthcare.

13.4.4 Energy Harvesting for Wearable Systems

Self-powered wearable electronics often utilize energy harvesting mechanisms converting physiological energy or environmental energy to electrical power, because reducing battery dependence is essential for missions in space [8].

- **Thermoelectric generators (TEGs):** Low-energy sensors can be powered by body heat gradients.
- **Piezoelectric and triboelectric nanogenerators (TENGs):** Utilize suit deformation or body motion to generate mechanical energy.
- **Flexible solar cells:** Used to strengthen the outer layers of the suit when engaging in extracurricular activities.
- **Energy storage:** Harvested energy is stored for use in solid-state micro batteries and printed thin-film super capacitors.

Figure 13.4 illustrates how this energy-autonomous method guarantees continuous biomedical monitoring without the need for frequent battery replacements and supports long-duration missions.

An important development in astronaut healthcare systems is the use of wearable and flexible electronics. These technologies, which combine wireless telemetry, e-textiles, biosensors, and energy harvesting, allow for independent, comfortable, and ongoing health monitoring in difficult settings. Next-generation space suits will transform from passive clothing into intelligent, adaptable health ecosystems by fusing nanomaterials, AI-driven analytics, and self-powered operation.

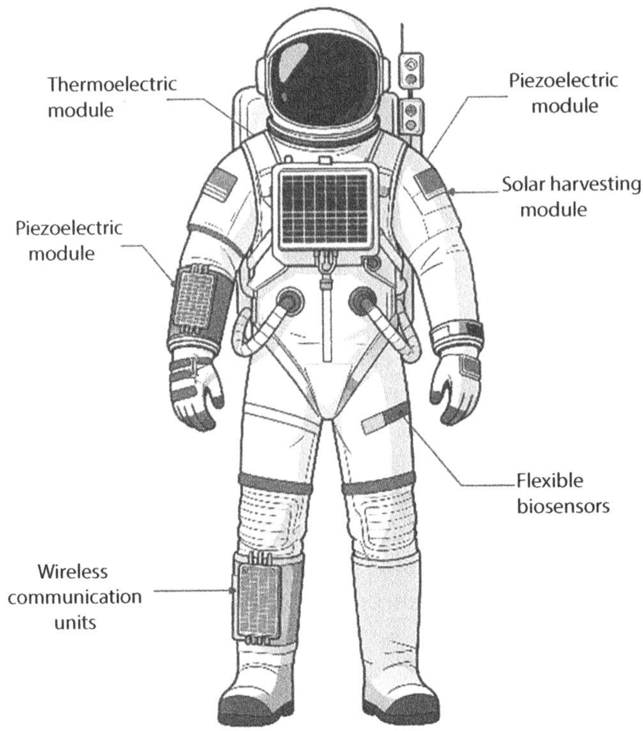

Figure 13.4 Energy-harvesting mechanisms for wearable systems.

13.5 Printed Sensors for Space Monitoring

In addition to astronaut health, printed and flexible sensor networks are essential for mission safety, environmental control, and spacecraft structural integrity. In order to monitor mechanical, thermal, and chemical parameters in real time across large and irregular surfaces, these distributed networks are incorporated into robotic systems, spacecraft panels, and habitats [9].

These lightweight, conformal, multifunctional sensors improve fault detection, streamline wiring, and enable self-diagnosis. This section presents the primary sensor types, design methodologies, and applications relevant to spacecraft and habitat monitoring systems.

13.5.1 Flexible Strain and Pressure Sensors for Structural Health

Structural health monitoring, or SHM, is crucial for detecting fatigue, delamination, and stress accumulation in satellites and spacecraft. As shown in Figure 13.5, printed strain and pressure sensors provide a conformal approach to measuring mechanical deformation across complex geometries [10].

- **Piezoresistive Sensors:** Make use of conductive inks with strain-dependent resistance, such as graphene, carbon nanotubes, or metallic nanoparticles. Perfect for ongoing solar panel and spacecraft hull monitoring.
- **Capacitive Pressure Sensors:** Measure changes in stress on spacecraft surfaces by using dielectric elastomer layers positioned between printed electrodes.
- **Flexible Piezoelectric Films (PVDF, ZnO):** Self-powered strain sensing is made possible by the generation of voltage under deformation.

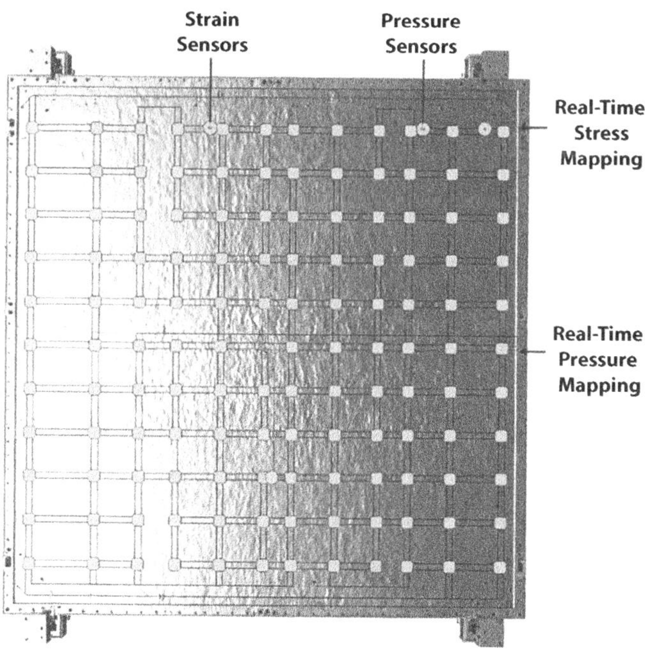

Figure 13.5 Printed strain and pressure sensors integrated on spacecraft panels.

- **Integration with Composites:** Smart composite panels can be created by laminating these sensors inside carbon fiber–reinforced polymer (CFRP) structures.

13.5.2 Printed Gas, Radiation, and Temperature Sensors

Multiple sensor nodes can be positioned on interior panels to integrate printed temperature and gas sensors into a network inside a space habitat. Lightweight, sensitive, and radiation-hardened sensors that are perfect for space conditions are made possible by these nodes' use of printed sensor technology with materials like PEDOT: PSS and nanomaterials like graphene or carbon nanotubes. Temperature variations and harmful gases are among the environmental parameters that the sensors can identify [11].

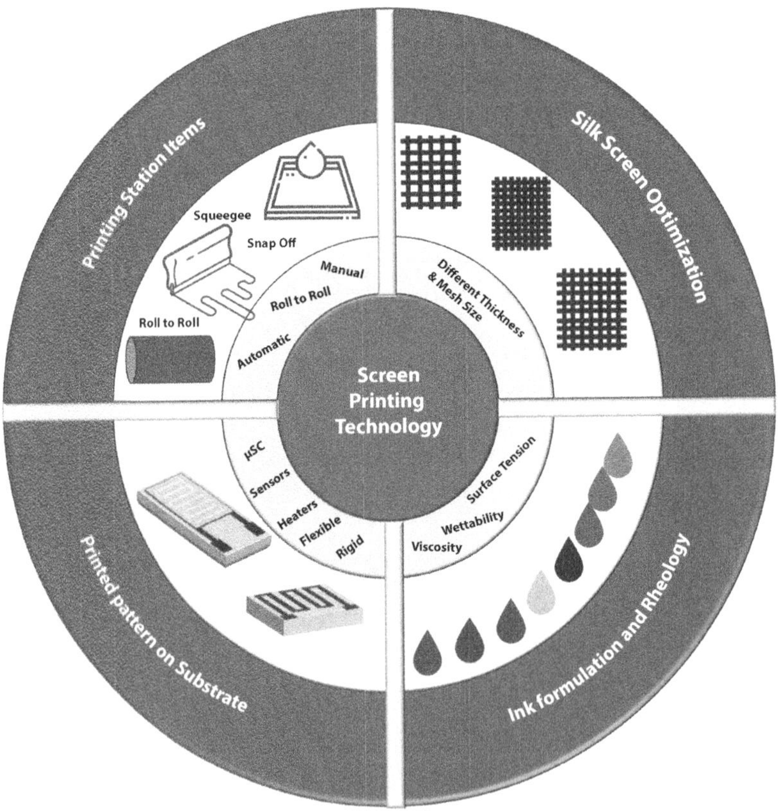

Figure 13.6 Network of printed gas and temperature sensors inside a space habitat.

Sensor nodes usually use wireless connectivity to send data to a central environmental monitoring system. Real-time environmental data collection, processing, and display are all done by this system. Because they simplify wiring and enable flexible sensor placement, these wireless sensor networks are beneficial in space habitats.

Several printed sensor nodes are incorporated into or positioned on habitat panels as part of the overall network architecture.

- Printed sensors combined with wireless communication modules (like BLE),
- Software for environmental monitoring for analysis and alerting,
- A central hub or gateway that gathers the sensor data.

High sensitivity, a broad temperature detection range, and compatibility with low-power wireless transmission platforms have been demonstrated by printed temperature sensors on flexible substrates, making them perfect for ongoing habitat monitoring.

As illustrated in Figure 13.6, a printed sensor network in a space habitat would be made up of sturdy, low-power printed temperature and gas sensors that are wirelessly connected to a central environmental monitoring unit to provide real-time data on the temperature and air quality of the habitat.

13.5.3 Distributed Sensors & IoT Integration

In large spacecrafts or space habitats, a distributed Internet of Things (IoT) system of printed sensors can create a high-level and integrated system of environmental measurements. The sensor patches are printed with necessary materials such as a microcontroller, a power module, and a wireless interface, such as the BLE, ZigBee, or Ultra-Wideband (UWB). The sensors can be used independently, which means that the information is processed at local level and the data is sent wirelessly to a monitoring unit [12].

The topology of the network is normally a mesh or star. Mesh topology also increases the redundancy since it allows nodes to connect with more than one neighbour, thus facilitating self- healing paths to communicate data. Star topology is an easy-to-manage topology because there are no switches or routers between all the nodes and a central point. Such designs guarantee real-time communication and sturdiness that is essential in the stability of monitoring within a habitat.

Localized signal processing and compression of data is made possible by edge processing at every node. This reduces the latency and reduces the bandwidth consumed by reducing the quantity of raw data sent. It enables the network to react fast to the changes in the environment and helps to minimize the load on core systems. The design makes use of self-healing conductive pathways *via* printed circuits and redundant sensor nodes to improve fault-tolerant functionality. These systems' constant operation, even in the event of possible damage or sensor failure, is an essential safety feature in the harsh and constrained space habitat.

13.5.4 Case Study: Printed Sensor Integration on Satellite Panels

One case in point of the recent advancements in small satellite technology is a CubeSat prototype that utilizes a flexible printed sensor network. The satellite structure integrates arrays and interconnects that are printed directly onto flexible substrates that fit the CubeSat frame. These printed sensors—environmental, pressure, and temperature sensors—are attached to the interior surfaces of the satellite in order to constantly obtain important data [13].

Each printed sensor module is integrated within telemetry systems and onboard microcontrollers for wireless communication and real-time data acquisition. Sensor data are sent to the CubeSat's main onboard computer over communication protocols, such as I2C or SPI, where it is processed and stored. Remote monitoring of the environment and satellite health is facilitated by the telemetry system which relays the data from the sensors to ground stations.

The sensor networks provided by this prototype concept are suitable for the very practical size, weight, and power (SWaP) requirements of CubeSat's, due to its compactness, light weight and power efficiency reported in Figure 13.7. Because printed electronics are such multifunctional materials, they can be effortlessly and bulk-free integrated into complex satellite geometries. Moreover, printed interconnects reduce mechanical stress points, which increases reliability and simplifies wiring.

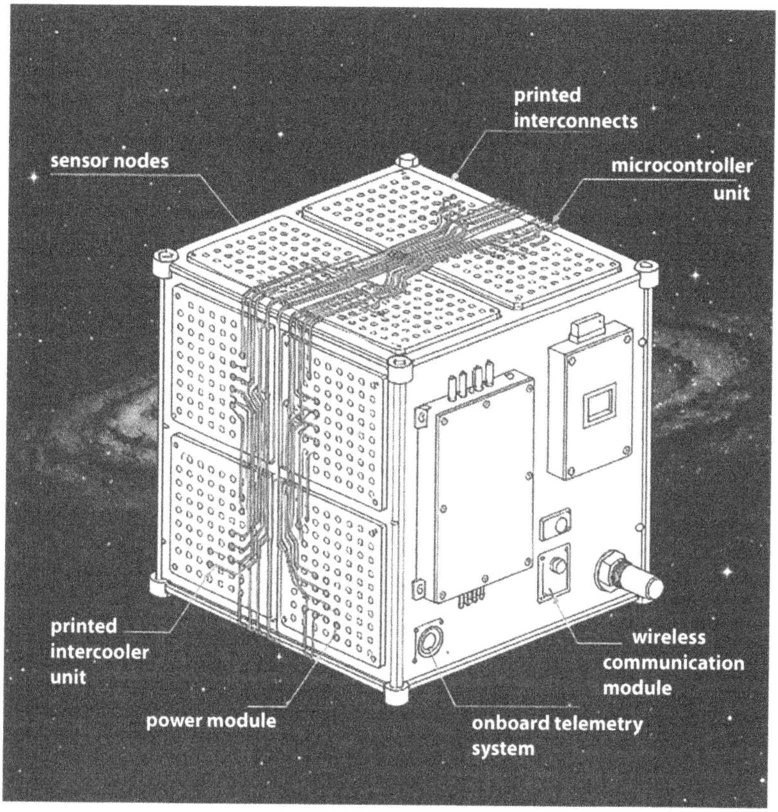

Figure 13.7 CubeSat prototype with integrated flexible printed sensor network.

13.6 Flexible Power & Communication Modules

Power generation, energy storage, and communication are the three primary roles of any space system. In the field of printed and flexible electronics, these modules need to be lightweight, radiation-hardened, and flexible enough to conform to curved or dynamic surfaces, like deployable arrays, spacecraft skins, or wearable technology.

Engineers can create compact, multipurpose architectures that allow for highly integrated and self-sufficient space platforms by utilizing wireless energy transfer circuits, thin-film photovoltaics, and printed antennas. The design techniques and materials utilized to create flexible power and communication modules appropriate for space applications are highlighted in this section.

13.6.1 Flexible Solar Cells and Energy Storage Devices

Thin, stretchable photovoltaic materials are deposited onto curved or adjustable substrates to produce flexible solar cells for spacecraft skins. These solar cells are highly efficient and can be installed in curved spacecraft panels. They are generally based on advanced thin-film materials and are composed of crystalline silicon or flexible backings of elastic polymers. Although constrained by shapes, this flexibility enables designs of spacecraft to have their surfaces ideally exposed to sunlight [14].

The printed supercapacitor modules are directly integrated with the flexible solar arrays to provide on-board energy storage. The extended cycle life, high power density, and wide operating temperature range of these 3D-printed or additively manufactured supercapacitors make them ideal for the space environment. They create a tiny, low-power subsystem by effectively storing and releasing the energy collected by the solar cells.

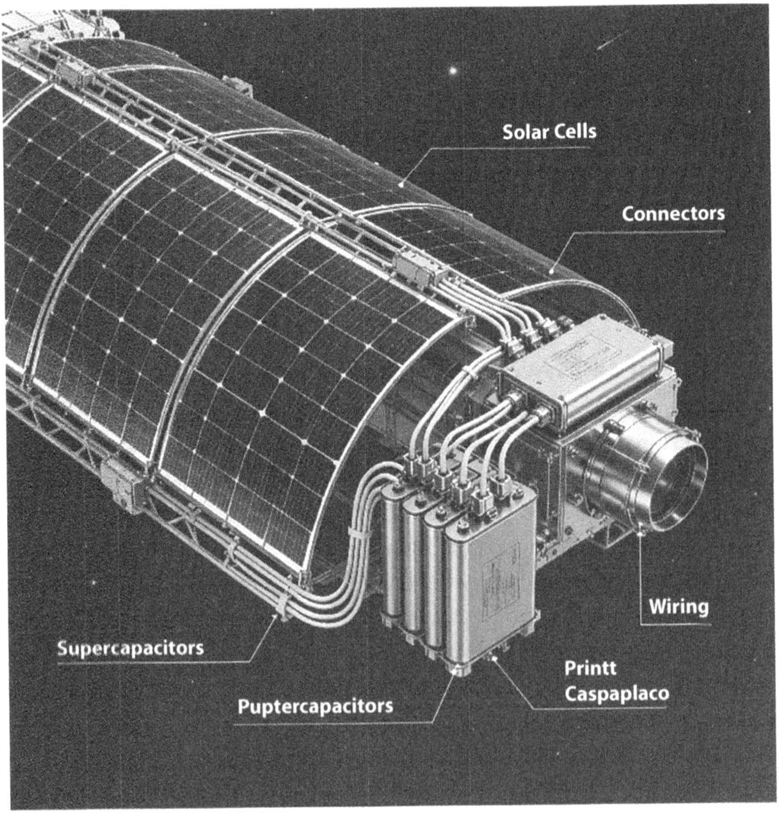

Figure 13.8 Flexible solar and energy storage integration for spacecraft skins.

As seen in Figure 13.8, a typical schematic has flexible solar cells uniformly covering the curved spacecraft panels, which are connected by printed conductive traces. The power produced by these traces is transferred to nearby printed supercapacitors that are part of an energy management circuit. In order to minimize power fluctuations during orbital day–night cycles, the combined subsystem can continuously power spacecraft instruments and systems. By decreasing weight, increasing surface area, and offering dependable energy storage appropriate for space missions, the integration of flexible solar cells and printed supercapacitors on spacecraft skins improves the design of power subsystems.

13.6.2 Printed Antennas for Space Communication

The importance of printed antennas and radio frequency circuits for space communication is growing because of their low profile, lightweight, and flexible form factors that satisfy the precise size, weight, and power requirements of spacecraft. Conductive inks are used to create printed antennas on flexible substrates like PET or polyimide films. As a result, antennas can be seamlessly incorporated into flexible skins or structural panels, or they

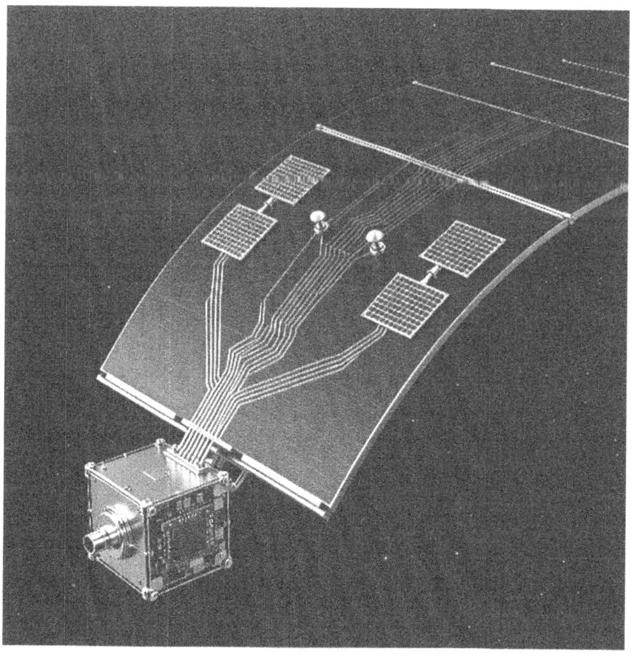

Figure 13.9 Printed conformal antenna array integrated on a CubeSat panel.

can adapt to the curved surfaces of spacecraft. Patch antennas, dipoles, and fractal designs that are tuned for frequencies between UHF and Ka band are common printed antenna types used in space. They minimize mass and volume while offering dependable signal transmission and reception [15].

These antennas are enhanced by printed RF circuits, as depicted in Figure 13.9, which combine passive and active parts through the use of printing methods like screen, inkjet, or aerosol jet printing. Filters, matching networks, amplifiers, and switches are among the components that make it possible for RF front-ends to be completely integrated and compact. The printed RF components are made to withstand vibration, radiation, and extreme temperatures found in space. Compactness, simplicity of integration into spacecraft architecture, and the possibility of low-cost mass fabrication are the main benefits of printed antennas and radio frequency circuits in space communication. They facilitate telemetry, command, data relay, and intersatellite links, among other space communication requirements.

13.6.3 Wireless Power Transfer and Energy Management

Wireless power transfer (WPT) removes heavy cabling and mechanical connectors that are prone to failure in space from wearable systems and distributed sensor networks.

- **Inductive and capacitive coupling:** Turn on mid-range power transfer from base stations to sensors or between spacecraft modules.
- **Resonant magnetic coupling:** Improves the efficiency of transfers for flexible receivers that are incorporated into spacecraft skins.
- **Rectennas (rectifying antennas):** Printed flexible rectennas are perfect for self-sustaining IoT nodes because they transform RF power into usable DC energy.
- **Energy management circuits:** Stable operation under a range of loads and temperatures is guaranteed by flexible voltage regulators and printed power management integrated circuits (PMICs).

Future reconfigurable space systems can be assembled more easily and with greater energy autonomy thanks to wireless energy frameworks.

13.6.4 EMI Shielding in Lightweight Structures

Signal integrity in closely spaced electronic systems is seriously threatened by electromagnetic interference (EMI). Lightweight and flexible EMI shielding materials are used to combat this.

- **Conductive polymer composites:** Adding CNTs, graphene, or MXenes to polymer matrices offers high shielding efficacy and flexibility.
- **Metal-coated fabrics:** Used in wearable technology to protect delicate biomedical sensors from radiofrequency radiation.
- **Multifunctional layers:** In a single printable coating, combine mechanical protection, heat dissipation, and shielding.
- **Space qualification:** EMI shields must remain stable without delamination or conductivity loss when exposed to radiation and heat cycling.

In intricate electronic systems, effective EMI mitigation guarantees accurate sensing, dependable communication, and system safety. Next-generation space electronic ecosystems are primarily made possible by flexible and printed power and communication modules. Thin-film photovoltaics, printed antennas, and wireless energy systems come together to give spacecraft and wearable technology mechanical adaptability, high functional density, and energy autonomy. Lightweight EMI shielding materials are used to support the shift to fully integrated, intelligent, and self-powered space platforms by ensuring signal integrity and system resilience.

13.7 Reliability, Testing, and Space Qualification

One of the most difficult operating environments for any electronic system is space. Even though they are lightweight and versatile, wearable, flexible, and printed electronics need to pass stringent reliability tests and qualification processes before being used. Following these procedures ensures that the systems will continue to function electrically and structurally throughout the mission, despite exposure to radiation, high temperatures, vacuum, micrometeoroid impacts, and mechanical wear and tear [16, 17]. This section highlights the testing protocols, material reliability plans, and qualification frameworks required to approve printed and flexible electronics for use in space.

13.7.1 Environmental Challenges in Space

Wearable and flexible electronic components in space face unique challenges that are much harsher than those on Earth. These devices are subjected to intense physical stresses and extreme levels of radiation, making it essential to understand their environment to enhance their materials and designs.

- **Thermal Extremes:** Spacecraft can swing between incredibly low temperatures of -150°C and scorching highs of +120°C, depending on how much sunlight they receive.
- **Radiation Exposure:** In space, components are bombarded by protons, gamma rays, and heavy ions, which can damage semiconductors, polymers, and conductive inks.
- **Vacuum Environment:** The vacuum of space leads to issues like outgassing, delamination, and moisture absorption, all of which can compromise insulation and adhesion.
- **Mechanical Stresses:** The stresses from vibrations during launch, the fatigue caused by microgravity, and the forces experienced during deployment all contribute to the long-term stability of flexible substrates.

Understanding these factors is crucial for developing robust electronic components that can withstand the rigors of space exploration.

13.7.2 Mechanical Durability and Fatigue Testing

Reliability of flexible electronics is highly influenced by its mechanical robustness. Devices must be strong enough to resist repeated folding, stretching, and bending without breaking.

- **Bending and Cyclic Fatigue Tests:** Assess resistance to electrical deterioration and micro crack formation under repeated stress.
- **Dynamic Flexural Endurance:** Allows for the simulation of flexible circuit deployment and operation on curved spacecraft structures.
- **Vibration and Shock Tests:** Replicating launch and orbital vibration environments in accordance with NASA GEVS or ESA ECSS standards.
- **Adhesion and Peel Strength Tests:** Check the integrity of the substrate-ink bonding while it is being heated.

13.7.3 Radiation Tolerance Evaluation

In flexible materials, radiation exposure can cause polymer degradation, displacement damage, and ionization. In the case of prolonged missions, radiation hardness testing ensures electronic stability.

- **Total Ionizing Dose (TID) Testing:** Evaluates the effects of cumulative radiation on dielectric layers and transistors.
- **Single-Event Effect (SEE) Testing:** Evaluates printed integrated circuits for transient phenomena such as latch-ups and bit flips.
- **Material Shielding Assessment:** Studies the absorption or deflection of ionizing particles by graphene, MXene, or polymer composites.
- **Degradation Monitoring:** Measures mechanical, optical, and electrical changes that occur post radiation.

13.7.4 Thermal Cycling and Vacuum Testing

Vacuum testing and thermal cycling evaluate how flexible devices react to outgassing phenomena and the temperature gradients that are very common on deep-space or low Earth missions.

- **Thermal cycling tests:** Compare extreme cold to extreme hot temperature to measure material expansion mismatch and delamination rates.
- **Vacuum outgassing tests:** Assess the rate of volatile release and how it affects adhesion or sensor performance.
- **Combined stress testing:** Simulates actual orbital situations by simultaneously exposing samples to thermal, vacuum, and radiation conditions.
- **Material compatibility analysis:** Guarantees the stability of the encapsulant, adhesive, and ink across all mission temperature profiles.

13.7.5 Self-Healing and Damage-Tolerant Materials

Self-regenerating materials are also a novel approach for long-term autonomously reliable space electronics. They allow devices to recover from minor fractures, radiation damage, or interruptions in conductive pathways.

- **Intrinsic self-healing polymers:** Include dynamic covalent bonds that reform when heated or exposed to ultraviolet light.
- **Microcapsule-based inks:** When damaged, release conductive fillers to restore the electrical continuity.
- **Graphene–polymer hybrids:** Show outstanding resistance to the spread of cracks and offer EMI protection.
- **Application areas:** Perfect for inflatable habitats, wearable suits, and deployable antennas where access to maintenance is restricted.

13.7.6 Space Qualification Protocols

Flexible and printed electronics are required to comply with strict space qualification regimes set by global organizations to ensure mission success.

- **NASA Technology Readiness Levels (TRLs):** States the maturity levels, which run the gamut from lab-scale prototypes (TRL 3–4) to hardware that is appropriate for flight (TRL 8–9).
- **ESA ECSS Standards:** Assist European missions with material selection, testing, and documentation. JAXA and ISRO Standards: Adapted to address launch vehicle-specific limitations and microgravity.
- **Flight Demonstration and Validation:** For instance, CubeSat deployment, suborbital evaluation, and/or integration of the product into the International Space Station (ISS) platform for in-orbit testing.

For flexible, printed, wearable space electronics, reliability, and qualification testing are the building blocks for trust. Engineers check these systems' mechanical endurance, radiation tolerance, thermal-vacuum resilience, and self-healing verification to make sure they attain flight-level robustness. By systematically adopting a transition trajectory based on flexibility for new electronic technologies, this approach speeds the path from the lab to mission-critical deployment, paving the way for future resilient, flexible space architectures.

13.8 Case Studies and Future Prospects in Space Missions

The way that astronauts, spacecraft, and habitats interact with electronic systems has changed dramatically as a result of the incorporation of flexible, printed, and wearable electronics into space technology. These technologies address important problems in long-duration missions, deep space exploration, and astronaut health monitoring with their lightweight design, modular deployment, and high functional density.

This section gives an overview of new opportunities and research directions that will impact the next generation of space electronics and includes significant case studies, both conceptual and experimental.

13.8.1 Flexible Sensor Arrays for Spacecraft Health Monitoring

Environmental sensing and structural integrity are critical for spacecraft maintenance and mission safety. Flexible sensor networks can be directly laminated onto spacecraft surfaces for distributed, real-time monitoring [18].

- **Applications:** Assess spacecraft panels for strain, temperature, radiation levels, or the development of microcracks.
- **Materials:** Sensors based on carbon nanotubes (CNTs) and graphene for low mass and high sensitivity.
- **Implementation:** Wireless data transmitters combined with sensors printed on polyimide or Kapton substrates.
- **Case example:** Conformal sensor skins that can identify structural fatigue during thermal cycles were demonstrated by NASA's "Printable Spacecraft Electronics Initiative."

13.8.2 Printed Antennas and RF Systems for CubeSats

Compact and low-mass communication systems are required for small satellites like CubeSats. For high-gain, multi-band communication in limited form factors, printed antennas provide a scalable solution [19].

- **Design types:** Microstrip and patch antennas printed using inkjet technology on flexible polyimide substrates.
- **Advantages:** Lightweight, able to be reconfigured in orbit, and compatible with CubeSat panels.

- **Example:** The ISRO Nanosat Program investigated printed antennas for low-Earth orbit (LEO) telemetry and data links.
- **Performance metrics:** Broadband tunability, robustness under launch vibration testing, and antenna efficiency >85%.

13.8.3 Wearable Biosensors for Astronaut Health and Performance

In space, astronauts endure tremendous physical and mental strain. Continuous, non-invasive health monitoring in microgravity environments relies heavily on wearable, flexible biosensors [20].

- **Measured parameters:** Stress biomarkers, muscle fatigue, skin temperature, heart rate, and hydration.
- **Device types:** Sweat-analysis sensors, flexible temperature arrays integrated into clothing, and printed ECG patches.
- **Power source:** Wireless power harvesting systems and integrated printed thin-film batteries.
- **Case example:** Flexible bio patches have been tested on ISS crew members by the European Space Agency (ESA) to track stress and sleep patterns.

13.8.4 Flexible Photovoltaics and Energy Harvesting Modules

For long-duration missions, planetary bases, or lunar habitats, deployable and conformal energy sources are crucial. High power-to-weight ratios and improved deployability are provided by flexible photovoltaic arrays that can adjust to uneven surfaces [21].

- **Technologies Used:** CIGS, perovskite, and organic thin-film solar cells.
- **Demonstrations:** The feasibility of flexible PV systems in orbit was demonstrated by NASA's Roll-Out Solar Array (ROSA) project.
- **Integration:** Hybrid systems that combine solar arrays and printed supercapacitors to allow for autonomous operation during eclipses.
- **Future Expansion:** Power modules for Mars habitats that use flexible, radiation-resistant photovoltaics.

13.8.5 Smart Materials and Self-Healing Components in Orbit

By using self-healing and smart-response materials, electronic systems can automatically repair themselves after being damaged by thermal cycling or micrometeoroids [22].

- **Mechanisms:** Conductive inks and polymer composites based on microcapsules that regain conductivity after breaking.
  - **Demonstration Missions:** The Materials International Space Station Experiments (MISSE) program of NASA tested self-healing polymers under vacuum and ultraviolet light.
  - Developing printed electronics that can withstand damage for reconfigurable satellite skins was the main goal of ESA's SHIELD program.
- **Potential Impact:** Increases mission longevity and lowers deep-space system maintenance needs.

13.8.6 Future Trends & Emerging Opportunities

Integration, intelligence, and interconnectivity—combining materials science, artificial intelligence, and advanced manufacturing—are the next frontiers for flexible and wearable electronics in space.

- Artificial Intelligence Integration: On-chip AI modules for sensor network adaptation and predictive diagnostics.
- On-demand 3D printing of flexible electronic modules on board spacecraft or lunar bases is known as "Additive Manufacturing in Orbit."
- Future flexible processors that use quantum tunneling or memristive components for edge computing are known as quantum and neuromorphic systems.
- Human-machine interfaces: haptic feedback arrays and printed neural sensors allow astronauts and robotic assistants to communicate easily.
- Interplanetary Communication Systems: Flexible, lightweight, deployable antennas for relay networks on Mars and the Moon.

It is clear from these case studies that space mission architectures are being redefined by flexible, printed, and wearable electronics. These technologies, which range from health-monitoring bio suits to self-healing spacecraft skins and conformal antennas, allow for intelligence, adaptability, and resilience in harsh environments.

13.9 Conclusion

The future of space systems is being redefined by flexible, printable, and wearable electronics, which provide lightweight, adaptable, and multipurpose alternatives to traditional rigid hardware. This chapter has discussed how developments in printed fabrication, additive manufacturing, and flexible materials allow intelligent habitat management in harsh space settings, astronaut biomedical sensing, and spacecraft health monitoring. Continuous, non-invasive astronaut health monitoring is made possible by wearable biosensors and intelligent e-textiles, while printed sensor networks integrated into spacecraft and shelters improve environmental safety and structural integrity assessment. Conformal communication modules, energy harvesting systems, and flexible power sources are all integrated to further encourage autonomous and energy-efficient operation. Ongoing research in self-healing materials, in-orbit manufacturing, and AI-enabled analytics is gradually tackling these constraints, even though issues with radiation tolerance, long-term durability, and microgravity fabrication still exist. All things considered, wearable and flexible electronics have the potential to become fundamental technology for human-centered, robust, and sustainable space exploration missions.

AI disclosure statement: The author confirms that no artificial intelligence (AI)-generated content or AI-assisted writing tools were used in the preparation of this chapter (including summaries and the abstract). All text, analysis, and conclusions are the result of the author's original work.

References

1. Kim, M., Mackenzie, D.M.A., Kim, W., Isakov, K., Lipsanen, H., All-parylene flexible wafer-scale graphene thin film transistor. *Appl. Surf. Sci.*, 551, 149410, 2021.

2. Dutta, T., Chaturvedi, P., Llamas-Garro, I., Velázquez-González, J.S., Dubey, R., Mishra, S.K., Smart materials for flexible electronics and devices: Hydrogel. *RSC Adv.*, 14, 12984–13004, 2024.
3. Huang, Q. and Zhu, Y., Printing conductive nanomaterials for flexible and stretchable electronics: A review of materials, processes, and applications. *Adv. Mater. Technol.*, 4, 2019.
4. Wiklund, J., Karakoc, A., Palko, T., Yigitler, H., Ruttik, K., Jantti, R., Paltakari, J., A review on printed electronics: Fabrication methods, inks, substrates, applications and environmental impacts. *J. Manuf. Mater. Process.*, 5, 2021.
5. Zocca, A., Wilbig, J., Waske, A., Günster, J., Widjaja, M.P., Neumann, C., Clozel, M., Meyer, A., Ding, J., Zhou, Z., Tian, X., Challenges in the technology development for additive manufacturing in space. *Chin. J. Mech. Eng.: Addit. Manuf. Front.*, 1, 1, 100018, 2022.
6. Bellisle, R., Bjune, C., Newman, D., Considerations for wearable sensors to monitor physical performance during spaceflight intravehicular activities, in: *Proc. 42nd Annu. Int. Conf. IEEE Eng. Med. Biol. Soc. (EMBC)*, Montreal, QC, Canada, pp. 4160–4164, 2020.
7. Newman, E., Huang, J., Pomerantz, M., Sellin, J., Multi-project telemetry-based digital twin environment for space-mission development, analysis, and operations, in: *Proc. IEEE Aerospace Conf.*, Big Sky, MT, USA, pp. 1–12, 2023.
8. Alhawari, M., Mohammad, B., Saleh, H., Ismail, M., Energy harvesting for self-powered wearable devices, 2018.
9. Al-Amri, A.M., Printed sensors for environmental monitoring: Advancements, challenges, and future directions. *Chemosensors*, 13, 2025.
10. Mu, C., Guo, X., Zhu, T., Lou, S., Tian, W., Liu, Z., Jiao, W., Wu, B., Liu, Y., Yin, L., Jian, X., Song, Y., Flexible strain/pressure sensor with good sensitivity and broad detection range by coupling PDMS and carbon nanocapsules. *J. Alloys. Compd.*, 918, 165696, 2022.
11. Abi-Ramia Silva, T.E., Burisch, F., Güntner, A.T., Gas sensing for space: Health and environmental monitoring. *TrAC, Trends Anal. Chem.*, 177, 117790, 2024.
12. Alattabi, K., Integration of sensor array techniques for efficient tracking of approaching individuals and objects: A technological perspective. *J. Inf. Syst. Eng. Manag.*, 10, 2025.
13. Shemelya, C., *et al.*, Multi-functional 3D printed and embedded sensors for satellite qualification structures, in: *Proc. IEEE SENSORS*, Busan, Korea, pp. 1–4, 2015.
14. Verduci, R., Romano, V., Brunetti, G., Yaghoobi Nia, N., Di Carlo, A., D'Angelo, G., Ciminelli, C., Solar energy in space applications: Review and technology perspectives. *Adv. Energy Mater.*, 2022.
15. Akan, V. and Yazgan, E., Antennas for space applications: A review, 2020.

16. Safdari, N., Monette, G., Beydoun, J., Martin, A., Varano, R., Laforce, F., Reliability testing of 28 Gbps/channel fiber optics transceivers for space applications, in: *Proc. Int. SpaceWire & SpaceFibre Conf. (ISC)*, Pisa, Italy, pp. 1–8, 2022.
17. Olawade, D.B., Ijiwade, J.O., Wada, O.Z., Toward net-zero in space exploration: A review of technological and policy pathways for sustainable space activities. *Sci. Total Environ.*, 972, 179145, 2025.
18. Duan, Y., He, S., Wu, J., Su, B., Wang, Y., Recent progress in flexible pressure sensor arrays. *Nanomaterials*, 12, 2022.
19. Abulgasem, S., Tubbal, F., Raad, R., Theoharis, P.I., Lu, S., Iranmanesh, S., Antenna designs for CubeSats: A review. *IEEE Access*, 9, 45289–45324, 2021.
20. Wang, Y., Dong, D., Zhou, W., He, J., Flexible wearable device applications for monitoring astronaut health: Current status and challenges. *Wearable Electron.*, 2, 77–84, 2025.
21. Jahandar, M., Prasetio, A., Lee, C., Kim, H., Kim, A.R., Heo, J., Kim, Y., Kim, S., Lim, D.C., Highly efficient flexible organic photovoltaic modules for sustainable energy harvesting under low-light condition *via* suppressing voltage-drop by metal-mediated cross-linkable polymer interfacial layer. *Chem. Eng. J.*, 448, 137555, 2022.
22. Pernigoni, L., Lafont, U., Grande, A.M., Self-healing materials for space applications: Overview of present development and major limitations. *CEAS Space J.*, 13, 341–352, 2021.

14

3D Printing and Additive Manufacturing for On-Demand Space Electronics

G. Loganathan[1]*, S. Sivasakthiselvan[2], M. Palanivelan[1] and Pham Chien Thang[3]

[1]*Department of Electronics and Communication Engineering, Rajalakshmi Engineering College, Chennai, Tamil Nadu, India*
[2]*Department of Computing Technologies, SRM Institute of Science and Technology, Chennai, Tamil Nadu, India*
[3]*Faculty of Journalism and Communication, TNU-University of Sciences, Thai Nguyen, Thai Nguyen Province, Vietnam*

Abstract
Due to the increasing length and complexity of space missions, pressure is growing to find a more flexible, quick, and sure technique of manufacturing electronic parts in space conditions. This chapter examines how the 3D printing and additive manufacturing technologies can be used to enable on-demand production of space electronics. It describes different printing methods including inkjet, aerosol jet, and fused deposition modeling (FDM) and the choice of materials that can be used in printable electronics in space like conductive inks, polymers, and metal alloys. Such issues as vacuum compatibility, thermal stress, and radiation resistance are identified, and some possible solutions are offered. Application of printed electronics in structural and functional units of satellites, landers, and lunar habitats is also discussed. Case studies and real-life examples of activities conducted by space agencies and even commercial organizations are given to demonstrate the existing capabilities and the future perspectives. This chapter ends by underlining the disruptive nature of additive manufacturing in making self-sustaining and sustainable in-space electronic fabrication systems.

Keywords: Additive manufacturing, deposition modeling, additive electronics, printable electronics, mEMS

**Corresponding author*: logeshece57@gmail.com

N. Sathish, Prasanth Aruchamy, Rajesh Kumar Dhanaraj, Nithya Rekha and Karthik Palani (eds.) Space Electronics for Deep Space Exploration, (327–350) © 2026 Scrivener Publishing LLC

14.1 Introduction

The 3D printing, or additive manufacturing (AM), is a revolutionary technology in various fields of engineering and science [1]. With regard to space exploration, it provides the possibility to produce mechanical and electronic parts on-demand, either in orbit or on alien surfaces [2]. The conventional space production involves a lengthy lead time and is also an expensive launch procedure. However, additive manufacturing facilitates making of complicated geometries, optimized structures as well as functional electronic systems without large inventories [3]. AM can be used to create spacecraft capable of making *in situ* repairs, and spending human time in deep space by combining materials science, robotics, and digital design [4]. It allows the production of components only when it is required through the elimination of reliance on supply chains on Earth and through the alleviation of payload limitations [5]. This is one of the key benefits of AM, as it can produce lightweight and, nevertheless, structurally sound components lowering the costs of the launch and enhancing fuel efficiency [6]. Furthermore, AM enables one to integrate several functionalities into single components, which have structural, thermal, and electronic attributes [7].

Recent developments, e.g., selective laser sintering (SLS), fused deposition modeling (FDM), and aerosol jet printing allowed processing a greater number of materials [8]. Alongside mechanical and electronic parts, AM has potentials of developing life support systems, habitats, and radiation shielding [9]. As a case in point, *in situ* resource utilization (ISRU) on the moon or Mars can make use of local materials, e.g., lunar regolith, to produce structural components. Spacecraft can automatically detect malfunctions, fix parts, and reconfigure subsystems by integrating AM with robotic and artificial intelligence (AI) technologies [10]. In terms of electronics, AM also makes it possible to create multi-layered, flexible, conformal circuits that are directly integrated into structures. This tactic improves performance in challenging space situations while conserving mass. Specific mission requirements can be made (adaptive antennas, special sensors, energy storage systems). Notwithstanding such advantages, there are certain issues of material properties in microgravity, quality control, and reproducibility.

Nevertheless, additive manufacturing models, such as the International Space Station with Additive Manufacturing Facility, have proven to be practical and consist entity. The integration of AM with digital twin in the future is likely to make space manufacturing processes real-time monitorable and optimizable. This would enable anticipatory control and error correction in the fabrication enhancing dependability in off-Earth production. Hybrid systems could also appear, that is, additive/subtractive systems that will integrate precision machining with high speed 3D deposition in order to improve the accuracy of the parts [11]. The other prospective direction is the multifunctional printing with nanocomposites, which have enhanced conductivity, radiation tolerance, and self-healing properties. Besides, close loop recycling of materials in the space habitats would contribute greatly to sustainability of missions. Debris and scrap metals can be recycled into new structures, and this will decrease the waste and supply requirements. Printing of the modular systems also ensures flexibility in long-term missions whereby tools, instruments or spares parts can be tailored whenever needed. In the future, it can be further developed to bioprinting tissues or even food on other planets. Lastly, because it enables resource economy, autonomy, and adaptability, additive manufacturing is essential to the future of space systems. With the development of the materials science, machine learning, and the robotic control, AM will become a crucial part of the sustainable space exploration and the off-Earth production.

14.2 Principles of Additive Manufacturing for Electronics

Electronics additive manufacturing is a technique that is used to deposit structural and functional materials in layers to create finished systems [6]. In comparison to the traditional fabrication, AM incorporates mechanical and electrical capabilities in single structures.

Inkjet printing: Interconnects and sensors are created by depositing conductive inks.

Aerosol jet printing (AJP): Fines conductive traces are applied to non-planar surfaces. Additive printing of polymer materials using conductive wire and filament is known as fused deposition modeling, or FDM. RF and microwave dielectric complex structures in stereo lithography (SLA) applications. Recent innovations in hybrid additive manufacturing (Hybrid-AM) combine multiple printing technologies in order to create multi-layered circuits, antennas, and embedded sensors with a very high spatial resolution [8]. The dielectric, magnetic, and conductive materials can be deposited sequentially in order to create functional electronic assemblies without the conventional processes of soldering or etching. These methods reduce parasitic effects, improve mechanical resilience, and allow electronics to be conformally deposited on flexible or curved substrates. Optimizing ink rheology and sintering procedures to provide high conductivity with low thermal budgets is another top concern in printed electronics, especially for polymer-based substrates [7]. Application of nanoparticle- and nanowire-based inks allows

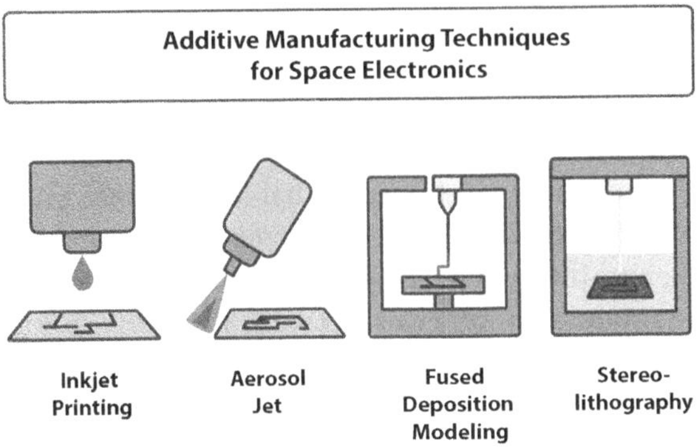

Figure 14.1 Overview of additive manufacturing techniques for space electronics: inkjet, aerosol jet, FDM, and stereolithography methods.

the creation of electrical, thermal, and optical properties that are tune able in space and aerospace. Also, direct printing of semiconducting layers and thin-film transistors (TFTs) are leading to fully printed communication and control modules [10]. New literature is also discussing *in situ* characterization and self-healing of printed circuits on the basis of embedded sensors and AI-controlled feedback control [1].

This will provide real-time health information of the circuit and adaptive repair processes in operation. Compact microwave filters and phased-array antennas can be created thanks to AM's precise control over dielectric constants and substrate shape in high-frequency systems. [9]. Very soon, additive manufacturing will help autonomous production of electronic modules in the alien surroundings. Figure 14.1 shows the structural–electronic hybrids that serve as both functional circuitry and mechanical supports could be printed using robotic technologies. Together with *in situ* resource utilization (ISRU), AM will transform the manufacturing of radiation-tolerant, lightweight, and adjustable electronic systems and space missions of the next generation [2].

14.3 Material Considerations for Space-Grade Additive Electronics

The choice of a material defines electrical, thermal, and mechanical stability of printed electronics in space [7]. Silver, copper, and graphene inks can be used as conductive materials and polyimide and PEEK dielectrics can be used as insulating materials in extreme temperatures. The electromagnetic shielding is enhanced by radiation resistant inks and polymer composites containing nanoparticles within [8]. The ISRU methods are designed to transform lunar regolith printable media [9]. Sintering behavior, surface roughness, and particle size distribution are very important factors in determining the electrical performance of the conductors printed. Nanoparticle-based silver inks are highly conductive but can easily get oxidized in a vacuum and, hence, require encapsulation strategies. Copper inks, on the other hand, are inexpensive. However, they must be stored in reducing atmospheres or sintered with the use of a laser to prevent degradation. Instead, graphene and carbon nanotube (CNT) networks are flexible and stable toward thermal cycling, ensuring that they are suitable in conformally electronics in spacecraft interiors [6]. Dielectric materials are of vital importance in the provision of insulation and signal integrity on high frequencies. Polyimide, PDMS, and PEEK polymer are some examples of polymers that have low dielectric losses and a high level of dimensional stability in the wide range of temperature variations experienced in space missions. New advancements in ceramic-polymer composites like alumina or a boron nitride-filled resin, improve on both mechanical stiffness and thermal conductivity at the same time printable [1]. In the design of space-grade electronics, another prominent issue concerns thermal management. Harmonic manufactured heat dissipators with coppergraphene composites and phase change materials are used to dissipate localized heat in power electronics.

Light lattice structures with built-in cooling channels can be additively manufactured to realize the efficient system thermal control without adding mass to the system [10]. Radiation tolerance is also resolved by material engineering, both on microlevel and nanolevel. Radiation shielding is increased by the addition of cerium oxide nanoparticles or boron enhancing polymers into the matrix. These types of hybrid composites help eliminate charge buildup as well as shield the sensitive electronic circuits against single-event effects (SEEs). Moreover, carbon-based nanolayer coating offers resistance to erosion by atomic oxygen in the low-Earth orbit conditions [2]. In the case of substrate materials, additive manufacturing can be used to utilize flexible and rigid platforms depending on the mission needs.

Deployable structures are made using flexible substrates like Kaplon and Ultem, whereas high-power and RF circuits are made using ceramic substrates, like alumina. New methods are also to investigate multifunctional substrates that can be used as both structural supports and electrical interconnect layers [3]. Material research Future studies concentrate on printable alloys and conductive polymers, which remain stability in vacuum, UV sunlight, and low temperature. *In situ* monitoring and machine-designed materials that are based on AI are able to streamline print parameters to manufacture identical performance. Moreover, by adding locally available regolith with metallic or polymeric binders, sustainable and self-reliant production of electronic and structural parts can be done on the Moon and Mars [5]. Finally, a combination of advanced materials, additive processes, and intelligent control will provide an opportunity to develop space quality electronics not only reliable and lightweight but also flexible enough to respond to a changing mission environment.

14.4 On-Demand Manufacturing in Orbit and Beyond

The future space missions have on-demand manufacturing as their core part. The Made in Space program by NASA had shown 3D-printed tools in the ISS [4]. Follow-up experiments were done on printing circuit traces and connector with multi-material printers [11]. On-demand manufacturing is not limited to simple tools but can be applied to complex structural, electronic, and electromechanical components. This is essential in long duration missions, where it is not possible to resupply the missions in space. Mission specific parts can be produced with automated feed- stock handlers and part checking modules in the additive manufacturing systems and with minimum crew action. These systems are in progress with NASA In-Space Manufacturing (ISM) project and the European Space Agency program (ESA) MELT program [3]. Autonomous manufacturing platforms in microgravity are composed of robotics, machine vision, and AI-driven decision-making to conduct print operations, inspection, and assembly.

Such systems include the use of digital twins of the physical printers and materials, which are virtual representations of physical items that are continuously updated through sensor feedback throughout the fabrication process. This will enable predictive print quality, void/defects identification, and on-site correction of geometrical errors without human supervision. Adaptive design strategies are also made possible by on-demand fabrication. Real-time optimization of components can be done based on mission conditions like load distribution, vibration response or even thermal expansion. For example, a structural panel which is printed on Mars could have additional layers of material added dynamically, depending on the environmental stress factors measured by embedded sensors. This adaptive model reduces the amount of waste materials and increases the life of the system [10]. Hybrid additive-subtractive manufacturing is an important field of development. Figure 14.2 shows the systems are built by linking precision milling or laser ablation units to the 3D printing heads and allow, therefore, surface finishing and fine tuning of features

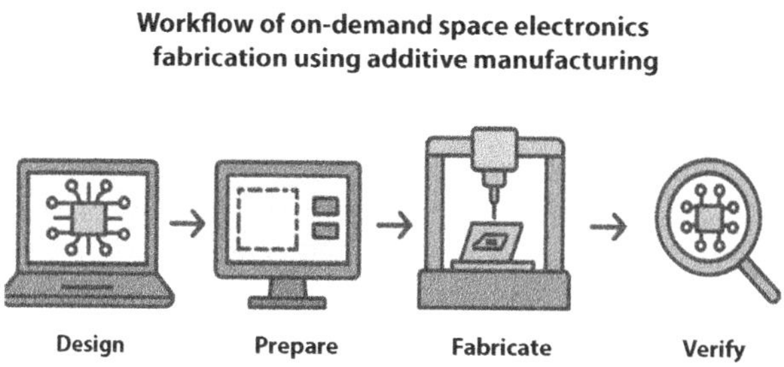

Figure 14.2 Workflow of on-demand space electronics fabrication using additive manufacturing.

immediately after deposition. Such hybridization gives both dimensional accuracy and electrical continuity to the printed electronic components.

Robotic arms with multiple axes are used to develop conformal equipment on curved surfaces of spacecrafts including antennas or thermal radiators [6]. Recycling feedstock is a vital part of sustainable manufacturing according to demand. Obsolete or damaged parts can be reused as a raw material through mechanical reprocessing and reuse of polymers and metals. Recycling systems Closed-loop recycling to regenerate filament by printing parts has been demonstrated in the ISS to lower the resupply of materials. *In situ* recycling of lunar regolith to create construction materials to be used to build surface habitation is a vision of future missions [9].

On-demand production is used in the case of electronics to print flexible circuits, embedded sensors, and light-weight antennas in mission frequencies. As multi-material printers can lay down conductive, dielectric, and structural materials in the same building cycle, they make possible the functional integration of mechanical and electronic systems [7]. Electrical conductivity can be high (e.g., laser-aided sintering or photonic curing) and the polymer substrates do not receive high amounts of heat. Techniques on how to manage thermal conditions of printed devices in space are integrated heat spreader and radiative coating. Traces run through structural layers in order to carry power electronics waste heat. Coatings of printed phase change material and graphene too enhance thermal stability with steep temperature gradients in orbit or surfaces of planets [8]. One of the issues of autonomous manufacture of space is quality assurance and reliability. Non-destructive methods of evaluation (NDE), e.g., X-ray tomography, thermography, and ultrasonic sensing, are undergoing modification to be applied *in situ* to probe printed parts. Artificial intelligence. which have been trained on print data on earth are capable of detecting defects and suggesting remedial measures in real time. Closed loop control systems permit repeatability of performance in repeated print cycles despite the instabilities of the fluids under the influence of microgravity [2]. Another way of realizing logistical optimization is through on-demand manufacturing coupled with spacecraft control systems. It can print certified design files, stored in digital part libraries, on demand *via* an onboard fabrication unit. This conserves on mass, volume, and use of physical spare parts.

The components, such as connectors, housings, and brackets, can be produced at the site with shorter lead-time. Wholly autonomous fabrication ecosystems would enable modular spacecraft to be supported in the long term, and would also be capable of being self-repairing and self-configurable. This would enable adaptive strategies to maintain systems, and would also increase missions and reduce the operational risk that would be integrated with AI diagnostics.

Future bases on the moon or Mars could resemble distributed manufacturing nodes linked together with digital networks with the tools, electronics, and building parts somehow being generated in a coordinated fashion [5]. Lastly, additive manufacturing on demand is an opportunity on sustainable, resilient, and self-sufficient space operations. As the material science, robotics, and machine intelligence converge, spacecrafts will be able to both fabricate, repair, and modify their systems to suit the mission, and will begin to reflect a change of the hardware as not a fixed system but a dynamically changing space infrastructure.

Omega In-Orbit Manufacturing: It enables a real-time repair of components [4]. Planetary Surface Fabrication: Local material is used in the Moon and Mars on the surface [3]. Structural Integration: Mechanical and electrical functionality [10].

14.5 Design Integration and Digital Workflow

Additive electronics are grounded on a digital design-to-manufacture process [8]:

1. Theoretical modeling of mechanical and electrical layers in 3D.
2. Specific parameterization of material and slicing of its layers.
3. Multi-material synchronization.
4. *In situ* monitoring.
5. Upon processing and validation.

AI-based optimization is used to improve space predictions and deposition prediction of failure to maximize accuracy of deposition. These electronic processes employ a combination of CAD and CAE to guarantee structural and electrical compatibility among layers. Digital twins mirror the conditions of printing, their model parameters are actively updated based on sensor data on temperature, viscosity, and deposition rate. Past print data is then analyzed with machine learning models to detect whether there are anomalies, and automatically adjust toolpaths to eliminate defects. Cloud-based collaborative systems in space enable production on-demand to permit the in-orbit manufacturing and ground control facilities to execute design iterations at the same time that the mission can be tailored to meet the needs of the customers [2].

The simulation design assists in optimizing the print structure, thermal support, and distribution of the print structure to minimize warping and residual stress. Digital quality metrics are an indication of the post-processing activities, like sintering, surface finishing, and conductive testing, and linked to mission assurance databases. Lastly, AI, digital twins, and additive workflows are integrated in such a way that, in a loop-based manufacturing ecosystem, it can make autonomous decisions and correct errors in microgravity environments.

14.6 Challenges in Space Additive Electronics

The key ones are compatibility with vacuum, thermal stress, radiation-degradation, and accuracy of deposition [2]. The aerospace electronics qualification standards are in a state of development [8]. Conductive inks have volatile binders, which outgas because of the vacuum condition leading to ineffective curing and performance of the conductive inks. Mechanical fatigue and micro-cracking of the printed traces in the multi-material composite materials and especially the printed traces are brought about by thermal cycling between the high-temperature and low-temperature regions. Radiation promotes scission and oxidation of the polymer chains, hence dielectric strength and dielectric conductivity degrades with time [7]. The concept of microgravity is more complex, because no one force is pushed on by the principle of buoyancy, and thus the deposition will be different. liquid or molten fluid flow. The closed-loop feedbacks required in the implementation of the correct layer registration and in the interfacial adhesion include the vision and thermal sensors. Also, space has limited *in situ* calibration standards that complicate the process of repeatability and certification of the processes.

Mechanical reliability and vibration tests, impact, and outgassing tests must be changed in order to simulate the conditions in space. The NASA, ESA, and ISO certification systems are being extended to additive manufacturing procedures and materials. Efforts are being made to come up with standardized evaluation measures of the electrical conductivity, adhesion strength, and radiation tolerance. Multi-dimensional computational methods are in use now to determine the probability that printed components are likely to fail, as a result of the development of microstructure during printing. More to the reliability and traceability of manufacturing cycles is provided by *in situ* diagnostics and predictive maintenance relying on machine learning. These difficulties will finally result in the long-term sustainability of additive manufacturing in deep-space mission-critical electronic systems.

14.7 Emerging Technologies and Innovations

Hybrid additive–subtractive systems, self-healing polymers, AI-controlled processes, *in situ* recycling are the latest trends in order to preserve the sustainability of the resources [10]. Hybrid: There is the additive–subtractive hybrid, which involves a combination of precision milling and layered deposition to achieve RF and microwave circuits accuracy to micron levels. Microcapsules in autonomous repair of cracks or electrical discontinuities can be made with polysaccharides, which can then be used in space to increase the durability of components. To dynamically identify defects through adjusting print parameters and find them in real time, AI-based monitoring systems are a combination of computer vision and forecasting analytics [8]. Recycling technologies would use *in situ* polymers and metals to produce feed-stock, which is utilized and reduce the dependency of the earth on material with Earth-based resupply missions. The new ones also include multi-material co-deposition systems which can print conductive, dielectric and structural materials with ease. It has been demonstrated that conductivity and radiation resistance of nanomaterial-enhanced inks café on the basis of carbon nanotube and MXenes are greater [7]. In autonomous production of future lunar and Martian habitats, a downsizing of printheads and robotic manipulators to being in microgravity are being tested. The innovations collectively bring the additive manufacturing to autonomous, intelligent, and sustainable space electronics ecosystems in full.

14.8 Cryogenic Testing, Qualification, and Validation for Space Missions

The other critical step in the development of the electronic systems in lunar and deep-space missions is the intensive cryogenic testing, qualification, and validation. Any cryogenic subsystems to be used in space should be able to service severe thermal, mechanical, and radiometric stresses over long durations in value of the mission, and that they cannot be repaired in-flight. As a result, a comprehensive ground-based test and qualification regimes are invaluable in any case of a successful mission and reducing the risks associated with flight failure.

Cryogenic testing aims at simulating the conditions of the environment that a system would experience in an orbit as closely as feasible. This includes ultra-low temperature, vacuum, and radiation exposure and mechanical loads connected with launch and maneuvering [6]. Because it is not feasible to reproduce all the conditions needed to simulate the whole spectrum of the space environment at the same time on Earth, the test methodologies are integrated with environmental simulation, accelerated life testing and analytical modeling of their robustness.

The basis of qualification is thermal–vacuum (TVAC) tests. Electrical assemblies are then placed in a TVAC chamber which is under low pressure conditions with temperature variations across the intended working and survival regions. TVAC removes convective heat transfer and so only conduction and radiation dissipation is possible, as in space. Electrical parameters, such as stability of bias of devices, leakage currents, noise performance, signal integrity, and others, are measured continuously in TVAC to identify temperature-dependent anomalies. In cryogenic electronics, TVAC may be continued long past the nominal operating temperatures in order to test functional margins, and to test the behavior of materials and interconnections at very low temperatures [7].

Another required component is thermal cycling qualification. The deep-space missions also require the repeated cryogenic and high-temperature outliers of the electronics because of the eclipses, instrument duty cycles, and maneuvers of the spacecraft. These are mechanical stresses that can lead to fatigue, delamination, and interconnect failure. Thermal cycling tests also hasten these effects by again and again testing assemblies by repeatedly cycling them between extreme temperatures to expose latent faults and confirm long-term reliability under mission-lifetime conditions.

Burn-in testing at high cryogenic temperatures is important to show early failures in life especially among semiconductor devices and interconnects. In burn-in, the components are subjected to constant operating conditions of cryogenic temperature and under an electrical load, accelerating defects associated with poor manufacturing, weak interconnections, and fringe elements [4]. The process is of particular concern to cryogenic electronics, where carrier freeze-outs or packaging stress fractures can only occur after long periods of low temperature operation.

Cryogenic stress and launch vibration qualification is an exception. During the launch process, electronic systems should be able to endure high mechanical vibration and shock and maintain operations at low temperatures. The cryogenic temperature stiffens and brittle nature of materials. Traditionally, therefore, qualification is done by the vibration and shock testing under ambient conditions, and subsequent cryogenic functional testing, or where possible, vibration testing at reduced temperature. This hybrid implementation approach will provide an insurance that the mechanical integrity and electrical performance remain intact in all phases of the mission [5].

In the development of space electronics, a distinct separation is made between the qualification testing and the acceptance testing. Qualification tests of representative units show that the design met the mission requirements and that the design is robust in worst-case operating conditions; acceptance tests of flight hardware show that it was properly made and that it works and that it does not have any manufacturing defects. Cryogenic systems: In cryogenic systems, limited-thermal cycling cryogenic functionality tests typically proceed the acceptance tests to ensure that the actual hardware is as per the qualified design [7].

The most complicated problem of cryogenic validation is the G2F correlation. Although the ground tests are extensively done, there may be disparities between the ground-test and in-flight operation in regard to thermal gradient, radiations, and the operation schedules. Engineers solve this by using described thermal and electrical models that reconcile a test data with the expected flight conditions, improving the models with the heritage data of the prior missions, and improving the predictive capabilities. The constant telemetry monitoring of the first phase of the mission offers a secondary point of verification, allowing adjustments of calibration according to the actual data on the operation.

14.9 Cryogenic Power Management and Energy Efficiency

Cryogenic power management is considered to be one of the least developed and hardest topics in design of space electronics. Power on missions to the Moon and deep space is extremely limited, and it is often provided by solar arrays, radioisotope thermoelectric generators, or little nuclear sources. At the same time, the cryogenic systems using electronic systems must have very high stability and reliability over long mission durations.

At low temperature, the electrical characteristics of power semiconductor device is by no means similar to those at room temperature. The carrier mobility, threshold voltage, and the parasitic resistance change in a fashion that directly affects the converter efficiency and the stability of the controller [2]. Consequently, power management architectures which have been designed to work at ambient temperature can be described to have lower effectiveness, or even break down, when operated at cryogenic temperature management. It is only possible by creating power management circuits which are now highly streamlined to run with low temperature rather than simply modifying existing ones.

Cryogenic DC to DC converters are required to bring the primary spacecraft bus voltages down to the low and constant supply is required by the cryogenic electronics and sensors. Switching devices such as MOSFETs at cryogenic temperature would be more likely to raise the threshold voltages and alter the transconductance, which makes it harder to design gate drive and predict switching control. Other passive elements, such as inductors and capacitors, alter also their magnetic permeability and dielectric constants, which affect energy storage and filtering properties. To reduce such problems, cryogenic DC-DC converters often incorporate fewer topologies with simple switching rate and low temperature components with low thermal properties [5].

Low-dropout regulators (LDOs) are significant in enabling noise-sensitive cryogenic circuits (particularly analog front ends and sensor readout electronics) to be provided with a clean and stable supply voltage. LDO behavior at cryogenic temperature depends on variations in the reference voltage generation, pass device behavior, and pass device stability of the feedback loop. Low temperatures tend to reduce leakage currents therefore can increase efficiency although higher threshold voltages may reduce dropout margin. These biasing networks and reference circuits are then to be designed with care in order to give stable regulation over a broad temperature range.

The power loss mechanisms in cryogenic systems are different as compared to those at room temperature and must be investigated to determine the most energy-efficient mechanisms. Despite the fact that the reduced thermal agitation tends to lead to reduced resistive losses, alternative loss mechanisms may be more important, such as switching losses, gate drive inefficiencies and magnetic core losses. Besides, the cryocooler energy consumed is also a significant percentage of the total energy budget. It is such that, not only does minimizing heat dissipation in cryogenic electronics increase electrical efficiency, but also reduces cooling needs, and this has a positive feedback effect on the power consumption of a system [6].

The stability of bias and efficiency trade-offs is a primary concern in cryogenic power management. In making sure that device performance is stable and in minimizing noise, particularly in analog and mixed-signal circuits, a stable condition of bias is required. However, active biasing to favor performance may introduce additional dissipation of power and thermal load to criogenic cooling system. To balance performance with energy efficiency, however, with care, designers are often relying on adaptive biasing schemes to adjust the operating point with temperature and workload.

The cryogenic networks of power delivery must supply a steady supply of electrical energy across both broad temperature ranges between hot spacecraft buses and cold components of the payload volume. This includes cryogenic compatible cabling, connector, and distribution architecture design, which minimize the voltage drop, electromagnetic interaction, and thermal conductivity. This particular isolation of power lines is of particular importance to thermal isolation in order that unwanted heat may be conducted into colder regions and ruin cooling characteristics. At the same time, its electrical integrity should be maintained in order to enable the stable to operate under the conditions of dynamic loads [8].

In summary, the space long-duration missions cannot be realized without the cryogenic power control and energy efficiency. Direct benefit has been given to the reliability of systems, cooling requirements, and mission life to new cryogenic DC to DC converters, low-noise voltage regulation, and power delivery architectures with optimized power delivery. As the missions in space continue on their way to more autonomy and exploration, invention of efficient and effective systems in the management of cryogenic power would keep on being at the center of research and engineering efforts.

14.10 Sensors and Instrumentation Enabled by Cryogenic Electronics

A major component of the preparation of the high-performance sensors and science instrumentation of lunar and deep space missions is cryogenic electronics. The highest space science ambitions require the measurements of the very weak signals, such as infrared radiation, minute temperature variations, or electromagnetic reflections of underground objects. At normal temperatures of operation which is much lower, thermal noise, and leakage currents limit such measurements and as such, inhibit the sensitivity and accuracy of such measurements to a large degree. The sources of noises are significantly reduced by utilizing cryogenic operation and sensors can operate near their physical limits.

Co-location of cryogenic electronics and sensors is also close to enhance performance through decreasing signal degradation during the transmission of a signal between the sensor and readout circuitry. Co-location of front-end electronics in the cryogenic sensor space minimizes parasitic capacitance, resistive loss, and electromagnetic interference. This is a trend typical of the modern space instrumentation particularly of the astrophysical, planetary, and subsurface exploration missions [9].

Infrared sensors are also among the most prominent businesses that have benefited by use of cryogenic electronics. The IR radiations that come out of distant heavenly bodies or ice-covered planetary surfaces are so minimal and can be easily lost in the thermal noise created inside the detector itself. Crying out is reduced by infrared cameras Crying out infrared detectors by cryogenic temperatures reduces dark current and thermal noise and enhances sensitivity and spectral resolution.

Space applications in bolometers and radiometers are based on the thermal and radiative measurements in space. Bolometers capitalize on the variations of temperature of an absorptive element to detect incident radiation, in a way that they are highly sensitive and highly sensitive to thermal noise. It can be increased significantly by cryogenically operating bolometers to reduce thermal variations and thereby increase the sensitivity of bolometers. Constant biasing, low-noise amplification, and precise readout, these small temperature-induced signals are given by cryogenic electronics. Radiometers, which sense the intensity of electromagnetic radiation at some frequency range, also benefit by the cryogenic low-noise amplifiers and reference circuits. Such instruments are also used in climatic studies of planets, measurements of cosmic background in the microwave, and thermal mapping of the surface of the moon and planets.

The other recent technological advancement in cryogenic electronics that has enabled an added new line of instrumentation is known as the cryogenic micro-electro-mechanical systems (MEMS) sensors. Acceleration indicators, gyroscopes, pressure sensors, and micro-mirrors can be employed at low temperature since there are lower levels of mechanical damping and quality factors. However, the problem of the material stiffness, the unremaining stress, and precaution measures are also valid with cryogenic operation. Cryogenic electronics are important in providing precise control signals, reduced noise readout, and consistency of MEMS sensors in those settings. Such sensors are now considered to be applicable to spacecrafts, planetary landers, navigation, vibration, and structural health [10].

Ice-penetrating radar systems, particularly to icy moons and polar caps of planetary bodies, are also another area of importance that may require potentially extremely critical usage of cryogenic electronics. These radar systems transmit or direct electromagnetic pulses to ice or regolith and infer the reflected pulses to ascertain the structure of the bottom of the surface and its composition. Cryogenic low-noise amplifiers and high-stability oscillators are necessary to observe weak return signals after a significant amount of propagation loss. Penetration depth and spatial resolution are also improved by cryogenic electronics because the thermal noise of the receiver chain has been reduced. These capabilities are needed so that spacecrafts, such as Europa Clipper, can study underground oceans and assess the habitability of icy worlds.

Planetary subsurface sensing does not only involve radar but also entails such techniques as electromagnetic sounding, thermal probing, and seismic measurements. Cryogenic electronics enhance such senses to achieve more signal fidelity and stability over time. Cryogenic amplifiers are used as an example because they are needed to detect very small seismic vibrations, which can be produced by a tectonic activity or collisions of meteorites that can be used to give the interior structure of a planet. Similarly, the cryogenic thermal measurements can be used to detect minimal quantities of heat that are connecting to geological processes or underground water bodies [8]. The necessity of such measurements is to learn more about the evolution of planets and define what resources may be utilized in the exploration in the future.

14.11 Communication and Signal Processing at Cryogenic Temperatures

The communication and signal processing subsystems are important in the data acquisition, transmission, and interpretation of space missions. These subsystems possess both special problems in cryogenic conditions, because the nature of devices has changed, there are very steep temperature distributions as well as very stiff noise limits. At the same time, the cryogenic operation possesses significant advantages, in particular, the low-thermal noise level and high signal fidelity. This has consequently seen cryogenic electronics have a greater role to play in facilitating high performance communications and signal processing in deep space and their use in planetary explorations.

Cryogenic low-noise amplifiers (LNAs) are considered one of the most crucial components of a cryogenic communication system. LNAs are usually located at the receiver chain in order to amplify the weak signals before any severe noise is introduced. Low temperature operation of

LNAs further practically eliminates thermal noise that would cause lower noise figures and increased sensitivity. This may be especially applicable in situations where the strength of received signals may be extremely low, e.g., in deep-space communication, planetary radar and radio astronomy. The most popular technologies used to make cryogenic LNAs are high-electron-mobility transistors (HEMTs) or special CMOS technology with nice noise properties at low temperatures.

Cryogenic RF front-ends must be carefully designed in order to be capable of coping with the variations in device behavior and passive component behavior. There are other parameters such as gain, bandwidth, impedance matching and linearity that may vary significantly due to temperature decrease. Passive components like capacitors, inductors, and transmission lines change in terms of dielectric constants, magnetic permeability, and conductor resistance. These manipulations can influence frequency response, matching conditions, and they can have a de-meriting effect on the system performance unless considered. It is possible to trust a large amount of characterization and modelling of the elements in the operating temperature range and strong margins on design, to allow stable operation in the cryogenic RF front-end design [7].

High-speed digital and analog signals through these gradients may be influenced by impedance discontinuities, mechanical stress of interconnects and variations in material properties. Signal reflections, crosstalk, and timing skew are problems that designers of cryogenic electronics must contend with, and these can become serious at lower temperatures. Care routing, controlled-impedance interconnects, and differential signaling are some of the techniques normally implemented to ensure signal integrity in cryogenic systems.

The other group of difficulties is in high-speed data connection through thermal gradients. The tools used in the contemporary space trigger volumes of data that are necessitated to be transmitted to either warm processing systems or onboard memory by the cryogenic payload of a satellite. These data interfaces should be characterized by large bandwidth and low error rates and cross interfaces between cryogen temperature and ambient temperature systems.

Signal processing with low noise may also be carried out in cryogenic temperatures. Cryogenic operation allows the possibility of carrying out very sensitive analog signal processing operations (filtering, mixing, amplification, etc.) in the immediate vicinity of the sensor or antenna [9]. This proximity lowers signal loss, noise pick-up even prior to digitizing. In other applications, the blocks of digital signal processing are placed in cryogenic regions to reduce the latency as well as to improve the efficiency of the entire system. This approach, however, should consider the dissipation of power since digital circuits can generate a significant amount of heat thus it will need to be cooled by the cryogenic cooling system.

The choice of optical or RF communication technology is also another trade-off of the design of cryogenic communication system. RF communication systems have been well proven and good experience in space missions. They are less strict to misalignment and variability of the environment but less bandwidth and more prone to noise. Comparatively, optical systems of communication possess considerably large data rates and enhanced security however, they must be extremely accurate and they are prone to environmental interference [8]. The optical elements operating within cryogenic conditions, e.g., lasers, modulators, and photodetectors, may have altered performance characteristics, which ought to be properly controlled. Cryogenic electronics play critical roles in driving, controlling and stabilization of such components in order to ensure that they would be reliable to work.

14.12 Thermal–Electrical Co-Design Methodology

Cryogenic electrical functions are very thermal sensitive. There are some parameters such as carrier mobility, which are highly temperature dependent, and leakage current and noise parameters. Although lower temperatures have been realized to enhance performance by lowering thermal noise and resistive losses, it can also lead to bias instability, device variability as well as tolerance to fabrication variability. Local heating may occur due to any power dissipation and this may lead to an unbalanced thermal balance that is required to sustain stable operation in the cryogenic regime [6]. In that way, electrical designs directly affect thermal behavior and the other way around which demands the co-design form rather than the design cycle.

A typical thermal–electrical co-design experience follows the system level starting with mission requirements defining what temperature, power and performance requirements are allowable. Thailand: This is based on thermal isolation considerations, and signal integrity considerations, and early-stage architectural decisions, such as the placement of the electronics with respect to sensors or antennas. Circuit level optimization can be done of the size and biasing and topology of devices not only in electrical aspects (gain, bandwidth, noise figure, etc.) but also by reduction of power dissipation and thermal gradients [7]. Design: It requires thermal and electrical analysis to be given feedback.

Delivering co-designing is centered around the simulation of electro-thermal. The advanced simulators are electrical circuit models and thermal solvers to predict the distribution of temperatures, paths of heat flow, and variation of performance under the realistic operating conditions. Such simulations are founded on temperature-dependent material characteristics, non-uniformity in the power dissipation, and cryogenic cooling system boundaries. The heat conduction models are usually the finiteness element methods (FEM) to model substrates, interconnects, and packaging structures and compact thermal models on the circuit and system level to help in the fast design cycle. Adequate knowledge of the electro-thermal modeling reduces the likelihood of the mission runaway or poor performance despite any unpredictable occurrences in the mission.

One of the significant problems of thermal–electrical-trade-off between cooling and power and system performance are usually incorporated in the thermal–electrical co-design. Cryogenic cooling facilities are expensive and few in quantity and volume and energy. To improve the electrical performance, cooling capacity may be more intensified to maintain the temperatures lower and more constant, but at the expense of extra complexity and power usage in the system. Conversely, extreme reduction of cooling power may have an impact on the performance or reliability [5]. The co-design methods enable the designer to quantitatively evaluate these trade-offs, to identify operating points that result in a scientific maximum, and be within the mission constraints.

Thermal–electrical co-design is the one that gives rise to the optimization of a system at the larger scale i.e. optimization of multiple subsystems. As an example, the signal processing may be moved nearer to the cryogenic sensor to reduce the noise and refine the quality of the data, yet the dissipation of power locally also increases, and it should be cooled. Similarly, the decisions concerning redundancy and fault tolerance influence power consumption and heat generation. As these interactions are considered in the initial phases of the design, co-design methods eliminate expensive redesign and as thermal and electrical subsystems, co-design techniques can be applied to ensure that subsystems grow concurrently.

14.13 Comparison between Terrestrial and Space Cryogenic Electronics

Cryogenic electronics finds application in terrestrial applications and space applications and these are quite different in the design philosophy, design constraints, and operating expectation of the two domains. This understanding of these differences is very important to both the understanding the special challenges of space cryogenic systems and the avoidance of direct solution transfer on earth to the space environment without the necessary adaptation. Such comparative perspective adds to the educational value of the chapter and the practical experience related to the design of cryogenic electronics [2].

The first significant difference is the restrictions that the operating environment does impose. Cryogenic electronics terrestrial cryogenic electronics, such as particle accelerator, medical imaging, and quantum computer laboratories, are generally housed in controlled locations with stabilized conditions. Space cryogenic electronics on the other hand must be able to operate in vacuity, extreme temperature gradient, and must have the ability to resist radiation, microgravity, and mechanical stress during launch and deployment. These are environmental factors that have a great influence in the selection of material, packaging solution, and system structure in space applications.

The other point of significant divergence is the availability of cooling. Onshore systems normally utilize large and long-lasting cooling systems, including liquid helium, liquid nitrogen, or large-scale closed-cycle refrigeration units supplied by external electric systems. This kind of system can be repaired, upgraded, or even replaced. Space missions involve the use of mass, volume and power as key factors and these elements have a heavy burden on cooling resources [6]. Very efficient active or passive cooling systems must be optimized to meet the mission requirements over a long duration of time. Space cryogenic systems must be able to work without refilling their cryogen or external support when deployed and therefore cooling efficiency and reliability is the major factor.

The cryogenic electronics in the earth and the space are also distinguished by the expectations of reliability. Even though high reliability is extremely essential in the systems on the ground, the failures can be overcome in several cases by repairing, changing the components or reconfigurating. In comparison, space cryogenic electronics must be capable of working to very high reliability levels, as once a spacecraft is in operation it is often impossible to repair it, or replace it. This demands a very slim design margin, numerous tests and numerous degrees of fault tolerance and redundancy. The focus on the long-term, maintenance free operation has a great impact on the design consideration of space cryogenic systems.

The other significant difference is that of maintenance as well as free operation. Direct human control, real-time control, manual control and scheduled maintenance are the common control ways in earthside cryogenic electronics systems. Space cryogenic electronics must be able to operate years, even decades, and in most instances miles distant on earth. Space systems then have very fundamental aspects in independent health management, fault identification and recovery functions [9]. The capabilities complicate the electronics but are needed to continue with the mission without human intervention.

Cost and qualification requirement of space and terrestrial application is also considerably different. Cryogenic systems are not habitually as stringently qualified (as space hardware) and may often be extremely expensive on earth. Space cryogenic electronics must be harshly tested and validated to demonstrate launchworthiness, radiation, thermal cycling, and extended operation. Such qualification activities substantially lengthen the development time and cost in order to minimize the mission risk. As a result, space cryogenic electronics are often focused on proven technology and incremental development rather than rapid development in the utilization of state-of-the-art components that are generally available only in terrestrial research laboratories.

14.14 Standardization, Space Qualification, and Design Guidelines

The effective functioning of the cryogenic electronics in the space missions does not only concern the development of technology but also the high standards and qualification processes. The space agencies and international bodies have established vast standards and guidelines in order to determine the capability of electronic systems to resist the harsh conditions of the launch and prolonged operations in space. These standards are key in risk reduction, interoperability and reliability of executing missions in different platforms and environments in the case of cryogenic electronics [5].

The design, testing, and qualification of space electronics have elaborate ones prepared by both major space agencies such as the National Aeronautics and Space Administration(NASA) and the European Space Agency (ESA) and the Indian Space Research Organization (ISRO). Problems are addressed as per the NASA requirements, e.g., NASA Technical Standards System such as thermal control, radiation hardness, material selection and reliability assurance. ESA standards are prepared in the European Cooperation of Space standardization (ECSS) system that provides harmonious guidelines in the selection of electronic components, environmental testing, as well as quality control. The ISRO specifications are also the requirements of space grade electronics with emphasis on robustness, indigenous qualification and mission requirements. Nevertheless, because these standards may generally be applied to space electronics, the cryogenic systems typically require particular interpretations and additional tests in order to consider the extremely low temperatures of use [8].

Technology Readiness Level (TRL) assessment is one of the most significant ways of space qualification of cryogenic electronics. TRLs provide a systematic way of determining the degree of maturity of a technology at the most primitive concepts that can be observed at TRL 1, to systems that have been flight-tested at TRL 9. There are many cryogenic electronic components and subsystems at intermediate TRL, particularly associated with new materials or new cooling systems or superconducting equipment.

14.15 Applications in Space Missions

CubeSats printed antennas, lunar module strain sensors and embedded Mars habitat wiring are some examples [5]. With additive manufacturing, the parts of satellites can be swiftly tailored to the specific mission payload and communication requirements. Antenna and waveguides inflated into the spacecraft structures reduce the mass and enhance the electromagnetic performance [3]. Lunar gateway systems: It is projected that lunar gateway systems will be provided with printed thermal management networks to manage electronics on board. Planet exploration could involve replacement of tools or sensor casings with *in situ* fabrication units in a rover [9]. AM can be used in Mars habitat to implement electrical circuits on wall panels to aid system integration, and minimize complexity related to cabling. Manufacture on-orbit facilitates the development of deployable facilities such as solar arrays and communication reflectors [11]. In the future, AM can also be used to enable fully automated power grid construction, sensor networks and robotic infrastructure on alien surfaces. These applications demonstrate how additive electronics can be applied to NextGen space mission light weight, resilient, recoverable systems.

14.16 Conclusion

Additive manufacturing and 3D printing are beginning to become mission-critical enablers of space electronics. In spite of the problems that remain, interdisciplinary research is heading towards sustainability, autonomous in-space fabrication. Such technologies significantly reduce the dependency of the supply chain on the Earth, and it is possible to produce components in real-time during missions that last long. This is due to the fact that the additive manufacturing together with the modern material aids in the production of light weight and radiation hardened electronic systems. Automation of design optimization based on AI is merely applied to ensure that scarce resources are utilized effectively in extraterrestrial environments. The future will be to make multi-material printing more precise, reliable and thermally stable in the extreme conditions [2]. The aerospace industries, space agencies, and academic institutions have been collaborating more toward accelerating the technological readiness. As research has matured, it will be common to print the antennas, sensors and circuit boards on demand in orbital and planetary operations. Lastly, additive manufacturing is the future of space exploration because it will allow a new generation of self-sufficient and adaptable space systems with the capability to enter the deep space. Thus, 3D printing may be regarded as among the pillar technology of the resilient and intelligent space infrastructures of the next generation.

AI disclosure statement: The author confirms that no artificial intelligence (AI)-generated content or AI-assisted writing tools were used in the preparation of this chapter (including summaries and the abstract). All text, analysis, and conclusions are the result of the author's original work.

References

1. Zocca, A., Wilbig, J., Waske, A., Günster, J., Widjaja, M.P., Neumann, C., Clozel, M., *et al.*, Challenges in the technology development for additive manufacturing in space. *Chin. J. Mech. Eng.: Addit. Manuf. Front.*, 1, 1, 100018, 2022.
2. Li, W., Zhang, C., Lan, D., Ji, W., Wang, Y., Solution printing of electronics and sensors: applicability and application in space. *Adv. Eng. Mater.*, 24, 9, 2200173, 2022.
3. Georgantzinos, S.K., Papadopoulou, E., Kostopoulos, G., Voulgaris, S., Tseni, A., Bakalis, P., Gkara, M., Mitropoulou, C.C., Lagaros, N.D., A Comprehensive Review of Additive Manufacturing for Space Applications: Materials, Advances, Challenges, and Future Directions. *Adv. Eng. Mater.*, e202501082, 2025.
4. Dunn, J.J., Hutchison, D.N., Kemmer, A.M., Ellsworth, A.Z., Snyder, M., White, W.B., Blair, B.R., 3D printing in space: enabling new markets and accelerating the growth of orbital infrastructure, in: *Proc. Space Manuf.*, vol. 14, pp. 29–31, 2010.

5. Georgantzinos, S.K., Papadopoulou, E., Kostopoulos, G., Tseni, A., Gkara, M., Mitropoulou, C.H., Lagaros, N.D., Additive Manufacturing for Space: Key Requirements, Applications, and Future Directions, in: *International Conferences on Engineering and Applied Sciences Optimization*, Springer Nature Switzerland, Cham, pp. 33–43, 2024.
6. Fu, X., Lin, Y., Yue, X.-J., Ma, X., Hur, B., Yue, X.-Z., A review of additive manufacturing (3D printing) in aerospace: Technology, materials, applications, and challenges, in: *Mobile Wireless Middleware, Operating Systems and Applications: 10th International Conference on Mobile Wireless Middleware, Operating Systems and Applications (MOBILWARE 2021)*, Springer International Publishing, Cham, pp. 73–98, 2022.
7. Zhang, Y., Zhang, G., Qiao, J., Li, L., Design and *in situ* additive manufacturing of multifunctional structures. *Engineering*, 28, 58–68, 2023.
8. Mao, M., Meng, Z., Huang, X., Zhu, H., Wang, L., Tian, X., He, J., Li, D., Lu, B., 3D printing in space: from mechanical structures to living tissues. *Int. J. Extreme Manuf.*, 6, 2, 023001, 2024.
9. Espera Jr, A.H., Dizon, J.R.C., Chen, Q., Advincula, R.C., 3D-printing and advanced manufacturing for electronics. *Prog. Addit. Manuf.*, 4, 3, 245–267, 2019.
10. Clinton, R., Prater, T., Werkheiser, N., Morgan, K., Ledbetter, F.E., NASA Additive Manufacturing Initiatives for Deep Space Human Exploration, in: *International Astronautical Congress*, no. MSFC-E-DAA-TN61658, 2018.
11. Schlake, E., Verma, S.K., Kandadai, N., Laser sintering of silver-palladium ink for on-demand manufacturing of electronics for space applications. *Sci. Rep.*, 15, 1, 27412, 2025.

15

Case Study on Electronics in the James Webb Space Telescope: Innovations and Challenges

Ravendra Kumar K.[1], Sapna[2*], Shilpa V.[3] and Girish H.[4]

[1]*CISCO Systems, Bangalore, Karnataka, India*
[2]*Manipal Institute of Technology Bengaluru, Manipal Academy of Higher Education, Manipal, India*
[3]*Department of Computer Science Engineering, Cambridge Institute of Technology, K R Puram, Bangalore, Karnataka, India*
[4]*Department of Electronics and Communication Engineering, Cambridge Institute of Technology, K R Puram, Bangalore, Karnataka, India*

Abstract
The James Webb Space Telescope (JWST) represents subsequent advancement in space exploration following the Hubble Space Telescope (HST) and Spitzer Space Telescope. Astrophysics posed substantial enquiries that demanded resolution, necessitating a large-aperture telescope specifically engineered for the infrared (IR) spectrum, which functions optimally when positioned away from Earth's heat and light interference. JWST is built to see the universe in infrared light. It has never been more sensitive or high-resolution for a wide range of astrophysical research. The primary reflection is made up of 18 hexagonal segments and is 6.5 m wide. It has a gold coating and a collecting area of about 25.4 m^2.

Keywords: JWST, galaxy, lagrange point, hubble deep field image, light rays

15.1 Introduction

Since Hubble's Deep Field image of Hubble made it possible to see a galaxy that was originally unseen, Hubble began at the time of Hubble [1].

**Corresponding author*: sapna.ise@cambridge.edu.in

N. Sathish, Prasanth Aruchamy, Rajesh Kumar Dhanaraj, Nithya Rekha and Karthik Palani (eds.) Space Electronics for Deep Space Exploration, (351–366) © 2026 Scrivener Publishing LLC

Therefore, the identification of that object was based not only on the location of the light, but on its color [2]. The red color indicated that the galaxy had been approximately 12.6 billion years old [3]. Because light of this color has shifted into an infrared spectrum, astronomers cannot detect it using optical telescopes (like Hubble) [4]. Hubble has successfully detected the earliest light from galaxies up to 500 million years after the formation of the universe (the Big Bang). JWST will enhance astronomers' ability to study galaxy formation and evolution over larger distances and over greater times than has previously been feasible by also collecting infrared emissions of galaxies [5].

The JWST would provide more and better opportunities for studying the universe than any telescope has before [6]. There have been very large advancements made in recent times in both the design of the mirror system on the JWST as well as the technology that is being used to create sensors, which are the devices that allow JWST to capture infrared images of objects outside the visible spectrum in the universe [7]. In addition to having segmented mirrors and infrared capabilities, JWST has been designed to include a sun shield to minimize the temperature of all the components of the telescope. Infrared instruments are generally much more susceptible to temperature than traditional imaging systems, so having an effective sun shield will allow all components of the JWST to maintain lower temperatures, which is critical for proper functioning of the telescope. The sun shield itself will be roughly the size of a tennis court.

The way the sunshield functions is that it blocks light and thermal radiation from Earth, sun and Moon from hitting the telescope [8]. There are five layers of the sunshield that do this which means that the tools on board JWST will be kept around −233°C (40°K) and it is an extremely cold environment. It helps make sure that when JWST makes infrared observations they will be as sensitive as possible. Another unique quality about the telescope(s) is that they are located at the second Lagrange point in the solar system which is located at around 1.5 million km from Earth (1.1 million miles) [10]. Because the second Lagrange Point would be the stable point it will help ensure that the JWST sunshield will always have an ability to work and block thermal heat, so it will not ever have a thermal disturbance while performing its job of observing the universe and reporting back data to scientists on Earth to help learn more about space. Also, because of the location of the telescope being at the second Lagrange Point there will be little to no disruptions created by any other objects in outer space when looking back at the universe [11]. Due to the combined size of its segmented primary mirror and its sensitivity to infrared wavelengths, the JWST offers a unique combination of capabilities not available from earlier telescopes. It also requires very low temperatures to maintain stability over long periods of time. That setup lets scientists' peek at the very first stars and galaxies out there. It is kind of amazing for figuring out where planets come from too. And then there is checking out atmospheres on worlds far away. It seems like those features just work together to open up all this early universe stuff.

15.1.1 The JWST Architecture

The JWST Mission designed as successor for the incredibly successful HST in mid-1990s. Finding "first light objects," analyzing the development of galaxies at that time period to the current, and researching star formation and the formation of solar-systems within our galaxy were the main goals of the exploration of the early universe. These four scientific objectives were later elaborated to include the study of roughly 5000 exoplanets since the system design was judged adequate without modifications [1].

The JWST is intended for a life of at least five years of operation from its position at the second Lagrange Point (L2) located between the Earth and Sun. The observatory was designed with an Ariane 5 Launch Vehicle to launch it directly into a 180-day orbit around L2. Since the observatory's deployed configuration is greater than the available volume of the Ariane 5 fairing or any other launcher fairing, the observatory's configuration is stowed for launch, and upon reaching the L2 point, it will undergo a complex series of deployments. The STScI oversees the management of both scientific and mission operations. The DSN enables communication with the observatory, handling both low rate of data S band commands and telemetry, as well as high rate of data Ka-band scientific data transmissions. Figure 15.1 illustrates an JWST Architecture.

I Driving Requirements

The JWST mission is structured hierarchically to document how lower-level "children" requirements are allocated across different system segments, including the Ground Segment, Launch Segment, and Observatory Segment. The JWST Project (JWST_PLAN_000633) gives a description of how the activities above will allow us to fulfil the JWST Mission objectives. The JWST Mission Level 1 Requirements can be found in Table 15.1 of the JWST Project. However, as the JWST Mission Level 1 utilization was further detailed with Level 2, there are other documents associated with JWST that are also relevant to completing the Observatory's mission. One of these documents is JWST_RQMT_000634, which contains a complete list of JWST Mission Requirements and can be referenced for purchasing JWST items including the Observatory.

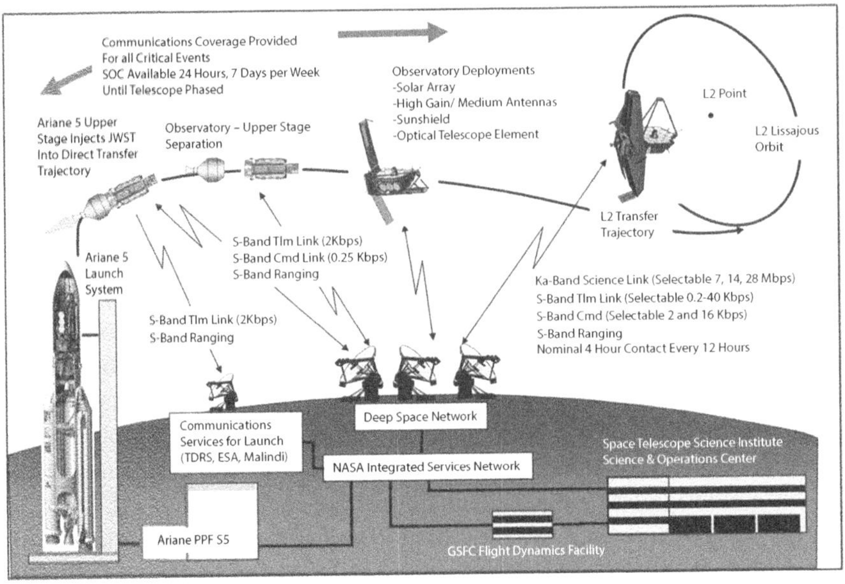

Figure 15.1 The JWST mission system architecture [1].

Table 15.1 Driving level 1 requirements.

Reqt #	**Requirement**
L1-1	Measure space density of galaxies to 2 μm flux density of 10^{-34} W m^{-2} Hz^{-1}
L1-2	Measure spectra of at least 2500 galaxies with R between 100 and 1000 to a 2 μm line flux density of 5.2×10^{-22} W m^{-2}
L1-3	Measure physical and chemical properties of young stellar objects, circumstellar disks, extrasolar planets, and solar system objects
L1-21	Data Volume: JWST shall deliver and process 229 Gbits per contact
L1-10	Data Availability: JWST shall deliver 95% of all Real Time (RT) and Stored Data
L1-8	Observing Time: 5×10^7 s for 5 yr
L1-9	Mission Life: 5 yr
L1-12	Orbit: L2 orbit

Table 15.2 Level 2 requirements driving the observatory design.

Reqt #	Requirement
MR-51	Wavelength Coverage: The observatory spectral coverage shall extend from 0.6 to 27 μm.
	Sensitivity: The observatory shall achieve the following signal-to-noise ratios (S/Ns) for the following source with a spectral resolution of *R*:
	$\lambda = 1.15\ \mu m$ (NIRCam): 1.10×10^{-31} W m^{-2} Hz^{-1} S/N = 10 in 10^4 s or less at $R = 4$
	$\lambda = 2\ \mu m$ (NIRCam): 6.25×10^{-21} W m^{-2} Hz^{-1} S/N = 10 in 10^4 s or less at $R = 4$
	$\lambda = 1.4\ \mu m$ (NIRISS): 1.26×10^{-33} W m^{-2} Hz^{-1} S/N = 10 in 10^4 s or less at $R = 150$
	$\lambda = 3.0\ \mu m$ (NIRSpec): 1.32×10^{-33} W m^{-2} Hz^{-1} S/N = 10 in 10^4 s or less at $R = 100$
	$\lambda = 2.0\ \mu m$ (NIRSpec) : 5.72×10^{-22} W m^{-2} S/N = 10 in 10^5 s or less
	$\lambda = 10\ \mu m$ (MIRI Imager): 7.0×10^{-33} W m^{-2} Hz^{-1} S/N = 10 in 10^5 s or less at $R = 5$
	$\lambda = 21\ \mu m$ (MIRI Imager): 8.7×10^{-32} W m^{-2} Hz^{-1} S/N = 10 in 10^5 s or less at $R = 4.2$
	$\lambda = 9.2\ \mu m$ (MIRI Spec): 1.0×10^{-20} W m^{-2} S/N = 10 in 10^5 s or less at $R = 2400$
	$\lambda = 22.5\ \mu m$ (MIRI Spec) : 5.6×10^{-20} W m^{-2} S/N = 10 in 10^5 s or less at $R = 1200$
MR-121	Maximum NIR stray-light levels in MJy $Ster^{-1}$:
	$\lambda = 2\ \mu m$: 0.091
	$\lambda = 3\ \mu m$: 0.070
MR-122	Maximum MIR stray-light levels in MJy $Ster^{-1}$:
	$\lambda = 10\ \mu m$: 3.9
	$\lambda = 20\ \mu m$: 200

(Continued)

Table 15.2 Level 2 requirements driving the observatory design. (*Continued*)

Reqt #	Requirement
MR-211	OTE Area * transmission >22 m^2 for wavelengths greater than 2 μm
MR-110	NIRCam Strehl greater than 0.8 over the NIRCam FOV
MR-116	MIRI Strehl at λ = 5.6 μm greater than 0.8
MR-113	24 hr Encircled Energy (EE) Stability: Without ground commands there shall be less than 2% rms variation in EE stability within a 0″.08 radius at 1 = 2 μm over a 24 hr period
MR-114	Conditions for Stability: The 24 hr stability requirement shall be met for any combination of target locations within the FOR including those separated by a worst-case thermal slew of 10°
MR-115	EE Long-term Stability: The EE within a radius of 0″.08 at a λ = 2 μm shall not change by more than 2.5% in less than 14 days following a worst-case slew
MR-104	The observatory field of regard shall be at least 35% of the celestial sphere

In general, requirements in Table 15.2 of the JWST_PLAN_000633 document have an influence on how well the observatory is designed and impacts the scientific capabilities. To meet the scientific mission of observatory, it is crucial to produce a stable platform. The parameters used in describing the field of regard (FOR) for the JWST Observatory are among these basic components of producing a stable platform for the observatory.

15.2 Related Work

I The authors [2] provide a comprehensive parametric cost model relevant to both space-based and terrestrial telescopes.

Model Structure: The cost model is expressed as a multivariable equation involving:

- Diameter (D) of the optical telescope assembly (OTA)
- Wavelength (λ) performance
- Operating temperature (T)
- Duplication factor (Y)

Findings:

- The cost of space telescopes is 50–100× more than ground telescopes.
- Diameter is the primary factor affecting costs.
- Technological advancements lead to a roughly 50% reduction in costs every two decades.
- Duplication helps only when manufacturing identical systems (e.g., interferometers).

II The authors [3] conducted a 2-year review of their original database, correcting inconsistencies and updating entries. They applied statistical regression techniques to telescope cost data, focusing on aperture diameter as the primary variable.

Key Findings:

- The size of the aperture continues to be the primary factor influencing costs.
- Telescopes with low areal density and decreased stiffness often has more cost per kg compared to those that are more rigid.
- The cost of a telescope accounts between 5% and 30% of the total mission expenses, revising the previous average estimate of 20%.

III The JWST [4] features a 6.5m segmented mirror and is engineered to function at very low temperatures in space.

- **Challenge:** Traditional cryogenic optical testing needed a huge, complicated, and expensive test tower inside a cryo-vacuum chamber, which made it hard to set up and expensive.
- **Innovation**: The authors came up with a new testing method that
    - does away with the required of cryogenic optical and test tower.
    - Puts important test equipment at outer space of the chamber so that it is easier to get to during testing.
    - Uses interferometry and wavefront sensing to check how well optics work in very cold temperatures.

This method reduces the costs and made things less complicated, it also lowered the risk by a lot, all while still making sure that JWST's optics were tested to a high standard.

IV The objective of this research work [5] was to reformulate the optical testing process for JWST was improved through reengineering the system in light of changing engineering constraints, budget limitations, and previously acquired experience through other tests, as well as other successes.

To eliminate the need for conventional full system cryogenic optical testing, to transition toward a modularized and distributed testing approach. To create a greater emphasis on testing subsystems in both ambient and cryogenic environments rather than testing with large test chambers as the primary test environment. To perform optical performance verification to eliminate the need for performing full end-to-end cryogenic testing by utilizing an integrated approach of applying interferometry, wavefront sensing to measure optical performance, and modelling to validate optical performance

The advantages are:

- The ability to reduce overall system costs and complexity,
- The ability to allow greater system flexibility and improve fault isolation, and
- The increased confidence in the alignment of segmented mirrors and system performance.

V This study reviewed both the Technical and Financial Management of JWST from both a Technical and Financial Management perspective. The primary purpose of this review was to present an updated review of JWST for the benefit of JWST project managers to assist them in determining their current status; the second purpose was to give specific action items/recommendations regarding the technical, financial, and managerial issues identified during this study.

The following is a summary of some of the key findings identified by the Committee in their assessment of JWST:

- Technical—The committee found that JWST has put in place all of the elements necessary to conduct a successful scientific mission.
- Financial—The committee found that JWST is expected to exceed its budget by approximately $1.5 billion, and that JWST needs to develop and obtain a new/updated budget and schedule.

The following are examples of some of the key recommendations put forth by the committee in their deliberations: From the outset of a project, establish reasonable timeframes and budget amounts, and use the budgets as the basis for allocating money to a project. Additionally, JWST's project financial management needs to be improved through the provision of more detailed or accurate information about the status of JWST's project financial management and by developing methods to better use JWST's financial reserves to enhance visibility on JWST's project's financial status. The same time that improvements are made in JWST's financial management structure, the technical progress of the JWST project must continue to occur.

VI This document has two objectives in assessing the technical, financial and management elements of the JWST.

1. To give a current and complete overview of the JWST from the above basis (technical/financial/management).
2. To make recommendations on ways of addressing various identified issues.

Context of the Mission for this Paper. The mission of the Chandra X-ray Observatory (launched in 1999) was to explore the Universe's extreme energy outputs. These events include Black Holes, Supernovae, and Neutron Stars.

Engineering Lessons Learned

- Precision optics: the mirrors of Chandra were to be aligned and polished to better than sub arc seconds; this was extremely challenging.
- Thermal stability: It was critical to maintain the thermal environment and the cryogenic environment to preserve the optical performance of Chandra.
- Systems integration: The importance of continually integrating on a project and testing all components as early as possible in the design phase to avoid surprises has been confirmed.
- Risk management: Identifying technical issues before they arise allows timely resolution of issues before they incur high costs due to delays.

Programmanic Insights

- Team differences: Successful integration of the optics, structure, and systems of Chandra required cooperation and collaboration among the design and engineering teams.

- Documenting and maintaining traceability of designs: Maintaining a comprehensive record of Chandra's detailed design documentation over an extended period was instrumental to managing the design and engineering of the instrument.
- Legacy effects: The success of Chandra's high-precision instruments has directly impacted the construction of high-precision tools for NASA's current and future missions such as the JWST and ATHENA.

V The use of the JWST's structure [8] again, as well as the JWST's architecture, are being utilized to significantly reduce risk, expense, and time associated with the mission planning. As a result of extensive testing of hardware and subsystems on previous missions, the team has demonstrated that utilizing hardware and subsystems from JWST will have great advantages for conducting future missions. The following list includes candidate missions for use of the JWST hardware used in previous missions.

Candidate Missions

- Design of the JWST can be modified for missions such as LUVOIR and HabEx.
- Segmented mirror technology of JWST provides a technological advantage for large aperture telescopes.

Advantages of Reuse

- Lower cost, due to the reduced need for R&D and testing.
- Shorter schedules from eliminating the need to perform first time engineering.
- Lower risk due to the implementation of systems and processes that have been previously tested and successfully implemented.

15.3 Innovations

The James Webb telescope is rethinking the way we look at astrophysics; in particular, the formation of galaxies, stars and planets. The James Webb comes equipped with a new type of mirror, which consists of 18 segments. After launch, these segments will eventually unfold and shape themselves into the final mirror shape. These mirrors will be made of a lightweight material called beryllium [17]. In addition, the James Webb telescope has a cryogenic cooler that will cool one of its instruments MIRI to 7K cooler and allow for better performance from this instrument by lowering the temperature at which the instrument operates.

The mirrors of the Webb telescope were designed to be foldable in sections like a drop-leaf table so that they would be small enough to fit inside a rocket.

15.3.1 MIRRORS—Lightweight Optics

NASA launched an initiative to explore innovative methods for constructing telescope mirrors. The AMSD programme was a joint 4-year initiative that included NASA the US Air Force, aimed at exploring methods for producing lightweight mirrors.

15.3.2 MIRRORS—Cryogenic Actuators and Mirror Control

Lee Feinberg, who manages the WOT Element loated at NASA Goddard to make the mirror segments work as one large mirror, each segment must be aligned to within 1/10,000th of a human hair's thickness. Further astonishing is scientists involved in the Webb-telescope project was to actually create the technique to accomplish.

15.3.3 MIRRORS—Wave Front Sensing

Wavefront sensing as well as control systems were created to identify and rectify any optical imperfections in the telescope. This was essential to make sure that the 18 segments of mirror function collectively as single big mirror.

15.3.4 Webb's Sunshield

The sunshield is responsible for creating a temperature disparity of almost 600°F between the spacecraft's warm and cool sides, extending across its five-layer structure that stands 4.8 m tall. It lowers the temperature at around 185°F on the warm to about −388°F on the cool side.

15.3.5 Webb Telescope Sunshields

A tennis court shield developed for Webb will serve the purpose of shielding the telescope from light from the sun. Constructed of five layers of thin

Kapton E and covered using aluminum for durability the Webb telescope's sunshields will also use silicon to reflect solar energy back into space.

I Improved Detectors

The Webb telescope will use improved infrared detector designs that are much quieter, much larger and have longer lifetimes than their predecessors.

II Micro-Shutters

Each micro-shutter can be opened and closed to view multiple objects with high resolution simultaneously, this means that observations could be made quicker than with traditional methods.

III MIRI Cryocooler

The detectors used by the MIRI instrument have been designed to operate best when cooled to less than 7K. A specialized "cryocooler" was developed for the MIRI detectors to achieve this temperature because passive methods of cooling did not provide enough heat rejection.

IV Backplane

This holds hardware of 2400 kg (2 ½ tonnes) and needs to stay almost still so that mirrors could see deep space. It was made to stay stable within 32 nm, which is 1/10,000th the thickness of a human hair.

V Collecting Data at Cryogenic Temperatures ASIC

Webb had a technical problem with changing "A-to-D" at outset very low cryogenic temperatures using an ASIC. This idea is not new, though.

15.4 Technical Challenges

Requirements create the following design Challenges for the JWST:

15.4.1 Cryogenic

To meet MIR standards and get the needed radiometric sensitivity, it is important to lower the noise from the detector and the heat coming from the OTE optics. The OTE's optics need to work at temperatures as low as 55°K, and the NIR-SI detectors, which utilizes HgCdTe, need to as low as 45°K. The MIRI detector, on the other hand, requires cooling to 6°K. Given the 6.6-m diameter of the OTE, passive cooling is the only viable option. This method involves making sure that the OTE and NIR SIs can see the cold space without any obstructions and are protected from the sun by sun shield. The NIR detectors would be cooled by the MIRI detector and cryogenic-radiator-margin (CRM) would have its own cryocooler. To use such a wide range of passive cooling, you need a very efficient insulating sunshield, careful planning and testing of all unwanted heat paths enough CRM to account for unknowns.

15.4.2 Size/Deployment

The OTE needs an aperture that is more than 6 m wide to get the right radiometric sensitivity. The sunshield needs to be about 15 m wide and 21 m long, which is same size as tennis court, in order to properly protect the OTE within the field of view set by MR-104. Even the biggest launcher fairings, that has diameters of 5 m, cannot fit an observatory this big. Because of this, the observatory must be folded up for launch and then set up in a complicated way once it gets into orbit. These deployments are not very reliable because there are so many parts involved, and any one of them could fail. In addition, the deployment of lightweight flexible sunshield membranes and cables presents the risk of snagging or tearing during operations due to the non-deterministic nature of these items. For this reason, it is critical to manage and control their use effectively at every stage of deployment.

15.4.3 Mass Constraints

Initial mass testing of the observatory indicated that none of the available lift capabilities from launchers would accommodate the amount of mass needed for a direct injection transfer trajectory. In addition to this limitation, other design-related issues made the situation more complicated because mass is the "currency" for spacecraft engineers in dealing with their problems. Also, having tight margins in terms of mass, and the fact that the overall design is more integrated in terms of the interfaces between functional elements, means that it is not as easy to identify the interface points between them. When making decisions on how traditional techniques of dynamic electrical separation of components and thermal interfaces will affect mass, it was necessary to consider the coupling between these methods. The nature of this coupling means that more thorough integrated analyses are needed to evaluate performance, as an alteration to one subsystem will likely have a significant impact upon other subsystems.

15.4.4 Optical Stability

Through 14 days and 24 hours, the stability of the Energy Encircled (EE) of the NIR Point Spread Functions (PSFs) determined how stable they were. Because of the technical and programmatic limitations on using complex active control systems, it was decided to use the design of the observatory to be able to be structurally and thermally stable as it moved through various positions in the Field of Rotation (FOR). This added complexity to the design, testing, and evaluation of stability.

15.4.5 Performance Verification

Due to its large size, weight, and wide range of environmental temperature extremes (ambient temperature conditions on the sunshield hot side and less than 55°K on the cold side) , it is impossible to perform a physical test of how well the observatory will perform in space. This became evident during the early planning stages of the mission [18]. To ensure the observatory will perform well in orbit, we rely upon detailed analytic models, which are composed of the individual element/subsystem models that were developed and validated, respectively, for cryogenic margin, pointing accuracy and will incorporate all of the components' interactions *via* precise performance margins assigned to them in cases where specific tests cannot be performed. In addition to these performance margins, any performance reduction because of the sportiveness would be estimated and offset through proper margin assignments and or mitigated through workmanship testing completed at the observatory level. Integrated systems analyses will continue to play an important role beyond serving as a tool for trade studies/design evaluation because they provide an understanding of how interconnected systems can be affected by changes in other systems.

15.5 Conclusion

The JWST illustrates the importance of keeping the fluids in long-term machinery as clean as possible. The synergy between Ferrography and Patch Membrane Filters at higher temperatures (greater than 550°C) provides an abundance of data and hence much greater utility than using either one individually. Understanding how Patch Membrane Filters are made will provide answers to many of the questions and concerns that arose from this testing. Greater application of heat to heat-treated Membrane Patches provides a number of new options for use in other applications and research areas. There will be significant opportunities for discovering additional applications for such filters to be utilized at higher temperatures. In 2023, ASTM will receive requests to perform a new standard on Patch Testing. All participants in the NASA Patch Test Research will be given appropriate credit for this successful research initiative.

AI disclosure statement: The author confirms that no artificial intelligence (AI)-generated content or AI-assisted writing tools were used in the preparation of this chapter (including summaries and the abstract). All text, analysis, and conclusions are the result of the author's original work.

Bibliography

1. Menzel, M., *et al.*, The Design, Verification, and Performance of the James Webb Space Telescope. *Publ. Astron. Soc. Pac.*, 135, 058002 (42pp), 2023 May.
2. Stahl, H.P. and Henrichs, T., Multivariable parametric cost model for space and ground telescopes, in: *Proc. SPIE 9911, Modeling, Systems Engineering, and Project Management for Astronomy VI, 99110L*, 6 Sept 2016, doi:10.1117/12.2234088.
3. Stahl, H.P., Henrichs, T., Luedtke, A., West, M., An Update to Single Variable Parametric Cost Models for Space Telescopes. *Opt. Eng.*, 52, 9, 3 June 2013, doi:10.1117/1.OE.52.9.091805.
4. Feinberg, L., Hagopian, J., Diaz, C., New Approach to Cryogenic Optical Testing the James Webb Space telescope, in: *Proc. SPIE 6265*, 62650P, 2006.
5. Atkinson, C., Arenberg, J., Matthews, G., *et al.*, Architecting a revised optical test approach for JWST, in: *Proc. SPIE 7010*, 70100Q, 2008.
6. Casanni, *et al.*, JWST independent Comprehensive Review Panel, Final Report, 10/29/2010.
7. Arenberg, J., Matthews, G., Atkinson, C., Cohen, L., Golisano, C., Havey, K., Hefner, K., Jones, C., Kegley, J., Knollenberg, P., Lavoie, T., Oliver, J., Plucinsky, P., Tananbaum, H., Texter, S., Weisskopf, M.C., Lessons We Learned Designing and Building the Chandra Telescope, Invited Paper, in: *Proc. SPIE 9144, Space Telescopes and Instrumentation 2014: Ultraviolet to Gamma Ray*, 91440Q, July 24, 2014, doi: 10.1117/12.2055515.
8. Arenberg, J., Atkinson, C., Barto, A., Lightsey, P., Knight, J.S., Yanatsis, D., Feinberg, L., Advantages to Reusing the James Webb Design for 2020 Decadal Missions, this conference, in: *Proc. SPIE, UV/Optical/IR Space Telescopes and Instruments: Innovative Technologies and Concepts IV, SPIE*, San Diego, CA, USA, 2009.
9. Machine Learning Models: Powering the Future of AI. *J. Sci. Technol.*, 4, 1, 40–65, 2023.

10. Pulicharla, M.R., Hybrid Quantum-Classical Machine Learning Models: Powering the Future of AI. *J. Sci. Technol.*, 4, 1, 40–6, 2023.
11. Roylance, B.J., Ferrography – then and now. *Tribol. Int.*, 38, 2005.
12. Bowen, E.R., Westcott, V.C., Anderson, D.P., Wear Particle Atlas, Final Report to Naval Air Engineering Center, Lakehurst N.J., Report NAEC 92–163, July 1982.
13. Elhabib, O.A., Ayad, M., Elmasry, A.A., Omar, A.A., Heat treatment of Wear Particles of Automotive Engines to Determine their Materials from Temper Colours. *Int. J. Mech. Eng.* (IJME), 2, 2, May 2013.
14. Barwell, F.T., Bowen, E.R., Bowen, J.P., Westcott, V.C., The Use of temper Colors in Ferrography. *Wear*, 44, 1977.
15. ASTM International, ASTM D7690-24: Standard Practice for Microscopic Characterization of Particles from In-Service Lubricants by Analytical Ferrography, West Conshohocken, PA, USA: ASTM International, 2024, doi: 10.1520/D7690-24.
16. Loede, R. and Snelling, K., NASA Wear Particle Analysis Report, Technical Report, Fort Myers, FL, USA: Trico Corporation.
17. National Aeronautics and Space Administration (NASA). Webb innovations. NASA Science. NASA Science – Webb Innovations, 2025, April 28.
18. Menzel, M., *et al.*, The James Webb Space Telescope optical telescope element, in: *Proc. SPIE, Optical, Infrared, and Millimeter Space Telescopes*, vol. 6265, Art. no. 62650N, 2006, doi:10.1117/12.670419.

16

Case Study on Overcoming Extreme Conditions on Mars with High-Tech Electronics and Sensing Systems in Rovers

R. Nagaraj[1]*, J. Logeswaran[2], Sri Dhivya Krishnan[1] and P. Manjula[3]

[1]*Department of CSE (Artificial Intelligence and Machine Learning), Sri Sairam Engineering College (Autonomous) West Tambaram, Chennai, Tamil Nadu, India*
[2]*Department of ECE, Mepco Schlenk Engineering College (Autonomous), Sivakasi, Tamil Nadu, India*
[3]*Department of Health Information Management and Technology, College of Applied Medical Sciences, King Faizal University, Alahsa, Eastern Province, Saudi Arabia*

Abstract
Mars offers one of the harshest conditions in the solar system, with low gravity, a thin atmosphere with high levels of CO_2, extreme swings in temperature, coarse dust, intense radiation, and long delays in communication with Earth. Even in such severe conditions, Mars is one of the best targets to be searched in this endeavour to find life and surface processes, such as aeolian sand and dust transport, impact cratering, and seasonal carbon dioxide frost, have created an active environment where terrestrial extremophile-like life is possible. Planetary rovers, such as the perseverance of NASA, Pragyan of ISRO, and several others are using the best of high-tech electronics and sensing systems to explore and study this difficult terrain. These can be radiation hardened computing, autonomous navigation, multi-sensor fusion, and strong communication systems together with scientific instruments, which are multispectral and hyperspectral cameras, spectrometers, ground-penetrating radar, and ISRU sensors. The future of resilience and increased science capabilities of operational robots and humans is based on emerging technologies such as quantum sensors, neuromorphic vision, AI-driven autonomy, and swarm-based exploration, which will make future robotic and human missions possible.

**Corresponding author*: nagarajsrin@gmail.com; ORCID: 0000-0003-0342-6325

N. Sathish, Prasanth Aruchamy, Rajesh Kumar Dhanaraj, Nithya Rekha and Karthik Palani (eds.) Space Electronics for Deep Space Exploration, (367–390) © 2026 Scrivener Publishing LLC

Keywords: Mars surface, extreme conditions, sensing systems, sensors, rovers

16.1 Introduction

16.1.1 Mars Structure

Mars, known as the Red Planet, appears reddish-orange at night due to abundant iron oxide covering its surface. Its diameter is about 6,791 km roughly half of Earth's 12,756 km—and its gravity is only 38% that of Earth's, meaning a 100-pound object on Earth would weigh 38 pounds on Mars. Like Earth, Mars rotates once every 24 hours and 37 minutes and has an axial tilt of about 25° (which can vary between ~15° and 35° over long timescales). Mars internal structure has three layers: (1) a central core radius is estimated between 1300 and 1500 km having heavy iron-rich melts (iron, nickel and sulfur); (2) the Mantle is chondritic in composition, consisting iron-magnesium silicates. In addition, the Fe/Mg ratio is higher in the Martian mantle. (3) The crust is basaltic in composition, whose thickness is estimated from 5 to 100 km. The crust is thickest beneath the elevated, heavily cratered terrain of the southern hemisphere, while it is thinnest beneath the major impact basins, such as Isidis and Hellas [1].

16.1.2 Mars Surface

Mars is hyper-continental because it lacks oceans and has a thin atmosphere, so it cannot store or redistribute heat effectively. As a result, daytime can be mild (around 20°C), but at night temperatures plunge to about −100°C, creating extreme day–night and seasonal contrasts much stronger than on Earth. Also, the liquid water cannot persist for long due to this thin atmosphere. Olympus Mons, standing around 17 miles (27 km) high—roughly three times the height of Mount Everest—is a striking example of Martian volcanism (depicted in Figure 16.1). The vast Valles Marineris canyon system, discovered by the Mariner 9 spacecraft in 1971, plunges as deep as 6 miles (10 km) and extends for roughly 2,500 miles (4,000 km). It is formed as the crust stretched and fractured, and evidence suggests water may once have flowed there. The planet's surface also shows many other volcanic features, and minor eruptions may still occur today. Geological features such as channels, valleys, and gullies on Mars suggest that liquid water once flowed across its surface. Some of these channels are enormous, reaching up to 100 km (60 miles) wide and 2,000 km (1,200 miles) long [2]. Planetary information of Mars is provided in the Table 16.1.

Table 16.1 Planetary information of Mars.

Planetary information of Mars	
Mean solar distance	227,943,824 km (1.5 AU)
Orbital eccentricity	0.093
Ecliptic orbital inclination	1.85°
Martian orbital period	686.98 Earth days
Average synodic cycle	779.94 Earth days
Average orbital speed	24.1 km/sec
Equatorial mean radius	3,396.2 km
North polar radius	3,376.2 km
South polar radius	3,382.6 km
Surface area	1.44×10^8 km^2
Mass	6.417×10^{23} kg
Average density	3.93 g/cm^3
Average surface gravity	371 cm/sec^2
Escape velocity	5.03 km/sec
Rotation period	24 hr 37 min 22.663 sec
Mars solar day (sol)	24 hr 39 min 36 sec
Obliquity	25.2°
Mean surface temperature	210 K (−82 °F, −63 °C)
Typical surface pressure	0.006 bar
Number of known moons	2

Additionally, the discovery of ancient shorelines and beach-like structures in the northern plains, particularly in regions like Utopia Planitia, provides compelling evidence for the existence of an ancient ocean. Northern plains of Mars, particularly Vastitas Borealis, are among the flattest and smoothest regions on the planet. This vast lowland area lies 4 to 5 km below the mean radius of Mars and is centered near the North Pole. On Mars, the northern hemisphere is low and smooth, while the southern

Figure 16.1 Olympus Mons, Mars's tallest volcano (source: Mars Global Surveyor spacecraft on April 25, 1998).

hemisphere is higher, heavily cratered, including Hellas Planitia, the largest impact basin at 2,300 km wide.

Mars has two small moons, Phobos and Deimos, were discovered by American astronomer Asaph Hall in 1877. Phobos, the larger moon, is about 27 km (17 miles) wide and brown-gray in color, with a massive crater roughly 10 km (6 miles) across and a surface patterned with grooves, possibly from the impact that formed the crater (depicted in Figure 16.2). Deimos is smaller, about 15 km (9 miles) wide, with fewer craters; its shallower craters appear lighter in color than the deeper, darker ones. Both moons are irregularly shaped, composed of carbon-rich rock mixed with ice, and covered in dust and loose rocks. Phobos is slowly spiralling toward Mars at about 1.8 meters (6 feet) per century and could collide with the planet or break apart into a ring within 50 million years.

The polar region of Mars contains vast deposits of finely layered stacks of water ice and dust, extending from the poles to latitudes in both hemispheres. These layers were likely deposited by the water ice and dust, with year-round ice caps on top. In winter, seasonal frost of solid carbon dioxide, or dry ice, forms from the thin CO_2-rich atmosphere [3]. Mars has four seasons, just like Earth — spring, summer, autumn, and winter — but their lengths and intensity are very different because Mars has a similar axial tilt (25.2°) but a much more elliptical orbit. Seasonal transitions are defined by the areocentric longitude of the Sun (Ls), which marks Mars' position around the Sun. Table 16.2 provides the distribution of mars season with Solar longitude (Ls).

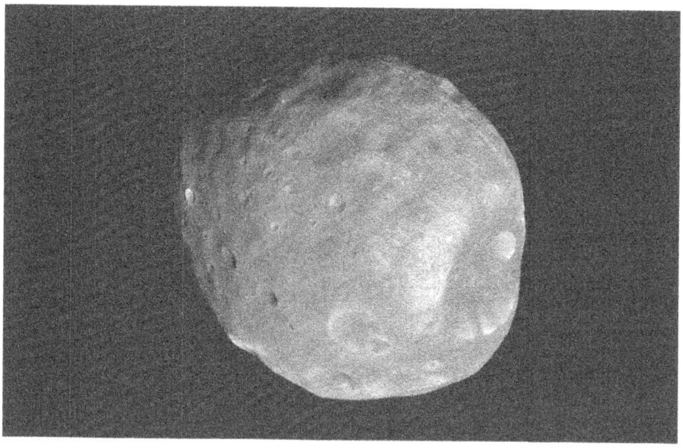

Figure 16.2 Phobos, the larger of Mars' two moons (source: image taken by NASA's High Resolution Imaging Science Experiment (HiRISE) camera on the Mars Reconnaissance Orbiter).

Table 16.2 Seasons of Mars with solar longitude (Ls) span.

Ls range	Northern hemisphere	Southern hemisphere	Duration (Mars days)	Duration (Earth days)
0°–89°	Spring	Autumn	199 sols	92.9 days
90°–179°	Summer	Winter	183 sols	93.6 days
180°–269°	Autumn	Spring	147 sols	89.7 days
270°–360°	Winter	Summer	158 sols	89.1 days

Its elliptical orbit means Mars is closest to the Sun when the southern hemisphere is tilted toward it, creating hotter, shorter southern summers and colder, longer southern winters. This extra solar heating produces stronger winds and atmospheric turbulence, helping generate the planet's largest dust storms. In contrast, northern summers are cooler and longer, because Mars is farthest from the Sun when the northern hemisphere is tilted toward it, resulting in gentler seasonal contrasts. During spring, the polar ice caps begin to retreat as carbon dioxide sublimates, temporarily thickening the atmosphere; in summer, an entire pole may lose its ice cover. As autumn and winter return, temperatures drop and the polar caps regrow, sometimes extending into mid-latitudes. Seasonal temperature gradients also drive cloud formation—especially around volcanoes during northern summer and in tropical cloud bands at other times of year—even though these clouds never produce rain. Despite occasional highs near 20°C, Mars remains an extreme environment, with winter lows near −140°C and highly variable atmospheric conditions shaped by its long, uneven seasons [4].

16.2 Extreme Conditions of Mars

16.2.1 Atmospheric Conditions

According to NASA, Mars' atmosphere is made up mostly of carbon dioxide (95.3%), with smaller amounts of nitrogen (2.7%) and argon (1.6%). Trace atmospheric constituents include 0.13% of oxygen, 0.08% of carbon monoxide, small amount of water vapor, nitrogen oxides, neon, krypton, xenon, and hydrogen–deuterium–oxygen compounds. The surface pressure of Mars is about 6.35 mbar, which is over 100 times less than Earth's pressure. Therefore, Martian air is not suitable for human survival, which makes oxygen is essential and critical requirement for any crewed missions. Because of thin atmosphere, the martian surface cannot retain heat effectively, leading to extreme temperature fluctuations between day and night. It also provides minimal protection from solar and cosmic radiation, posing a serious threat to both electronic systems and human health [5].

16.2.2 Temperature and Humidity Extremes

Mars is much colder than Earth because it lies farther from the Sun and has a thin atmosphere of about 95% carbon dioxide, which cannot retain heat effectively. As a result, temperatures on Mars vary widely across its surface. The average temperature is around −80°F (−60°C), but it can drop up to −200°F (−128°C) near the poles during winter and rise to a maximum of about 70°F (20–21°C) at midday near the equator. These extremes make a harsh and frigid world compared to Earth, where the lowest recorded temperature is −128.6°F (−88°C) in Antarctica and the highest is 134°F (56°C) in Inyo County, California, USA. The Curiosity rover recorded an air temperature of 43°F (6°C) during the daytime. Similarly, the Perseverance rover measured a high of 8°F (−13°C) and a low of −112°F (−80°C) on January 29, 2022. During nighttime, the frost forms on rocks as temperatures drop, creating periods of very high relative humidity, sometimes reaching 100%, before the frost turns to vapor at dawn. As the Sun warms the surface during the day, the air becomes much drier, with relative humidity dropping sharply [6].

16.2.3 Dust Storms and Wind

Martian dust storms are classified by scale into local, regional, and global events. Local storms can occur anywhere and, in any season, though certain regions—such as Acidalia, Utopia, Arcadia, and Hellas—show recurring activity, usually lasting under three days. Regional storms last persist more than 2 days and can transport dust across long distances ($\geq 1.6 \times 10^6$ km^2), and are most common in southern spring and summer (LS 180°–360°), except during the solstitial pause. Global, or planet-encircling, storms form when multiple local and regional storms merge and can persist for several weeks, making them rare and irregular events. Seasonally, regional dust activity is categorized into three phases: A, B, and C. Type A storms occur between LS 180° and 270°, last approximately four months, and cover most low- and mid-latitudes, sometimes evolving into global storms. Near the southern summer solstice, Type B storms develop over the south polar region and last about two months. Type C storms arise near LS $\approx$ 330°, last one to two months, and may also briefly reach global scale before conditions calm and a new cycle begins.

Dust storms begin with dust lifting driven by surface wind stress and convective vortices such as dust devils. Severe solar heating of the southern spring and summer increases atmospheric circulation and near-surface winds, establishing positive feedback that raises more dust. Storms are initiated and maintained by baroclinic waves, cyclonic systems, and saltation of sand-sized particles. The dusting also heats the atmosphere, thereby increasing meridional movements and enhancing the occurrence of storms both locally and globally. The interplay between the orbital and spin can also cause a reaction of the atmospheric circulation, which contributes to interannual variability. The dust storms alter the thermal structure, heat the mid-level atmosphere, increase global circulation, and change the surface through polar deposition and redistribution of materials. Such consequences are problematic for the rover when it is navigating, causing a decrease in visibility, damage to hardware, a reduction in the amount of energy produced by the solar panels, insufficient communication, and even the possibility of toxicity [7]. The dust of the Martians is very abrasive and electrostatic, extremely fine, and can readily be raised in storms. Consequently, it will be able to absorb space suits, machines, solar panels, and seals. The intensity of perchlorates is also dangerous to health when inhaled or absorbed into the body.

16.2.4 Radiation

The Martian environment experiences higher radiation than Earth's surface due to its thin atmosphere and zero global magnetic field. Also, the mars surface radiation averages about 240 to 300 mSv per year, about 40 to 50 times the average on Earth. The thin atmosphere allows more ultraviolet (UV) radiation to reach the surface. Mars experiences more exposure than Earth, which receives about 0.26 mSv per year. Curiosity measured roughly 230 mSv on Mars, with models predicting 156–274 mSv depending on solar activity. During solar minimum, cosmic radiation may reach 1.2 ± 0.5 mSv per day, including 0.4 mSv per day from albedo neutrons generated in the regolith. Much of the radiation on Mars originates from highly penetrating Galactic Cosmic Rays (GCRs), which are challenging to shield against, as well as secondary particles generated when the Martian atmosphere and surface interact with cosmic rays [8].

Radiation on Mars varies with elevation, latitude, and atmospheric pressure—deep basins and denser atmospheric periods offer slightly lower exposure. Solar Proton Events (SPEs) are high-energy proton particle bursts released by the Sun during solar flares or coronal mass ejections. These SPEs cause short-term spikes (increase radiation levels for short periods), with Curiosity observing about a 25% dose increase in a single day. Over time, radiation weakens materials, causing plastics and metals to become brittle and can interfere with or damage delicate microelectronics, making radiation-hardened systems essential for Mars operations.

16.2.5 Physical and Environmental Constraints of Mars

The risk of radiation on Mars poses a considerable threat to astronauts, habitats, and electronics, which raises the long-term health and equipment risks associated with the Mars object due to high-energy particles, solar storms, and galactic cosmic rays. A combined effect of its low gravity (approximately 38 percent that of Earth) is that it allows for human survival during long periods of stay, leading to muscle atrophy, bone wastage, cardiovascular and fluid imbalances. It also has a fragile atmosphere, which results in very low aerodynamic drag for parachutes, limiting their performance. Additionally, aerobraking is more challenging, so spacecraft must use heat shields, retro-propulsion, and sophisticated landing systems. As a result, safe landings are much more challenging than on Earth.

All these extreme conditions on Mars necessitate the use of advanced sensing and high-tech electronics for landers and rovers. Such sensing and other electronic systems must be able to withstand very high levels of radiation, coarse dust, and extreme temperature fluctuations, all while operating in a thin atmosphere. At the same time, they are required to maintain proper measurements and sound operations. Safe navigation, precision of data gathering, and durability on the Martian surface require sophisticated sensors and solid designs that are electronically balanced.

16.3 Electronics System

Mars rovers require advanced electronic systems that can ensure the rover operates effectively in hostile environments, allowing it to survive and conduct scientific exploration. Examples of these electronic systems include hardened embedded computers, autonomous navigation systems, power systems, scientific systems, high-reliability communications systems, and specialized thermal management systems, as described below.

16.3.1 Radiation-Hardened Computing

The application for exploring Mars requires electronic components that may withstand the severe environmental conditions. Perseverance is a kind of rover, subjected to the ionizing radiation, which is one of the most serious threats in Mars environment. The three primary sources of these ionizing radiations are galactic cosmic rays, Solar flares, and storms, which can cause high-energy particles and proton belts within the Earth's magnetic field.

Earth's atmosphere and the natural magnetic effects give protection for the electronic components from external radiation. Meanwhile, the Martian environment conditions are different from those on Earth, and it has less shielding due to its thin atmosphere and minimal magnetic field effect. It may cause damage to electronic materials and the external high-energy particles can disrupt the semiconductor operations, which leads to faults or outright hardware failure.

The semiconductor chip used in common consumer electronics applications cannot withstand the Mars environment. The energy particles in the Mars environment cause a disruption in the charge distribution in the semiconductor systems. For this condition, it is necessary to have a more specialized defense mechanism. The environmental factors also cause logical errors or flip in bit values, and these cumulatively result create the malfunction of the semiconductor over time. The interplay of various factors influences the probability and severity of such radiation-induced faults, including the extent of physical shielding encasing the chips, the intrinsic radiation tolerance of the chip materials—especially the silicon die—and the energy spectrum and flux of the incident ionizing radiation. One well-known cumulative effect is termed the total ionizing dose (TID) effect.

The continuous effect of the radiation causes the degradation in the semiconductor lattice structure that leads to shift in the threshold voltage level, increased current leakage, and finally functionality loss in the transistor [9]. These issues affect the lifespan and reliability of electronic components, making it primary to include the radiation-hardening strategies in the design and material chosen phase for spacecraft and rover electronics. In Mars rovers, a Radiation-Hardened (RAD-HARD) processor was preferred in an advanced class of computers. This processor was specifically designed to function reliably and withstand in extreme radiation and temperature conditions. In one model of NASA's Perseverance rover, a computer was created using the Rover Compute Elements (RCEs), which was served as the rover's main computing units. The various electronics components are depicted in Figure 16.3.

These RCEs are designed with an extra backup unit to protect from the fault tolerance, allowing the systems to detect errors and transfer important

Figure 16.3 Important components in the deployed rover.

tasks to a backup processor when in idle. The BAE RAD750 is a type of main processor which is used in recent Mars missions. This specifically featured PowerPC-based microprocessor can withstand high radiation and operate at speeds of up to 200 MHz. It can withstand extreme temperatures on Mars, and it also possesses better processing power with high memory capacity and computational throughput.

In recent era, the rover's systems are equipped with advanced memory and data-storage solutions in order to withstand the space challenges. To hold the huge data in Mars, this system uses larger size of non-volatile flash memory. In addition to that, the system contains some amount of volatile memory, especially Error-Correcting Code (ECC) RAM, which helps to detect and correct bit errors in real-time, ensuring the integrity of the data during processing. These memory technologies are work together to manage and compress the larger quantities of scientific data, which again sent back to the Earth through the allocated limited-bandwidth communication channels. In Mars, the temperature variation is often exceed the predicted operational limits of electronics components. Hence, the right decision is to shield these electronics system using temperature-controlled enclosures consist of heaters and thermal insulation to maintain optimal operating conditions. Shielding materials made using aluminium or any other specialized composites can further protect these delicate components from radiation penetration. Hence, these approaches can make space mission computers possible to work reliably with minimal faults, and also consider that repairs are not possible in space.

The radiation-hardened processors, error-correcting memory, redundancy system design, and physical protection features are an impressive implementation of space electronics design followed in Mars exploration. These requirements address the challenges of computational reliability, fault tolerance, and performance in the complex environment and the success of the mission depends on the autonomous task of the rover. The technology in the Mars rovers is therefore the best of aerospace engineering in electronics, involving the development of semiconductor physics, materials science, and system architecture to address the needs of interplanetary exploration. With such innovations, Mars rovers like Perseverance give humanity the ability to explore the solar system, performing groundbreaking research while withstanding the harsh conditions of one of the most hostile environments an engineered system has ever encountered.

16.3.2 Navigation and Autonomy

Rovers must travel over unpredictable terrain and avoid hazards with minimal human control. They are equipped with inertial measurement units

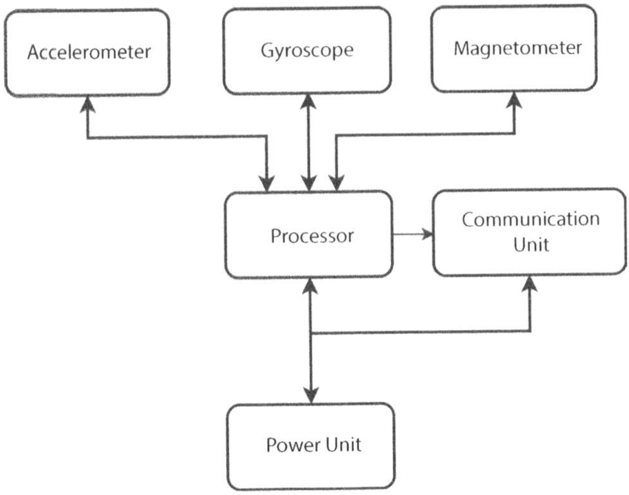

Figure 16.4 Block diagram of the Mars Rover IMU.

(IMUs) that provide three-axis orientation for precise movement and tilt estimation, as well as autonomous navigation algorithms utilizing onboard processors, perception modules, and motion planning to cover distances of up to 35 mph in rocky Martian landscapes [10]. The detailed block diagram of the Mars Rover IMU is depicted in Figure 16.4.

16.3.3 Power Electronics and Energy Storage

Mars rovers depend on durable power systems for keeping instruments functioning on Mars. Newer missions use Lithium-ion batteries and a Multi-Mission Radioisotope Thermoelectric Generator (MMRTG), while earlier rovers like Spirit and Opportunity were powered by high-efficiency triple-junction solar panels. Radiation-hardened regulators, converters, and MOSFET drivers manage power distribution and battery charging, keeping the rover running through cold nights and dust storms.

16.3.4 Space-Qualified Scientific Instruments

Specialized sensor electronics systems support scientific payloads on the rovers. These sensor electronics are High-resolution multi-spectral cameras such as Mast Camera Zoom (Mastcam-Z), Navigation Camera and Panoramic Camera, Laser-Induced Breakdown Spectroscopy (LIBS) systems that use custom spectrometers and onboard digital processors, X-ray and UV spectrometers such as Planetary Instrument for X-ray

Lithochemistry (PIXL) and Scanning Habitable Environments with Raman & Luminescence for Organics & Chemicals (SHERLOC), Subsurface radar modules namely Radar Imager for Mars' Subsurface Experiment (RIMFAX) and Atmospheric and environmental instruments are Mars Environmental Dynamics Analyzer (MEDA) and Mars Oxygen *In-Situ* Resource Utilization Experiment (MOXIE). Each of these modules relies on highly miniaturized, robust analog and digital front-end electronics for precise measurements in Mars' demanding environment.

16.3.5 Communications and Data Handling

Mars rovers communicate with lander and satellite through Ultra-High Frequency (UHF) radios and communicate directly with Earth using X-band high-gain antennas. Their communication system is engineered with radiation-hardened line drivers, transmitters, and error-correcting memory to support deep-space data rates while surviving harsh environmental conditions. Image Compression and Extraction (ICER) are commonly applied to compress imaging and sensor data and store it before transmission to maximize effectiveness and minimize transmission errors.

16.3.6 Environmental Protection and Thermal Management

The complete electronics system is enclosed in an insulated and heat-resistant Warm Electronics Box (WEB) with multilayer thermal insulation and heaters (including Kaplan strip heaters), to maintain the devices within the required operating temperature despite the harsh weather conditions on Mars. They can endure dust, vibrations, and severe changes of temperature because of their available flexible wiring, thermal straps and strong materials, which ensures their reliability in Martian environment, which is hostile.

16.3.7 Key Electronics Manufacturers

Many manufacturers, such as BAE Systems, Intersil (Renesas), Analog Devices, and Infineon, offer a variety of radiation-hardened electronics that are utilized across power management, sensing, and data handling systems. These parts have been carefully selected based on their qualification for shock, vibration, radiation, and temperature resistance. Mars rover electronics represent the highest form of engineering, designed to meet extreme reliability, where special designs must deal with the challenges of distance, radiation, dust, and temperature, so that each mission is able to explore the extremes of exploration on the Red Planet.

16.4 Mars Sensing System

Planetary rovers are also designed to travel through uncharted terrain, which studies the surface materials and sends valuable information back to their base. The success of missions, such as Perseverance on Mars and Chandrayaan-3 on the Moon, relies heavily on the reliability and smartness of their sensing systems. These two main interdependent activities are intended to obtain scientific information and an autonomous navigation procedure. The hostile and unpredictable climate on Mars presents a myriad of challenges for the rovers to overcome. It is a difficult task to ensure real-time control as the communication delay can reach several minutes [11], and the rover must act independently due to severe weather, dust storms, rough surface, and the absence of GPS. These sensors continuously scan the environment to control movement and to avoid risks, in addition to gathering the required scientific data [12]. Such sensor systems are of two broad types. The sensors in navigation and engineering enhance mobility and safety by facilitating obstacle detection, terrain mapping, localization, and system diagnostics. Scientific sensors monitor the rocks on the surface, the soil, the atmosphere, and other features of the planet's surface and environment to accomplish the mission objectives and investigate the geological history and potential for life on Mars. These, together with the navigation sensors, are essential for conducting an effective exploration of the planets.

16.4.1 Engineering and Navigation Sensors

The rovers utilize a combination of navigation sensors to navigate the rocky terrain on Mars safely, continuously monitoring their movements to map the uncertain terrain and determine their exact position. All these sensors help the rover to be aware of its surroundings, prevent danger, and choose safe routes to move on its own, even when the communications with the Earth are not instantaneous.

I. Wheel Encoders (Odometry Sensors)

Wheel encoders will help the rover to establish the distance it has travelled and its direction of travel. Adding up these small rotational vectors will enable the rover to determine its position within the travelled distance, called odometry or dead reckoning. However, odometry becomes inaccurate due to mechanical fatigue, rough roads, and wheel slip. To minimize this drift, the inertial measurement unit (IMU) data and optical sensors data are added to the wheel encoder data.

II. Inertial Measurement Unit

The IMU consists of gyroscopes that monitor the change of orientation, such as pitch, roll, and yaw, and accelerometers that monitor linear acceleration and movement. Combining the information from several sensors, the rover is constantly informed of its movement and stability, helping it to travel safely in rough areas and slopes. Moreover, the IMU ensures the rover's reliability in motion, even in extreme conditions, as it enables localization in the short term when the cameras lose their visual targets during dust storms or low-light conditions.

III. Vision Sensors

The pair of stereo cameras receives two images of slightly different perspectives- much like the vision of a human pair of binoculars. The subtraction of these images creates high-quality 3D depth maps of the local terrain using the onboard computers of the rover. The maps aid the rover in recognizing barriers, evaluating the surface roughness, classifying the terrain, and determining safe and effective paths. The Navigation Cameras (NavCams) are located at the top of the rover, providing wide-angle shots that are used in visual odometry, path planning, and mapping. The next-generation Mastcam-Z generates zoomable stereo imaging capability, providing high-resolution color images and panoramas to guide navigation and scientific research.

Hazard Cameras (HazCams) are positioned at a low location on the chassis. They are deployed to examine the surroundings of the wheels to detect steep drops, rocks, or loose soil that can trap the rover. All these camera systems enable the rover to maintain a continually updated 3D perspective of its surroundings [13]. The rovers apply Simultaneous Localization and Mapping (SLAM) algorithms through a combination of cameras, Light Detection and Ranging (LiDAR), and IMU to help in determining the motion as well as building terrain maps to navigate and act autonomously in complex terrain. Visual odometry tracks surface features in sequential images, and IMU measurements are employed to eliminate drift and enhance the accuracy of the motion. NavCam-Z provides the Autonomous Navigation (AutoNav) system of the rover with high-resolution, zoomable, stereo images that can be used to produce finer 3-D terrain images and identify hazards. All these combined functions enable the Perseverance rover to perform autonomously and require minimal ground orders, allowing it to cover long distances throughout the day [14].

16.4.2 Scientific Sensors

The spacecrafts used on the planets, including the MastCam-Z on Perseverance, have improved the imaging cameras, which shoot high-resolution color and stereoscopic images to facilitate analysis. With the help of these photographs, one can see the texture of rocks, layers of sediments, erosion patterns and other significant geological processes [15]. The majority of contemporary cameras are sensitive to the visible, infrared, and ultraviolet areas, and this implies that they are sensitive to minerals and moist materials that reveal past water activities. Besides having a scientific meaning, these visual data sets are also helpful in determining optimal sample locations and planning rover routes [16]. A superior understanding can be given by spectroscopic instruments, which determine the chemical structure of materials on the planet. Other X-ray spectrometers, like PIXL, utilize special fluorescence emission to determine the elemental composition. Mineral phases and chemical structures can be identified using spectrometers such as Raman and Laser-Induced Breakdown Spectrometers (LIBS), like SHERLOC and ChemCam, on Mars missions. Such sensors provide the necessary evidence for studying astrobiology, as they identify materials containing carbon and potential biosignatures [17]. In current missions, the techniques employed include drilling, sample collection methods, and ground-penetrating radar (GPR). Radio waves are directed to the Earth by GPR machines, such as the RIMFAX radar on Perseverance, which then penetrate the levels and structures under the surface and any potential ice.

Combined with mechanical exercises and spectroscopic analysis of samples, the study can be extended beyond the naked eye and reveal information about the distribution of water reservoirs and the geological history of the planet. Rovers are increasingly prevalent, and they use small laboratory equipment, which enables them to study materials on the ground. For example, the Sample Analysis at Mars package on Curiosity comprises an isotopic ratio and organic molecule detection package that utilizes mass spectrometry and gas chromatography. Robotic arms, sample-handling devices, and onboard analysis chambers can be used to assist rovers in collecting, processing, and analyzing samples with minimal input from Earth. PIXL is an X-ray fluorescence device that detects the distributions of the elements on a sub-millimeter scale and can be attached to the Perseverance robotic arm. They have a mechanism for deep probing of rock surfaces through chemical imaging, thanks to their precision positioning and micro-focused X-ray beam. The small-scale maps are used to raise awareness of the patterns related to settings impacted by water and potentially by former microbial activity. Finally, the scientific and engineering sensors will make a rover not only a remotely controlled machine, but a more complex autonomous investigator. As scientific equipment generates the data needed to achieve mission goals, engineering sensors provide motion and protection. Their integration provides a combined sensing structure that gives rovers the much-desired freedom to sense, comprehend, and analyze planetary environments.

16.4.3 Sensor Fusion and Data Processing

Planetary rovers also use sensor fusion to combine data sets from several instruments to form a transparent and credible image of their treacherous scenery. Such built-in processing is required for autonomous navigation, hazard detection, and scientific decision-making, particularly in cases where noisy or temporarily sensor data are involved.

I. Multi-Sensor Integration Concept

No single sensor can have a complete or reliable situational awareness on a different planet. Dust can block the cameras, the wheel encoders can drift due to slip, and the IMUs may be biased. Multi-sensor integration can resolve these weaknesses by making use of the information gathered by the vision sensors, LiDAR, radar, inertial units, among other onboard units, to create a more precise estimation of the rover state. This integrated awareness enables the rover to maintain localization, identify hazards, and plan safe routes despite the degraded state of one or more sensors or when they are temporarily unavailable. Sensor fusion principles are based on the concepts of redundancy and complementarity. Redundant sensors promote reliability, and complementary sensors complete information gaps. As an example, the combination of the stereo camera depth and the LiDAR measurements will produce terrain models with a combination of the terrain model and the detailed visual representation.

II. Methods of Filtering and Estimation

Raw sensor data is usually very noisy, and since rovers rely on coarse estimation, sophisticated estimation methods are employed to derive useful state information. Kalman Filters (KF), as well as nonlinear variants such as Extended Kalman Filters (EKF) and Unscented Kalman Filters (UKF), are used to estimate the position, velocity, and orientation of the rover, and then update these estimates when new data is received. Particle Filters (PFs) are particularly useful in nonlinear or complex environments because they can simultaneously describe multiple possible states. To maintain the accuracy of localization, modern systems often integrate IMU forecasts with visual updates or LiDAR data.

III. Fusion of Real-Time Data for Terrain Modeling
The sensor fusion is used to make rich 3D terrain models. The rover generates Digital Elevation Models (DEMs) which describes the surface shape and traversability through stereo depth maps, LiDAR point clouds, and radar derived elevation data. The models are applied in making decisions on the placement of the wheels, the choice of paths and the safety of driving. Rovers (e.g., Perseverance) run these data streams on board and update local maps as they move across the terrain.

IV. Function of Autonomy and AI on board
Long communication times between Mars and Earth require the rovers to make numerous decisions on their own. This autonomy is aided by the AI methods, including convolutional neural networks (CNN) to interpret images and reinforcement learning to optimize routes [18]. The AutoNav system of Perseverance utilizes computer vision and real-time hazard detection to select safe routes without requiring instructions from Earth. The use of low-power edge AI hardware will be incorporated into future missions to further enhance onboard decision-making.

16.4.4 Case Study: NASA's Perseverance Rover—Autonomous Navigation and Scientific Exploration on Mars

One of the most advanced missions to Mars is the NASA Perseverance rover, which landed in Jezero Crater in February 2021. It delivered highly sophisticated autonomy (ability to explore complex terrain with minimal instructions from Earth) and a substantial scientific payload to Mars with the aid of the exact Sky Crane landing system. Perseverance is equipped with complete instrumentation and the first sample collection and storage system for future delivery to Earth to study the geology of the crater and find evidence of past life [19].

I. Terrain Conditions and Mobility Challenges
The conditions of the mobility in Mars are highly different both on Earth and on the Moon. Having a gravity approximately 38 per cent less than that of the earth, and with a landscape full of dust and loose soil and sharp rocks and rough ground, safe passage requires a high degree of traction and detection of hazards. To manage rocky landscapes and inflict less damage than is generally experienced during missions, Perseverance employs six redesigned aluminium wheels with better tread patterns. Stereo vision, wheel odometry,

and detailed terrain modeling all form part of Perseverance's navigation system to navigate the multifaceted environment of the Jezero Crater, comprising ancient river deltas, sand ripples, and broken rock. It features a suspension system (Rocker-Bogie) that enables it to move easily over rocks and slopes while keeping its wheels in contact, enhancing control.

II. Navigation and Communication System

Perseverance is loaded with one of the most sophisticated autonomous navigation systems ever. Its AutoNav system allows the rover to examine the terrain and select safe routes automatically, an essential feature due to the five 20-minute communication lag between it and Earth. Applications of perseverance include Stereo Navcams to map the terrain, Hazcams to detect hazards in the near field, IMU and wheel encoders to estimate the state, and Terrain-relative navigation (TRN) in landing to accurately aim at safe areas. The orbiters which also facilitate communication include MAVEN and Mars Reconnaissance Orbiter, which send the data back to Earth using the high bandwidth UHF links.

III. Environmental Adaptation and Power Management

Mars is a very hostile planet, characterized by global dust storms, low temperatures of below −90°C, and huge daily temperature variations. To handle this environment, Perseverance relies on a radioisotope thermoelectric generator (RTG) to provide continuous power and heat its internal systems, which is an advantage over solar-powered rovers that are more exposed to dust and darkness. Its thermal system directs much of the heat generated by the RTG, insulation, and circulates warm fluid across essential parts of the vehicle, enabling it to operate in a stable manner throughout the year, even when it is cloudy and the high dust concentration.

IV. Scientific and Engineering Outcomes

Perseverance has yielded significant scientific and engineering achievements. It detected the presence of carbon-bearing materials in rocks of Mars, examined sedimentary layers indicating the existence of an ancient lake, retrieved and preserved the first rock cores which could be used in sample return in the future and measured weather patterns and dust in the atmosphere. On the engineering front, it not only demonstrated the capabilities of AutoNav and terrain-relative (or relative-stabilized) navigation in the landing phase but also served as the basis to support Ingenuity, the first aircraft to fly on another planet. These achievements combined outline the way high-level autonomy will make possible complex exploration missions to Mars and precondition the next generation of robotic and human missions.

16.4.5 Emerging Technologies and Future Directions

Since planetary exploration is shifting toward greater autonomy, efficiency, and scientific complexity, future rover systems will incorporate new advanced technologies that make them much more effective with respect to their sensing and analysis as well as their functioning. These innovations will be directed to resolve persistent problems, including low power, slow communications, and hostile environmental conditions, and, in due course, allow rovers to be more self-reliant and adaptive to other worlds.

I. AI-Driven Terrain Classification and Adaptive Navigation

AI plays an imperative role in helping the rovers to interpret the environment surrounding them and react accordingly. Deep-learning algorithms that have been trained on planetary images can automatically identify the different types of terrains, such as rocks, loose sand, dunes, and crater edges, to enable the rover to change its driving control according to its predicted stability and safety. More autonomy is added by means of reinforcement learning which allows the rover to learn its route-planning more efficiently by experience. Together, they allow lessening the need to have continuous human control, faster decision-making and allow the rovers to move longer distances per day [20].

II. Compact Hyperspectral Sensors for Mineralogy

The rovers of the next generation will use smaller hyperspectral cameras which can record hundreds of wavelengths at once rather than the multispectral cameras which can record only tens of wavelengths at the same time. Such instruments provide extensive mineralogical data, and can detect weak spectral signatures of hydrated minerals or salts, or potential organic compounds. Micro-optics and onboard processors have advanced to the stage where they can be placed into devices with very restricted size and power requirements. In combination with hyperspectral machine learning spectral analysis, hyperspectral systems will enable the rovers to detect scientifically significant targets in real-time, thereby reducing unnecessary sampling and data transmission.

III. Swarm Rover Systems for Distributed Exploration

Future missions may utilize teams of small rovers that work together as a coordinated swarm, rather than relying on a single large rover. Through wireless mesh networks, these smaller units can share information and divide responsibilities—such as mapping, sampling, or monitoring the environment—while collectively building a fuller picture of the terrain. Swarm systems naturally increase mission resilience because the failure of one unit does not end the mission. Algorithms inspired by insect behavior and self-organizing robotics are being developed to manage distributed navigation, task assignment, and rerouting in communication-delayed environments such as Mars canyons or the Moon's far side.

IV. *In Situ* Resource Utilization (ISRU) Sensing

Long-term robotic and human activity on other worlds depends on the ability to locate and use local resources. Emerging ISRU sensing technologies are being developed to detect and measure materials such as water ice, oxygen-rich minerals, and metallic cores. Instruments such as advanced spectrometers, neutron detectors, and compact drilling systems will enable rovers to assess resource abundance in real time. When paired with autonomous excavation or processing units, future rovers could help generate oxygen, fuel, or construction materials directly from the surface (Perseverance's MOXIE experiment successfully produced oxygen from the Martian atmosphere) [21].

V. Role of Quantum Sensors and Neuromorphic Vision Chips

Accurate innovations in sensing and computation will have a tremendous boost on rover autonomy. Quantum sensors are based on principles, such as atomic interferometry, which provide measurements of gravity, magnetic fields, and motion that can be very precise, eliminating the drift problem inherent to conventional inertial systems. Meanwhile, neuromorphic vision chips resemble the dynamics of biological neurons in that they can process visual data not in the form of complete image frames, but in fast event signals. This saves an enormous amount of power and allows quick reaction to risks or unexpected alterations in terrain. Combined, these technologies would provide future rovers with nearly human sensorial accuracy and still be lightweight, energy-efficient, and very radiation-resistant.

Planetary rovers are high-technology robotic systems that comprise mobility, perception, and scientific instruments to carry us beyond the Earth. The missions of Perseverance, Curiosity, Chandrayaan-3 demonstrate that sensing systems play a crucial role in the success of rovers, allowing them to sense the environment, navigate and perform their scientific experiment under the most severe conditions. This study characterizes the two-fold classification of sensors in rovers, namely scientific discovery and navigation and survival, and the interdependence of these sensors. With modern rovers, it is now possible to process data and act upon it nearly in real-time, defeating communication lag and terrain ambiguity thanks to sensor fusion, autonomous operation using AI, and the ability to process using radiation hardened processors. Two examples of case studies are Pragyan and Perseverance, which illustrate how different design philosophies can be combined to attain the same goal: how to maximize a scientific payoff and operational resilience.

Combining quantum sensing, swarm-based exploration, neuromorphic vision, and hyperspectral imaging in the future will significantly increase the capabilities of rovers on planets. These technologies will enable autonomous decision-making, contextual decision-making, shift mission tasking away from passive observation to adaptive exploration, and bring about long-term planetary exploration and human habitation. Rover sensing systems, in essence, are the foundation of intelligent and self-directed planetary exploration.

16.5 Conclusion

To explore Mars, people will have to overcome harsh environmental conditions, including low gravity, a thin atmosphere composed of carbon dioxide, a wide range of temperatures, coarse dust, and high levels of radiation. Rovers such as Perseverance and Pragyan, which feature high-tech electronics and sophisticated sensing systems, gain the independence to make autonomous decisions, gather scientific evidence, and remain reliable even when communication is delayed. Real-time terrain and resource analysis is possible due to the integration of various technologies, including radiation-hardened processors, multi-sensor fusion, hyperspectral cameras, spectrometers, ground-penetrating radar, and ISRU sensors. New technologies, such as AI-guided autonomy, quantum sensors, neuromorphic vision, and swarm exploration, will provide greater resilience and adaptive dynamic decision-making, which will form the basis of long-term presence on Mars for both robotic and human missions.

AI disclosure statement: The author confirms that no artificial intelligence (AI)-generated content or AI-assisted writing tools were used in the preparation of this chapter (including summaries and the abstract). All text, analysis, and conclusions are the result of the author's original work.

References

1. Golombek, M.P. and McSween, H.Y., Mars: Landing site geology, mineralogy, and geochemistry, in: *Encyclopedia of the Solar System (Third Edition)*, T. Spohn, D. Breuer and T.V. Johnson (Eds.), pp. 397–420, Elsevier, 2014.
2. Checinska Sielaff, A. and Smith, S.A., Habitability of Mars: How welcoming are the surface and subsurface to life on the red planet? *Geosciences*, 9, 9, 361, 2019.
3. Clark, B.C., Kolb, V.M., Steele, A., House, C.H., Lanza, N.L., Gasda, P.J., Martínez-Frías, J., Origin of life on Mars: Suitability and opportunities. *Life*, 11, 6, 539, 2021.
4. Almatroushi, H., AlMazmi, H., AlMheiri, N., AlShamsi, M., AlTunaiji, E., Badri, K., Young, R.M., Emirates Mars Mission characterization of Mars atmosphere dynamics and processes. *Space Sci. Rev.*, 217, 8, 89, 2021.
5. Amils, R., González-Toril, E., Fernández-Remolar, D., Gómez, F., Aguilera, Á., Rodríguez, N., Sanz, J.L., Extreme environments as Mars terrestrial analogs: The Rio Tinto case. *Planet. Space Sci.*, 55, 3, 370–381, 2007.
6. Pál, B., Kereszturi, Á., Forget, F., Smith, M.D., Global seasonal variations of the near-surface relative humidity levels on present-day Mars. *Icarus*, 333, 481–495, 2019.
7. He, F., Rong, Z., Wu, Z., Gao, J., Fan, K., Zhou, X., Wei, Y., Martian dust storms: Reviews and perspective for the Tianwen-3 Mars sample return mission. *Remote Sens.*, 16, 14, 2613, 2024.
8. Ehresmann, B., Zeitlin, C., Hassler, D.M., Guo, J., Wimmer-Schweingruber, R.F., Berger, T., Reitz, G., The Martian surface radiation environment at solar minimum measured with MSL/RAD. *Icarus*, 393, 115035, 2023.
9. Keys, A.S. and Howell, J.T., Technology developments in radiation-hardened electronics for space environments, in: *Proc. Int. Symp. Space Technol. Sci.*, 2008.
10. Molli, S., Durante, D., Boscagli, G., Cascioli, G., Racioppa, P., Alessi, E.M., Iess, L., Design and performance of a Martian autonomous navigation system based on a smallsat constellation. *Acta Astronaut.*, 203, 112–124, 2023.
11. Maimone, M., *et al.*, Autonomous navigation results from the Mars Exploration Rover (MER) mission, in: *Experimental Robotics IX*, Springer, Berlin, Heidelberg, 2006.
12. Tunstel, E., *et al.*, Mars exploration rover mobility and robotic arm operational performance, in: *Proc. IEEE Int. Conf. Syst. Man Cybern.*, vol. 2, IEEE, 2005.
13. Andolfo, S., Petricca, F., Genova, A., Precise pose estimation of the NASA Mars 2020 Perseverance rover through a stereo-vision-based approach. *J. Field Rob.*, 40, 3, 684–700, 2023.
14. Maki, J.N., *et al.*, Mars exploration rover engineering cameras. *J. Geophys. Res. Planets*, 108, E12, 2003.
15. Hayes, A.G., *et al.*, Pre-flight calibration of the Mars 2020 Rover Mastcam-Z multispectral, stereoscopic imager. *Space Sci. Rev.*, 217, 2, 29, 2021.

16. Rice, M.S., *et al.*, Mastcam-Z multispectral database from the Perseverance rover's traverse in the Jezero crater floor, Mars (sols 0–380), 2022.
17. Sharma, B., Mathur, A.C., Sharma, R., *et al.*, Beyond Earth's horizon: AI's contribution to space exploration, in: *The Intelligent Universe: AI's Role in Astronomy*, pp. 359–389, John Wiley & Sons, 2025.
18. Johnson, A.E., *et al.*, Implementation of a map relative localization system for planetary landing. *J. Guid. Control Dyn.*, 46, 4, 618–637, 2023.
19. Durga Prasad, K., *et al.*, Contextual characterization study of Chandrayaan-3 primary landing site. *Mon. Not. R. Astron. Soc. Lett.*, 526, 1, L116–L123, 2023.
20. Lv, F., Li, N., Liu, C., Gao, H., Ding, L., Deng, Z., Liu, G., Highly accurate visual method of Mars terrain classification for rovers based on novel image features. *Entropy*, 24, 1304, 2022.
21. Hoffman, J.A., Hecht, M.H., Rapp, D., Hartvigsen, J.J., SooHoo, J.G., Aboobaker, A.M., Eisenman, D.J., Mars oxygen ISRU experiment (MOXIE)—preparing for human Mars exploration. *Sci. Adv.*, 8, 35, eabp8636, 2022.

17

Future Trends in Space Electronics and the Road to Interstellar Exploration

V.C. Diniesh[1*], Prabhakaran Abraham[2], N. Sathish[3] and Mustafa Almahdi Algaet[4]

[1]Department of Electrical and Electronics Engineering, Amrita School of Engineering, Amrita Vishwa Vidyapeetham, Coimbatore, Tamil Nadu, India
[2]School of Computer Science and IT, JGI Knowledge Campus, Jain (Deemed to be University), Bangalore, Karnataka, India
[3]Department of Electronics and Communication Engineering, Sri Venkateswara College of Engineering, Sriperumbudur, Tamil Nadu, India
[4]Network Department, Faculty of Information Technology, Elmergib University, Al Khums, Libya

Abstract

Even the most advanced missions within our solar system can be accomplished using current scientific knowledge, without needing breakthroughs in fundamental science. The rapid evolution of space technology has significantly increased the demand for advanced and resilient electronic systems capable of withstanding harsh extraterrestrial environments. With the development of radiation-hardened and thermally stable components, spacecraft can now operate effectively in harsh environments such as deep space and lunar surfaces. High-speed optical communication is transforming data transmission, ensuring faster and more secure communication across vast distances. Such missions fall under the category of which relies strictly on established scientific principles. However, the situation changes drastically when it comes to interstellar exploration. Sending robotic flyby missions to nearby stars using nanoprobes is theoretically feasible with existing technology. Finally, this provides the next generation recent trends towards interstellar assessment and space electronics to go beyond realistic future technologies.

Keywords: Space electronics, deep space exploration, interstellar missions, autonomous systems, radiation-hardened electronics, emerging technologies

**Corresponding author*: vc_diniesh@cb.amrita.edu

N. Sathish, Prasanth Aruchamy, Rajesh Kumar Dhanaraj, Nithya Rekha and Karthik Palani (eds.) Space Electronics for Deep Space Exploration, (391–412) © 2026 Scrivener Publishing LLC

17.1 Introduction to Deep Space Missions and Interstellar Dreams

Journeys undertaken by humans beyond Earth have always attracted attention, from the initial moon landings to the robotic spacecraft currently exploring the solar system's outer regions. Our goals for the next several decades include traveling to other stars and doing space exploration. There is a smaller gap between dreams and reality as a result of developments in systems engineering, electronics, materials, and propulsion. In their explorations of interstellar space and the solar system's periphery, the Voyager 1 and 2 spacecraft contributed to the expansion of human understanding. Voyager 1 entered interstellar space in 2012, becoming the first artificial object to do so. In 2018, Voyager 2 was launched. These missions have provided us with unprecedented data about the solar system's outermost regions. Asteroid composition and planetary atmospheres are just two of the many topics explored by robotic explorers, while the Deep Space 1 mission demonstrated the effectiveness of ion propulsion systems in accelerating spacecraft over enormous distances. To start with this chapter lays out the case for why space electronics must advance lists the basic technological faults that hinder long-duration missions and then suggests a theoretical framework for an idealized plan for space travel.

17.1.1 Need for Advancing Space Electronics

The nervous system of spacecraft is space electronics, which sense, compute, command, and transmit communication and control. There are huge demands kept on these systems during voyages to deep space, let alone interstellar expeditions. The use and exploration of space have changed significantly over the past decade. This modification is known as "the newspace revolution" [1, 2]. Spacecraft electronics behave like a complex central nervous system, with dedicated computer systems that control every aspect of functioning. Modern spacecraft usually have more than one type of computer for backup and specific tasks. For example, the Computer Command System (CCS) is the main controller, the Flight Data System (FDS) processes data, and the Attitude and Articulation Control System (AACS) records where the spacecraft is pointing. These systems need to be able to interpret data packets in real time while working under severe limits on power and mass.

This constraint is specifically strong when missions go beyond our solar structure, when communication latency with Earth can last for years. In 2025, researchers suggest a novel active shielding or real-time mitigation system.

To prevent irreparable harm, particle detectors in this architecture selectively disable circuits that are susceptible to damage. Fault detection using ML models helps improve the in-satellite load system [5]. This design accomplishes an exposure accuracy of almost 95%. This approach gives great potential for utilizing marketable electronics systems with upgraded performance in challenging environments. Communication takes longer for missions that reach farther. Deep-space missions will eventually have to do more on their own, identifying issues, resolving them, and making informed judgments without waiting for Earth. From Figure 17.1 represent the combining underwater networks, terrestrial IoT/V2X systems, and satellite constellations makes it possible to connect to the internet from anywhere in the world. Underwater acoustic communication, RF/5G on land, and high-frequency/optical systems in space all need different kinds of technology, which can cause problems with interoperability, latency, and reliability.

In deep space and interstellar missions, where human involvement is impossible, AI and autonomous control systems are important to the spacecraft's electronic nervous system. Hence, for capturing the images, Hybrid machine learning models are needed to provide a proper solution to the world [6]. Spacecraft with modern AI systems can navigate, change course, and detect defects without ground control. The European Space Agency's (ESA's) Operations Satellite (OPS-SAT) employed camera images to manage attitude, while NASA's Deep Space 1's Remote Agent Experiment demonstrated early

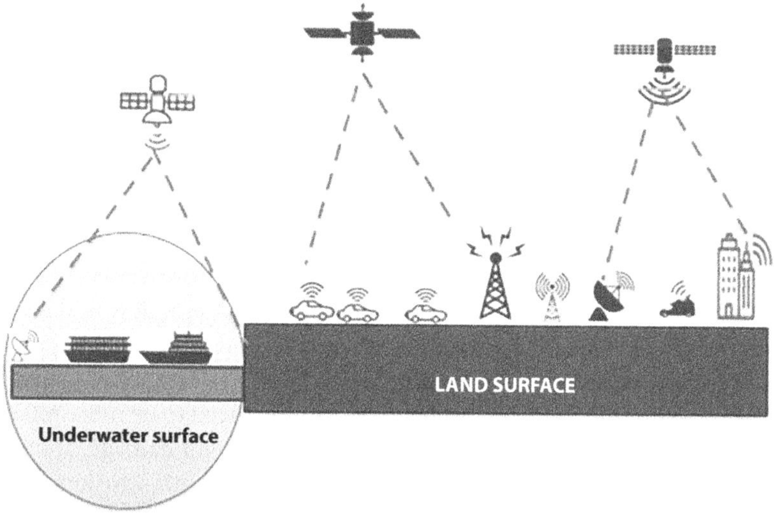

Figure 17.1 A holistic architecture for advancing electronics for sea, land, and space communication systems.

autonomy. The compatibility of modern interplanetary propulsion technology, like propellant-less [3], is still debated.The use of basic science ethics in these strategies is not necessary for exploring knowledge like fusion drives [4].

The rapid growth of missile threats, specifically hypersonic and ballistic weapons systems, has made it very vital to have advanced space-based electronics that can provide tracking and guide interceptions in a real-time environment. IR sensors, fast processors, and ultra-high-speed communication systems are required to detect, monitor, and classify real threats and decoys over long distances in recent missile defense satellite constellations. Beyond detection, interceptor missiles need accurate tracking data to make hits on the interceptions of targets travelling at several miles per second with accurate prediction necessary to predict or guide [8]. A mission with perfect interception is needed for the system. Space electronics must maintain continuous track custody from launch to interception and coordinate with multiple defensive systems across global networks.

17.1.2 Gaps in Current Technology for Long-Duration Missions

Even with major progress, there are still major challenges to be solved before deep space or interplanetary travel becomes possible. Propulsion efficiency, radiation protection, long-duration life support systems, sustainable energy sources, communication delays, astronaut psychological disorders, planetary entry/descent/landing technology, and the exorbitant expense of such missions are some of the most significant concerns. Extending solar power over large distances reduces it impractically. The supply and decay of isotopes are two of the limitations of radioisotope power. The main sources of electricity for spacecraft nowadays are radioisotope thermoelectric generators (RTGs) and solar panels. However, both things have big challenges in deep space. Solar power doesn't operate past the orbit of Jupiter, and RTGs only make a small quantity of electricity that gets weaker over time. Fusion reactors and antimatter propulsion are examples of revolutionary power sources that are still only ideas for interplanetary voyages that would have to last for decades or even centuries. There are alternate sources being driven on, like radioisotope Stirling converters and minor fission reactors, but they are hard to make bigger for extended missions that need to be more reliable.

Electronics and surfaces aboard spacecraft can sustain damage from dust grains or relativistic-velocity atoms. Along the flight path, erosion can be caused by even a thin layer of micrometeoroids. Spacecraft have to navigate on their own, adjust their trajectories, keep an eye on their health, deal with unforeseen circumstances, and choose which data to send because signals take years to reach their destination. However, current autonomy is still insufficient for missions that last for decades without human oversight. Spacecraft need non-volatile, radiation-hardened memory along with error correction and data cleansing systems to avoid radiation-induced bit flips, memory deterioration, and long-term data corruption. For missions anticipated to last hundreds of years, current memory architectures could not be adequate.

17.1.3 Defining the Vision and Requirements of Interstellar Exploration

A well-defined goal and a detailed outline of the project's functional, performance, and architectural specifications are necessary to transform engineering aspirations into space travel reality. Some of the challenging visions and requirements are the need for future design and development. An interstellar mission goal is to reach one or more target star systems, collect and return important scientific data, or deliver human or biologically relevant payloads in an acceptable amount of time. Table 17.1 shows key needs and hurdles for deep-space missions. High speed and safe deceleration are limited by energy and braking methods. Systems must be reliable, fault-tolerant, and resistant to radiation, heat, and cold over centuries. Power must remain efficient and data links need low delay with strong processing. Autonomous control, smart decisions and scalable designs are vital for future adaptability.

17.2 Evolution and Present State of Space Electronics

Space electronics have improved a lot over the years. Today's radiation-hardened circuits are small, powerful systems designed to use less mass and energy while resisting damage. In the early days of space travel, electronics were safer and simpler. As semiconductor technology on Earth advanced, space systems began using newer parts, such as radiation-resistant ASICs and later special radiation-hardened microprocessors. At the same time, miniaturization and lightweight materials allowed satellites and probes to carry stronger payloads without adding extra weight or power use. Over the years, new design methods have focused on making systems more reliable even in extreme radiation environments. Modern satellites now use fault-tolerant processors that can keep working even when some parts fail. The use of advanced packaging techniques helps protect delicate chips from space conditions. Energy-efficient designs ensure that long missions can operate on limited power. These changes

Table 17.1 Key requirements and challenges for deep-space missions.

Essentials	Key requirements	Challenges
Speed	Need maximum speed of light	Momentum and energy limitations dominate
Deceleration	Capability to sluggish down in destination system	Braking and reverse momentum
Reliability & Fault Tolerance	System-level tolerances for failures, self-repairing capabilities, and redundancy	Failures must not build on each other, and hardware must last for hundreds of years.
Resilience to radiation	Manage cumulative radiation dose, SEUs, and latch-ups.	Techniques for reducing damage, and design that can handle errors
Temperature management	In extremely cold or hot climates, maintain constant operating temperatures and eliminate internal heat buildup.	Thermal control that is either passive or active including systems that change phases
Power and energy management	Make sure that systems, operations, and communications have enough power all the time.	High efficiency power systems that break down and decrease over time
Data and communication management	High gain on chip processing, priority based scheduling	Less latency and reliable link
Autonomous control & routing	Self-center trajectory control corrections and decision	Intelligent deep learning, ML fault tolerance
Scalability	Future adaption with existing infrastructure and payload	Distributed systematic Architecture

have made missions to deep space safer, longer-lasting and more powerful than ever before.

Modern space electronics are becoming highly reliable and, in many cases, perform almost as well as commercial electronics. Many missions now use commercial off-the-shelf (COTS) parts that are specially adapted to handle radiation with the help of shielding, error correction, and active protection. Future systems are expected to be more flexible and innovative with research focusing on combining computing, memory, communication and sensors into compact, radiation resistant modules. This includes work on heterogeneous integration and modern 3-D chip architectures that can pack more capability into smaller and stronger designs. Quantum devices are also shaping the future. For instance, quantum accelerometers based on cold atom interferometry could provide much more accurate navigation in deep space missions. As exploration pushes farther toward interstellar travel, electronics must be modular, redundant and designed to degrade gracefully so that a single failure does not affect the entire system. These trends point to electronics that last longer, adapt to challenges and help the next generation of deep space and interstellar missions possible.

17.2.1 Historical Milestones and Technological Transitions

Space missions in the late 1950 and 1960, such as Sputnik (1957) and Apollo (1960–1970), relied on bulky transistor-based systems with significant shielding to withstand the harsh conditions of space. These early frameworks were standalone and rigid, offering limited with less flexibility. By the 1990 and 2000, the approach shifted toward modular avionics, which allowed systems to be more easily upgraded and reconfigured. Advances in fault detection and isolation techniques improved system reliability, while the introduction of ASICs and FPGAs enabled more compact, efficient and programmable hardware, giving mission planners greater control and adaptability. The significant important change in technology in the previous decade has been the merging of commercial electronics with space specific protection methods. Engineers are using high-performance terrestrial processors with extra layers of error correction, redundancy and shielding increasingly instead of depending solely on expensive rad-hard parts. NASA's Deep Space Optical Communications proposes a transition to laser-based, high-bandwidth communications utilizing advanced electronics. Modern 3D chip stacking, phase-change photonics materials and AI-driven autonomous fault management are all improving space electronics. It is replacing static, overdesigned systems with designs that are flexible, strong, and fast for future trips to deep space and other stars.

17.2.2 Current State-of-the-Art in Deep Space Electronics

Deep-space electronics are currently pushing the limits by using commercial parts that can handle radiation, innovative mitigation tactics, and strong structures. At the same time, the idea of active shielding or dynamic mitigation is becoming more popular. A new study presents a system of real-time particle detectors that selectively power cycles sensitive circuits to stop permanent damage, claiming ninety-five percent detection accuracy. Silicon is making progress in sensing, moving us closer to tiny, low-power, and high-efficiency detectors. Recent tests look at techniques to protect against radiation damage so that performance can be kept up in orbits that are not friendly. In space power electronics, radiation-hardened and radiation-tolerant technologies are becoming more popular. Component manufacturers make space-friendly microcontrollers and power controllers. The current frontier is highly advanced, but it's still basically a step towards the final design for electronics that can work in space.

17.2.3 Limitations in Scaling Existing Designs toward Interstellar Goals

Deep-space electronics have reached high levels of advancement, yet interstellar missions continue to face serious challenges. The most critical among them is radiation resistance, since long-term exposure to high-energy particles can gradually degrade or even destroy electronic components. Current systems use shielding, redundancy, and error correction to stay alive for years. Active shielding and mitigation techniques are still being studied, and physical shielding can't be raised forever without a huge penalty for mass. This means that electrical degradation over very long periods of time is still a problem that needs to be tackled. Another big drawback is that it doesn't last long and doesn't give off enough power. The solar panels on most spacecraft lose power over time and distance. Finally, problems with communication and scale are big problems. Even Deep Space Optical Communications can only maintain high throughput at global distances. Over long distances, beam divergence, aiming stability, and data classification become extremely challenging problems for humans to resolve.

17.3 Harsh Realities of Deep-Space and Interstellar Environments

17.3.1 Extreme Radiation, Temperature, Vacuum, and Microgravity

Space missions have to deal with radiation levels that are much higher than those on Earth. Spacecraft are continually subjected to galactic cosmic rays (GCRs), solar energetic particles (SEPs), and trapped radiation belts as they leave the magnetosphere's protective orbit. These can cause latch-ups, single-event upsets, and long-term material degradation, compromising electronic component stability. Devices that are resistant to radiation nonetheless break down over missions that last decades. Solar storms are challenging to forecast, which makes planning missions even harder. This means that designs need to be fault-tolerant, and shielding or active mitigation measures need to be more advanced. Lastly, long-term microgravity affects systems in ways that are less obvious but just as essential. Fluid activities has an impact on the effect on cooling management systems and fuel space. Moving parts need to be checked for reliability because they wear down differently in space than they do on Earth.

17.3.2 Reliability, Redundancy, and Fault-Tolerance Challenges

Reliability has always been a vital key component of designing space eco systems since setbacks in orbit are often permanent. Deep research space and interstellar goals require spacecraft to conduct for decades, which significantly extend the lifespan of many electronic components. Radiation destruction, heat process flow cycling, and the natural way of resources could affect problems years into the space mission system. To ensure the system is resilient, parts must be evaluated and scrutinized, and designs should handle wear and tear rather than prevent. Fault tolerance, in turn, requires spacecraft to detect, isolate, and repair unexpected problems autonomously. Future spacecraft will need AI-driven fault detection and self-repair, adaptive electronics reconfiguration, and technologies that reroute computation or load around damaged components. Thus, building systems that run with minimal ground intervention, adjust dynamically to faults, and survive wear and environmental stressors is the main problem.

17.3.3 Communication Delays and Autonomy Needs

One of the biggest problems in exploring other stars is how long it takes to communicate. In our own solar system, it can take minutes to hours for signals to get from one spacecraft to Earth. This delay slows scientific data flow and prohibits mission control from responding rapidly to defects or unexpected events. Building trust in these systems will take years of space testing and demonstration. But without advanced autonomy, interstellar probes will either fail too soon or yield very little useful data. Communication delays and autonomy are not two different concerns; they are strongly related and will be important for future trips to deep space and other stars.

17.3.4 Power and Thermal Constraints

Spacecraft operating in deep space require reliable power sources that do not depend on sunlight. Solar arrays perform efficiently in the inner solar system, but their output decreases with distance, leaving them ineffective for interstellar missions. Instead, long-duration power must come from alternatives such as radioisotope thermoelectric generators (RTGs), nuclear reactors, or other advanced energy systems that can sustain electronics far from the Sun. The issue of thermal management is closely linked to power. In orbit, where conduction and convection are impossible, spacecraft electronics and propulsion systems create heat that needs to be released. There is only radiative cooling, which uses panels and coatings, but when systems grow or waste heat builds up over extended missions, it loses efficiency. Future interplanetary missions must offer thermal and power flexibility. Some examples of this include phase-change materials that smooth out temperature changes, self-regulating radiators that change size dependent on the amount of heat load, and smart power distribution networks that assure that subsystems use energy efficiently. AI can aid spacecraft in controlling their power usage, operating under balancing vital tasks, placing survival initial in emergencies, and stepping up the most recent science work while using up limited sources. Energy and physical temperature limits, which exceed beyond the fundamental engineering items, affect the mission's effectiveness, threshold limit, and resolution. Monitoring the condition of Satellite Power system is challenging task to ensure its reliable operation and prevent any difficulties in the chronical. Hence if the system collapses, the satellite mission could break, so identifying anomalies is challenging and requires to be considered [7].

17.4 Breakthrough Hardware Technologies on the Horizon

Wide bandgap nano materials like gallium nitride (GaN) and silicon carbide (SiC) are instantly changing power electronics system and communication systems say transmission, reception system for space. These materials exhibit greater breakdown voltages, thermal resistance, and flipping speeds than silicon, making harsh-condition power conversion quicker and effectiveWide bandgap nano materials like gallium nitride (GaN) and silicon carbide (SiC) are instantly changing power electronics system and communication systems say transmission, reception system for space. These materials exhibit greater breakdown voltages, thermal resistance, and flipping speeds than silicon, making harsh-condition power conversion quicker and effective. Spacecraft propulsion motors, power conditioning, and communication amplifiers use GaN and SiC devices because they can withstand severe temperatures and radiation. Because they can perform at higher voltages and temperatures, they don't need as much cooling, which is very important in space.

3D IC stacking and chiplet-based solutions are desirable because spacecraft need small, powerful computers. Chiplets combine processors, memory, sensors, and communication parts in a way that is not uniform, unlike monolithic chips. This method eliminates interconnect delays, maximizes bandwidth, and allows malfunctioning sections to be routed. 3D integration not only makes deep-space missions work better per watt, but it also reduces space and weight and generates designs that are easier to shield from radiation. High-performance computing is already looking into disaggregated architectures here on Earth, so this move fits with that trend. Next-generation spacecraft will need memory that is both high-density and resistant to radiation and that can keep data safe for decades. Innovative non-volatile storage systems promise radiation-resistant storage.

In the future, interplanetary electronics must be capable of adapting and repairing themselves to survive decade's long missions. Advanced nanomaterial coatings and conductive polymers could enable circuits to detect faults, bypass damaged sections and restore functionality without external intervention. Reconfigurable hardware, similar to FPGA architectures, would allow systems to modify their operations in real time to address unexpected challenges. When combined with AI driven autonomy, these self-centric and adaptable electronics could significantly extend mission lifespans, ensuring that spacecraft remain functional even after aging or exposure to hazards like micrometeoroid impacts.

17.5 Intelligent and Autonomous Electronic Systems

AI and ML are needed for spacecraft that need to work on their own for lengthy periods of time or in places that are hard to reach. Single-event upsets and radiation damage can happen to traditional accelerators like GPUs and TPUs. Scientists are working on developing AI accelerators that can grip a lot of sensor data packets on board and are resistant to radiation. Because of this, it will be possible to identify potential dangers in real time, distribute resources, and rank the importance of important data. For instance, dust storms on Mars or micrometeoroid strikes in interstellar space are examples of conditions that are impossible to predict. With these specialized processors, spacecraft would be able to operate independently, which would reduce the amount of raw data that would need to be transmitted back to Earth. Additionally, it would make spacecraft more responsive to situations that are complex.

When conventional von Neumann computer systems are used for missions that continue for decades, there are significant concerns regarding the amount of power that is consumed, the speed at which they operate, and their reliability. The concept of cognitive computing, which is informed by the functioning of the human brain, has the potential to assist with perception, pattern recognition, and While requiring very little power while simultaneously processing a large amount of data in order to make decisions in space that are subject to change dependent on what it is that occurs. In a similar vein, bio-inspired designs, such as self-organizing networks or evolutionary computing models, could make it possible for spaceships to modify their behavior in real time. This would be comparable to the flexibility that is displayed by living organisms when their environment undergoes change.

Future missions might employ swarms of small tiny probes instead of one large spacecraft. Each probe has electronics component and communicate the work of detecting, processing, and communicating. This method not only decreases the risk of the mission, but it also lets intelligence develop, where probes work together to create a bigger picture of science. Inter-probe RF communications can be managed by swarm electronics. This enables redundancy, shared policymaking, and response to bad conditions.

Interstellar systems may value autonomous fault tolerance most. Spacecraft must diagnose, rectify, and identify problems without human assistance due to years-long communication delays. This includes rerouting damaged circuits, employing self-repair materials, or modifying FPGA-based hardware. Artificial intelligence helps spacecraft identify science targets, adjust itineraries, and prioritize energy use. These skills transform spaceships from simple machines that follow directions into smart, flexible explorers that can expand.

17.6 Future Communication and Sensing Paradigms

17.6.1 Delay-Tolerant Networking and Optical Communications

As humans continue to explore further into space, communication architectures must evolve past physical limits of distance, latency, and reliability.

Two paradigms in this context are delay-tolerant networking (DTN) and optical communications (laser-based links). Together, these two paradigms offer a scenario for continuous, efficient, and scalable data transfer in interplanetary and deep-space conditions [9].

I. Delay-Tolerant Networking

Unlike ground networks, where connectivity is taken for granted, communication in deep space is subjected to intermittent disconnections and high signal delays. Standard internet protocols (e.g., TCP/IP) fail in this environment because of timeouts and packet loss. DTN solves this problem by embracing a "store-and-forward" approach where data bundles are held up in intermediate nodes (e.g., satellites, planetary orbiters, or ground relays) awaiting re-establishment of a persistent connection. This makes DTN naturally robust against sporadic connectivity and extended delays of several minutes (Earth–Mars) to hours for outer planets.

NASA, ESA, and ISRO are already testing DTN in space missions. NASA's Disruption Tolerant Networking project [9], having been tested on the International Space Station and planet probes, shows how DTN can form a backbone to interplanetary internet. The dream is a Solar System Internet (SSI), with planetary nodes, spacecraft, and orbiters as relay points. DTN will increasingly be the key in the future not only for spacecraft telemetry, but also for quantum-secure communication protocols, which will probably need high-reliability hybrid architectures combining quantum-secure keys with bundle-based classical transfer.

II. Optical Communications

Aside from DTN, optical communications are emerging as the next-generation deep-space links. When compared to RF systems that need greater wavelengths and enormous antennas, laser-based links possess lower wavelengths, enabling 10–100× greater data rates, reduced power needs, and more concentrated beams [10]. These advantages make them ideally suited for data-intensive missions such as space telescopes and planetary imaging. Space telescopes have been proven by NASA's LLCD and LCRD [11, 12] proved optical links in near-Earth space, while the DSOC [13] experiment on Psyche is extending tests beyond Earth orbit. These efforts illustrate that laser systems can deliver gigabit-per-second downlinks, transforming deep-space data return. Eventually, delay-tolerant networking and optical communications represent the backbone of future deep-space missions. They provide the scalability, hardness, and efficiency needed to allow not only exploration and scientific discovery but eventually the creation of permanent human colonies beyond our world.

17.6.2 Quantum Communication and Quantum Sensors for Deep-Space Communication

Deep space communications, or the transmission of data from Earth to spacecraft beyond geostationary orbit, are faced with fundamental challenges: huge distances, high signal attenuation, high latency, and harsh environmental conditions. Traditional radio-frequency systems, while being trusted, are bandwidth- and noise-limited. Emerging quantum technologies specifically quantum communications and quantum sensors—can possibly improve or transform deep-space connections through higher security, precision, and sensitivity.

I. Quantum Communication in Deep Space Principles

Quantum sensors use superposition and entanglement to get over the restrictions of classical measurement. In deep-space missions, navigation as well as communication are facilitated. The following are (i). Atomic clocks: Offer ultra-stable timing for navigation and synchronization, demanded by QKD and entanglement distribution. (ii). Atom interferometers: Accurately measure accelerations and gravitational fields; already demonstrated on the ISS (Cold Atom Lab). (iii). Quantum magnetometers: Spin defect-based in solids for mapping planetary magnetic fields. (iv). Squeezed light and quantum dots sensors: Enhance detection sensitivity for weak signals or spectral lines. These systems are capable of stabilizing optical links, enhancing pointing accuracy, and locking synchronization for quantum protocols.

II. Experimental Milestones and Impactful Use Cases in Deep Space

Table 17.2: Quantum communication for space applications has come a long way, and this book, Universal and Interstellar Quantum Telecommunications Demonstrations and Studies, reviews the major milestones. Micius [14] satellite marked a chronical breakthrough by representing satellite to ground quantum key distribution and entanglement distribution across kilometers to millions, realistically laying the foundation for the first satellite space-based quantum network. Building on this, ISRO [15] successfully tested free space QKD between ground base stations, very high securely transmitting of images and video, aiding as concept for potential satellite missions. In parallel Thales Alenia Space and SpeQtral [16] are tailing entanglement-based satellite systems, aiming to show overall quantum secure communication networks. The theoretical research extends the vision further, suggesting that under good conditions, scale of interstellar travel Quantum communication may be possible, emphasizing the long-term possibilities of these technologies for deep space goals.

Table 17.2 Universal and interstellar quantum wireless communication studies.

Entity/project	Demonstration/ contribution	Scale/impact
Micius Satellite	Satellite-to-ground QKD and entanglement distribution	Thousands of kilometers, first space-based quantum network
ISRO	Free-space QKD between ground stations; secure image and video transmission	Proof-of-concept for future satellite QKD
Thales Alenia Space & SpeQtral	Collaboration on entanglement-based satellite systems for secure networks	Towards global quantum-secure communication infrastructure
Theoretical Studies	Models of interstellar quantum communication viability	Suggest feasibility under favorable conditions

There are many difficult problems with quantum communication in deep space. One of the biggest and the most challenging problems is that photons vanished and beams spread out over enormous distances between planets, which makes data signals much ineffective. This is compounded by decoherence potentiality since quantum states can be contaminated by space cosmic radiation and space background noise. Synchronizing is challenging in the face of Doppler shifts and relativistic phenomena that make timing and alignment problematic. Furthermore, technology is presently lacking mature quantum repeaters and memory, technologies used to maintain entanglement and long-distance communication. Lastly, advancement depends on making space-qualified single-photon detectors that allow for high efficiency and radiation resistance, making them reliable in the hostile environment outside Earth [17].

III. Quantum Sensors in Deep-Space Systems

Quantum sensors use superposition and entanglement to go beyond the constraints of classical measurement. In deep space, they help with both navigation and communication [18]. In space-based quantum communication systems, atomic clocks provide consistent timing for navigation, synchronization, QKD, and entanglement distribution. Atom interferometers

[19] can also measure gravitational fields and accelerations with great precision. The Cold Atom Lab has already done this on the International Space Station. Quantum magnetometers [20–22], which use spin defects in solids, are another powerful tool because they can map planetary magnetic fields very accurately. All of these quantum technologies make space exploration and navigation more precise, safe, and able to probe planetary environments. Quantum dots and squeezed light sensors: Enhance detection sensitivity for weak signals or spectral features. These devices can stabilize optical links improve pointing accuracy and ensure secure synchronization for quantum protocols.

IV. High-Precision Timing with Atomic Clocks

High-precision atomic clocks use the natural oscillations of atoms most notably cesium, rubidium, strontium and trapped ions as ultra-stable frequency references to define time with extraordinary accuracy, now reaching fractional uncertainties at the 10^{-18}–10^{-20} level in advanced optical lattice and ion clocks. These systems rely on laser cooling, vacuum isolation and quantum control to minimize noise and environmental shifts, allowing them to measure time so precisely they would drift less than a second over the age of the universe. Beyond setting the global time standard, such clocks enable breakthroughs in fundamental physics tests, GPS and navigation, telecommunications synchronization, geodesy and even dark matter searches, while ongoing work on portable and quantum-enhanced designs aims to push timing precision into practical, real-world environments.

V. Long-Range Interstellar Data Relays

Long-range interstellar data relays are envisioned communication systems that overcome the immense challenge of sending signals across light-year distances, where even powerful transmissions arrive on Earth as single photons per second after years in transit. By combining ultra-precise laser links, photon-counting detectors, nuclear or beamed power supplies and architectures such as distributed probe chains or solar gravitational lens relays that use the Sun's gravity as a massive amplifier, these systems could transform faint trickles of data into usable scientific streams from other star systems. While bound by light-speed delays, such relays are essential to return high-bandwidth images and spectra from missions like Breakthrough Starshot and represent the first steps toward a future interstellar internet connecting humanity with distant probes and, eventually, colonies.

17.7 Powering the Next Generation of Space Electronics

Every joule of energy is critically important for interstellar voyages. A lot of research is going on right now with electronics that use very little electricity. This kind of system lets computers, sensors and communication devices work on very low voltages and go into deep sleep mode while they're not being used. Using sub-threshold logic, approximation computation, event-driven processing and other methods can greatly cut down on power use. These methods, along with radiation-tolerant architecture, make it possible to carry out longer missions in places where a steady supply of power isn't always available. Solar arrays are the most common power source for missions in the inner solar system and near-Earth, although their efficiency drops as the distance increases. Dynamic isotope power systems and very small fission reactors are essential candidates for interstellar missions, since they can provide continuous, long-term energy for decades or even centuries. Stirling engine–based converters offer higher efficiency than static thermoelectric systems, making them attractive for deep-space applications. Compact fission reactors, on the other hand, deliver scalable power levels and can support both propulsion and high-demand instruments. Together, these systems represent practical pathways for reliable energy supply in interstellar spacecraft without reliance on solar input.

Solar sails and beamed laser power transfer are two samples of energy harvesting technology that missions may use to get more power on board. Solar sails can both provide a small amount of power for use on board and use the momentum of photons to move them. This is possible because they include integrated photovoltaic panels. Beamed power doesn't need local fuel or isotopes, but it does need very precise beam control and long-term stability over astronomical distances. This makes it a promising but difficult technology for the future. Even with sophisticated sources, it will still be very important to manage and store electricity well. Next-generation spacecraft will need smart power distribution networks that prioritize important components, redirect energy around defects on the fly and make charging cycles more efficient. Supercapacitors, solid-state batteries and advanced fuel cells are all examples of storage technologies that can handle radiation and have a high density. They can all provide reliable buffering during peak load times. These technologies will let spacecraft handle survival, science operations and communication on their own, even when energy levels alter. They will also have AI-powered energy controls. Better power electronics will change spaceships from passive energy users to adaptive energy managers that can live for decades in difficult situations.

17.8 Manufacturing, Packaging, and Sustainability

17.8.1 Additive Manufacturing and In-Space Fabrication

3D printing, often called additive manufacturing, is becoming a game-changing technology for space electronics and buildings. It lets spaceships make spare components, circuit boards, or even whole modules in space or on the surfaces of distant planets. This cuts down on the requirement to send extra hardware from Earth. In-space fabrication could also allow for designs that change based on conditions, like printing shielding layers as radiation levels rise or making new connectors and housings after micrometeoroid damage. Recent tests on the International Space Station (ISS) have shown that it is possible to 3D print polymers and metals in microgravity. This opens the door to making electronics on demand for long trips or missions to other stars.

17.8.2 Advanced Thermal and Mechanical Packaging for Harsh Conditions

Spacecraft electronics have to work in very harsh conditions, like radiation and very cold temperatures. Ceramics, multilayer shielding and rigid housings are used in current packaging. However, future designs are heading towards adaptive thermal coatings, phase-change encapsulants and composites that can withstand vibrations. Packaging will have to deal with both microgravity effects and mechanical stresses over time, making sure that parts stay stable for decades. Flexible substrates are also changing the way packaging works, making it possible to make lighter, stronger enclosures that can survive in space without needing repairs.

17.8.3 Modular, Designing for Longevity, Recyclability and Resource Efficiency

Once launched, traditional spacecraft are monolithic, which means they can't be changed. Interstellar missions, on the other hand, need modular architectures that let components be separated, fixed, or replaced by changing circuits or turning on dormant backups. Chiplet-based designs, avionics that are straightforward to install and electronics that are spread out over numerous probes. This modularity lets spacecraft improve their processing or sensor power in the middle of a mission by uploading new firmware, turning on circuits that aren't being used enough, or adding new parts that were manufactured *via* additive manufacturing. Adaptability is needed to resolve unexpected challenges over decades.

Finally, electronics of the future should be manufactured from materials that last a long time and are good for the environment. Along with being able to withstand radiation and ageing, systems will also need to break down in a way that doesn't interfere with important operations. If parts of the spaceship may be utilised for different things throughout time, the mission could last longer. This is possible if the spaceship is made of recyclable materials and has designs that use resources wisely.

17.9 Roadmap to Interstellar-Grade Electronics

17.9.1 Technology Readiness Level Progression

Pathfinder avionics will show radiation-tolerant COTS parts with active fault mitigation, optical downlinks and autonomous fault detection beyond simple resets soon (2025–2032). This will raise readiness to TRL 5–6 through small sat and lunar gateway trials. By the 2030s, endurance missions to the outer solar system will test long-lasting electronics, chiplet-based avionics, neuromorphic processors for onboard science triage, and better isotope power systems. These trips will happen with orbiters circling Jupiter, Saturn, or Uranus. Swarm demonstrations in the late 2030s and 2040s will show how coordinated probes can sense and process information in different places. Inter-probe communication and shared autonomy will make swarm electronics more advanced.

The next big leap, which is expected to happen in the future, will be in power and autonomy. For example, compact fission reactors or high-efficiency isotope systems will provide stable long-term energy and spacecraft will be able to determine their own science goals, manage their own energy and change their paths without any interference from Earth. The interstellar pathfinder will be a probe with electronics that can last for centuries, mend themselves and be put together in different ways. It will also be able to send scientific data back from a nearby star system, completing decades of slow progress.

17.10 Conclusion and Future Outlook

17.10.1 Convergence of Technological Breakthroughs

Space electronics are at a phase in their growth where a number of new technologies are coming together. Wide bandgap semiconductors, chiplet-based designs, neuromorphic processors, quantum-enabled sensors and self-healing circuits are all pieces of a broader puzzle. Because to this convergence, spaceships won't have to worry about their electronics breaking anymore. Instead, they will utilise smart, energy-efficient and flexible systems that can survive extreme environments.

The goal for interstellar exploration in the future is to make electronics that can work on their own for a hundred years or more, even when they're light-years away. These systems will act like the nervous system for interstellar probes, allowing them to think, adapt and prosper without having to be watched over all the time from Earth. Future spacecraft will include sustainable electricity, adaptive temperature control, swarm-based intelligence, and modular lifespans to make them more resilient in both technology and philosophy. Scientific progress will unveil the enigmas of exoplanets, cosmic evolution, and even extraterrestrial life, altering humanity's destiny as a species that transcends its home star.

17.10.2 Final Thoughts on the Path Forward

The road to interstellar-grade electronics is not straight or certain; it needs progress to be made in engineering, science, governance and ethics all at the same time. Space authorities, commercial companies and international regulatory bodies need to work together to make sure that exploration stays fair, responsible and long-lasting. With each achievement, from relativistic ancestors to long-lived planetary probes, Confidence grows in the ability to send humans on the first ever interstellar mission. It is not just about sending spaceships across the galaxy; it is also about expanding human imagination and ensuring that technology is as strong and flexible as the vision for reaching other worlds.

AI disclosure statement: The author confirms that no artificial intelligence (AI)-generated content or AI-assisted writing tools were used in the preparation of this chapter (including summaries and the abstract). All text, analysis, and conclusions are the result of the author's original work.

References

1. Denis, G., Alary, D., Pasco, X., Pisot, N., Texier, D., Toulza, S., From new space to big space: How commercial space dream is becoming a reality. *Acta Astronaut.*, 166, 431–443, 2020.

2. Pelton, J.N., *Space 2.0: Revolutionary Advances in the Space Industry*, Springer International Publishing, Berlin/Heidelberg, 2019.
3. White, H. and March, P., *et al.*, Measurement of impulsive thrust from a closed radiofrequency cavity in vacuum. *J. Propul. Power*, 33, 4, 2017.
4. Genta, G. and Kezerashvili, R.Y., Achieving the required mobility in the solar system through direct fusion drive. *Acta Astronaut.*, 173, 303–309, 2020.
5. Ganesan, M. and Lavanya, R., A deep learning approach to fault detection in a satellite power system using Gramian angular field. *Int. J. Eng. Syst. Modell. Simul.*, 12, 2–3, 195–201, 2021.
6. Swamy, P.A., Rithani, M., SyamDev, R.S., A hybrid model for disaster damage detection using satellite images, in: *Proc. 7th Int. Conf. Image Information Processing (ICIIP)*, IEEE, pp. 838–843, 2023.
7. Preetha, S.K., Sai, J.V.M., Raj, V.S., Sekhar, M.J., Lavanya, R., Anomaly detection in satellite power system using deep learning, in: *Proc. IEEE Int. Conf. Electronics, Computing and Communication Technologies (CONECCT)*, pp. 1–5, 2024.
8. Gokkul Nath, T.S., Sudheesh, P., Jayakumar, M., Tracking inbound enemy missile for interception from target aircraft using extended Kalman filter, in: *Proc. Int. Symp. Security in Computing and Communication*, Springer, Singapore, pp. 269–279, 2016.
9. National Aeronautics and Space Administration, NASA, ESA use experimental interplanetary internet to test robot from International Space Station, NASA News Release 12–391, 2012.
10. National Aeronautics and Space Administration, Delay/Disruption Tolerant Networking (DTN), NASA (Communications with Missions), 2023.
11. National Aeronautics and Space Administration (SCaN), Lunar Laser Communications Demonstration (LLCD), NASA, 2023.
12. National Aeronautics and Space Administration, Laser Communications Relay Demonstration (LCRD) overview, NASA, 2023.
13. National Aeronautics and Space Administration, Deep Space Optical Communications (DSOC) technology demonstration, NASA, 2023.
14. Austrian Academy of Sciences, Secure quantum communication over 7,600 kilometers, 2018.
15. Indian Space Research Organisation, Quantum key distribution (QKD): Free-space QKD, ISRO Press Release, 2021–2022.
16. Swayne, M., Quantum communications by satellite: SpeQtral and Thales Alenia Space launch new experimental phase, The Quantum Insider, 2025.
17. Berera, A. and Calderón-Figueroa, J., Viability of quantum communication across interstellar distances. *Phys. Rev. D*, 2022.
18. Wright, J., *et al.*, Could we eavesdrop on interstellar communications that pass through our solar system?, Penn State News, 2022.
19. Burt, E.A., *et al.*, Demonstration of a trapped-ion atomic clock in space. *Nature*, 595, 43–47, 2021.

20. NASA Science Editorial Team, Quantum sensing *via* matter-wave interferometry aboard the International Space Station, NASA Science Highlight, 2025.
21. NASA Science Editorial Team, Quantum scale sensors used to measure planetary scale magnetic fields, NASA Science Feature, 2024.
22. National Aeronautics and Space Administration / Jet Propulsion Laboratory, Superconducting nanowire single-photon detectors (SNSPD) for deep space optical communications, NASA JPL Microdevices, 2023.

Index

Also of Interest

Check out these other titles from Scrivener Publishing

Artificial Intelligence Applications in Aeronautical and Aerospace Engineering, Edited by K. Sathish Kumar and R. Naren Shankar, ISBN: 9781394268764. This book is a comprehensive guide for anyone in the aeronautical and aerospace fields who wants to understand and leverage the transformative power of artificial intelligence to enhance safety, optimize performance, and drive innovation.

The Intelligent Universe: AI's Role in Astronomy, Edited by Yogesh Chandra, Manjuleshwar Panda and Mahesh Chandra Mathpal, ISBN: 9781394355488. Uncover the universe's secrets with this essential guide that provides a comprehensive exploration of how artificial intelligence is revolutionizing modern astronomical research.

Internet of Vehicles, Edited by S. Neelakrishnan, Niresh Jayarajan, Malathy Sathyamoorthy, Rajesh Kumar Dhanaraj, and Archana Naganathan, ISBN: 9781394275083. Position yourself at the forefront of the transportation revolution with this guide to mastering computational intelligence that serves as the essential linchpin for the safe, sustainable, and hyper-connected Internet of Vehicular Things.

Autonomous Systems in the Internet of Vehicles, Edited by Balamurugan Balusamy, Sandeep Kumar Mathivanan, Prabhu Jayagopal, S.K.B. Sangeetha, and Ali Kashif Bashir, ISBN:9781394311699. Accelerate your expertise in the future of autonomous navigation by mastering essential fusion algorithms that enable vehicles to operate safely and reliably in complex, dynamic environments.

Autonomous Vehicles Planning and Control, By Yong Bai & Liang Zhao, ISBN:9781394355044. Master the future of marine exploration and technology with Autonomous Vehicles Planning and Control, which provides a comprehensive, interdisciplinary guide to the principles, control, and real-world applications of autonomous marine vehicles.

AI for Autonomous Vehicles: The Future of Driverless Technology, Edited by Sathiyaraj Rajendran, Munish Sabharwal, Yu-Chen Hu, Rajesh Kumar Dhanaraj, and Balamurugan Balusamy, ISBN: 9781119847465. This title presents the research prospects and security challenges present in the exciting new field of autonomous vehicles by comprehensively introducing advanced concepts of artificial intelligence (AI) in driverless technology.

AUTONOMOUS VEHICLES VOLUME 1: Using Machine Intelligence, Edited by Romil Rawat, A. Mary Sowjanya, Syed Imran Patel, Varshali Jaiswal, Imran Khan, and Allam Balaram. ISBN: 9781119871958. Addressing the current challenges, approaches and applications relating to autonomous vehicles, this groundbreaking new volume presents the research and techniques in this growing area, using Internet of Things, Machine Learning, Deep Learning, and Artificial Intelligence.

AUTONOMOUS VEHICLES VOLUME 2: Smart Vehicles for Communication, Edited by Romil Rawat, Purvee Bhardwaj, Upinder Kaur, Shrikant Telang, Mukesh Chouhan, and K. Sakthidasan Sankaran, ISBN: 9781394152254. The companion to *Autonomous Vehicles Volume 1: Using Machine Intelligence*, this second volume in the two-volume set covers intelligent techniques utilized for designing, controlling and managing vehicular systems based on advanced algorithms of computing like machine learning, artificial Intelligence, data analytics, and Internet of Things with prediction approaches to avoid accidental damages, security threats, and theft.

AI AND IOT-BASED INTELLIGENT AUTOMATION IN ROBOTICS, Edited by Ashutosh Kumar Dubey, Abhishek Kumar, S. Rakesh Kumar, N. Gayathri, Prasenjit Das, ISBN: 9781119711209. The 24 chapters in this book provide a deep overview of robotics and the application of AI and IoT in robotics across several industries such as healthcare, defense. education, etc.

Printed and bound by CPI Group (UK) Ltd, Croydon, CR0 4YY
19/09/2026
14965090-0001